Polymer Support Fluids in Civil Engineering

ice
Institution of Civil Engineers
publishing

Polymer Support Fluids in Civil Engineering

Carlos Lam
DPhil CEng MICE MHKIE

Stephan A. Jefferis
PhD CEng CEnv FICE CGeol FGS FRSA

Published by ICE Publishing, One Great George Street,
Westminster, London SW1P 3AA

Full details of ICE Publishing representatives and distributors can be found at: www.icebookshop.com/bookshop_contact.asp

Other titles by ICE Publishing:

Civil Excavations and Tunnelling – A Practical Guide, Second edition
R. Tatiya. ISBN 978-0-7277-6153-8
ICE Specification for Piling and Embedded Retaining Walls, Third edition
Institution of Civil Engineers. ISBN 978-0-7277-6157-6
Earthworks: A Guide, Second edition
P. Nowak and P. Gilbert. ISBN 978-0-7277-4116-5

www.icebookshop.com

A catalogue record for this book is available from the British Library.

ISBN 978-0-7277-5786-9

Commissioning Editor: Michael Fenton
Development Editor: Maria Inês Pinheiro
Production Editor: Madhubanti Bhattacharyya
Market Development Executive: Elizabeth Hobson

Typeset by Academic + Technical, Bristol
Index created by Laurence Errington
Printed and bound in Great Britain by TJ International, Padstow

FSC
www.fsc.org
MIX
Paper from responsible sources
FSC® C013056

This book is dedicated to the contractors, consultants, academics and material suppliers who have done so much over many years to advance the understanding of excavation support fluids in civil engineering. Without their unstinting help this book would not have been possible. We also recognise the special contributions of our families, teachers and students.

Contents

About the authors ix

01 **Introduction to fluid-supported excavation** **1**
1.1. What is a support fluid? 1
1.2. Terminology 1
1.3. A brief history of drilling and support fluids 2
1.4. How support fluids work 5
1.5. Critical functions of support fluids 7
References 13

02 **Constraints on the use of bentonite support fluids** **17**
2.1. Introduction 17
2.2. The steady decline in the reserves and the perceived reduction in performance of some bentonites 17
2.3. Mixing and hydration requirements of bentonite fluids 18
2.4. Requirement of surface treatment plant 21
2.5. Disposal of used fluids: environmental regulations and on-site treatment 22
2.6. Environmental impact 25
2.7. Soil dispersion 26
References 27

03 **Introduction to polymers and polymer fluids** **29**
3.1. Introduction 29
3.2. What is a polymer? 29
3.3. Basic polymer science 30
3.4. Basic classification of polymers of interest for use as support fluids 34
3.5. Dissociation of polymers in solution 41
3.6. Key properties of polyacrylamides 45
References 53

04 **Commercial polymer systems** **57**
4.1. Introduction 57
4.2. Natural biodegradable polymers 59
4.3. Polyacrylamide-based systems 64
4.4. Blended polymer systems 83
References 86

05 **Site equipment and procedures** **89**
5.1. Introduction 89
5.2. Polymer fluid mixing 89

	5.3.	Pumping	92
	5.4.	Storage and self-cleaning	96
	5.5.	Managing excavations in soils with high groundwater pressures	100
	5.6.	Excavation tools	102
	5.7.	Cleaning in excavations	105
	5.8.	Recycling and rejuvenation for reuse	107
	5.9.	Disposal	109
	References		110
06	**Fluid properties and field testing**		**113**
	6.1.	Introduction	113
	6.2.	Dissolved solids and electrical conductivity	113
	6.3.	pH	114
	6.4.	Water hardness	118
	6.5.	Viscosity	120
	6.6.	Density	131
	6.7.	Sand content	137
	6.8.	Residual polymer concentration	138
	References		142
07	**Applications of polymer fluids to the construction of deep foundations around the world**		**145**
	7.1.	Introduction	145
	7.2.	Europe	145
	7.3.	Africa	157
	7.4.	Asia	159
	7.5.	The Americas	170
	References		182
08	**Detailed analysis of case studies**		**187**
	8.1.	Introduction	187
	8.2.	Glasgow	187
	8.3.	Stratford, London	201
	References		226
09	**Rheological properties of polymer support fluids**		**229**
	9.1.	Introduction	229
	9.2.	Fundamentals of non-Newtonian fluid flow	229
	9.3.	Laboratory techniques	232
	9.4.	Flow behaviour	235
	9.5.	Transient viscoelastic properties	252
	References		261

10 .... **Specifications and associated controls** **265**
10.1. Introduction 265
10.2. Polymer types 265
10.3. Sampling locations 266
10.4. Test procedures commonly employed 267
10.5. Excavation practice 271
10.6. Review of existing polymer fluid specifications 271
10.7. Concluding remarks 283
References 283

Index **285**

About the authors

Carlos Lam DPhil CEng MICE MHKIE
Carlos Lam holds degrees from The Hong Kong Polytechnic University, Imperial College London and the University of Oxford, all in civil engineering. He was formerly a lecturer in geotechnical engineering at The University of Manchester and served on the editorial advisory panel of the *Proceedings of the Institution of Civil Engineers – Geotechnical Engineering*. He currently works for the Geotechnical Engineering Office in Hong Kong. He has received several awards, including the Fugro Prize from the Geotechnical Division of the Hong Kong Institution of Engineers. He has published extensively on the use of polymer fluids for geotechnical construction in international journals and conferences. He gratefully acknowledges the support of the ICE Research and Development Enabling Fund in the preparation of this book.

Stephan Jefferis PhD CEng CEnv FICE CGeol FGS FRSA
Stephan Jefferis is a director of Environmental Geotechnics Ltd, a visiting professor at the University of Oxford and an emeritus professor of the University of Surrey, and was the chairman of the British Geotechnical Association from 2015 to 2017. He has over 40 years of experience in research and consultancy on geotechnical processes. He has worked extensively and internationally on the application of bentonite, bentonite–cement and polymer slurries in piling, diaphragm walling, tunnelling and cut-off walls. He is the author of over 100 papers, and is regularly invited as a keynote speaker in international conferences.

Polymer Support Fluids in Civil Engineering
ISBN 978-0-7277-5786-9

http://dx.doi.org/10.1680/psfce.57869.001

Chapter 1
Introduction to fluid-supported excavation

1.1. What is a support fluid?

In civil engineering a support fluid is a liquid that can be used to support an open excavation in the ground. Support fluids are used in a wide variety of civil engineering processes including piling, diaphragm walling, slurry tunnelling, horizontal directional drilling, oil and water well drilling, drilling site investigation boreholes, and the formation of reactive barriers and deep drainage walls. In each of these applications, the role of the fluid is to keep the excavation open until a permanent element (concrete or other backfill, a lining, instrumentation, etc.) is installed. In processes where the support fluid is also used to transport the cuttings back to the surface and to cool and lubricate the drill tool, the fluid may be called a 'drilling fluid' as it does more than simply support the excavation.

Although the concept of a support fluid is very simple, specifying, achieving and maintaining the required fluid properties and the associated equipment can be complex and vary significantly between applications. This book will focus on the use of polymer support fluids in piling, diaphragm walling and deep trenching, as these are currently the principal applications for these fluids. Where appropriate, information will also be given on other applications to cover, so far as possible, the full spectrum of applications of polymer fluids in civil engineering. The use of polymer fluids in applications such as concrete mix formulation and pipe lubrication in pipe-jacking are, however, not covered in this book.

1.2. Terminology

Since the first use of polymers in excavation support, many different terms have been used rather loosely to refer to them, including 'mud', 'slurry' and 'fluid'. The term 'mud' is clearly derived from the petroleum industry where a drilling fluid engineer on an oil rig is often known as a mud engineer. According to the *Oxford English Dictionary*, the word 'slurry' means 'thin sloppy mud or cement. Also in extended use, any fluid mixture of a pulverised solid with a liquid (usu. water).' Therefore, the term 'slurry' should properly be used only to describe suspensions of solids in a liquid. Although both terms, 'mud' and 'slurry', can be used interchangeably to describe bentonite and other clay-based suspensions, it is incorrect to use them to describe polymer solutions, as they are solutions not suspensions. To further confuse the situation, it may be noted that even water is referred to as 'water slurry' by the American Association of State

Highway and Transportation Officials (AASHTO, 2010). Clearly this is an inappropriate usage, at least to the purist. As the term 'fluid' can be used to describe both solutions and suspensions, it is therefore used in this book for both slurries and polymer solutions.

The terms 'drilling', 'boring', 'excavating/excavation' and 'support' have all been used to describe the function of the fluid. Although most of them can describe the action of digging, as in 'digging fluid', only the terms 'support' or 'excavation-support' focus on the main function of the fluid; that is, supporting an excavation. As a result, in this book the term 'polymer support fluid' (as used in the book title) or simply 'polymer(s)' will be used to describe the material. This usage aligns with that by the Federation of Piling Specialists and the Institution of Civil Engineers (FPS, 2006; ICE, 2017), which both use the term 'support fluid'. It is worth emphasising that the main difference between a drilling fluid and a support fluid is that the former is required to perform multiple tasks simultaneously, including hole cleaning, cooling, lubrication and borehole stabilisation, while for the latter the first three functions are not specifically required, although they may be useful.

The word 'polymer' is used in this book to refer to any water-soluble polymer, and specific descriptive terms such as 'natural', 'biodegradable' and 'synthetic' will be added as and when necessary to indicate particular features of a polymer. Currently, some of the most commonly used support fluids are based on polyacrylamides. In the polymer science literature, polyacrylamides are usually referred to as 'PAMs' or 'PHPAs', depending on whether or not the material is partially hydrolysed. The abbreviation PHPA, which is short for 'partially hydrolysed polyacrylamide', originated in the oil industry and is commonly used in civil engineering texts. In this book, the abbreviation 'PHPA' will be used to denote the partially hydrolysed form of polyacrylamide, and the abbreviation 'PAM' to denote the unhydrolysed form.

1.3. A brief history of drilling and support fluids

1.3.1 Water

The use of water to assist in drilling boreholes can be traced back thousands of years. According to Brantly (1971) and Darley and Gray (1988), Confucius wrote of wells a few hundred feet deep being drilled for brine in China during the early part of the Zhou dynasty. Water was poured into these wells to soften the rock and to help remove the cuttings by a bailer. Similarly, the ancient Egyptians are known to have used water to help remove cuttings from holes drilled in quarries. However, the use of circulating water to remove cuttings was not developed until the early 19th century in Europe. In 1833, a French engineer, Pierre-Pascal Fauvelle, observed water spouting from an artesian well during boring. Inspired by what he had seen, Fauvelle went on to invent a system involving pumping water down a hollow boring rod to bring the drilled particles to the surface (Fauvelle, 1846). At about the same time in England, Robert Beart filed a patent for essentially the same system, involving the use of a rotating hollow drill rod and flush water (Beart, 1845). A detailed account of the history of well drilling techniques is given by Brantly (1971), and is not repeated here. Although water is still occasionally used for the construction of deep foundations today, it is now widely appreciated that water can cause swelling and softening of water-sensitive clays, mudstones and shales, thus causing loss of excavation stability and/or reduced load-bearing capacity of the completed foundations (ICE, 2017).

1.3.2 Bentonite

The concept of creating a deep excavation using clay slurry appears to have been developed separately in Europe and in the USA at about the same time in the mid-20th century. The first deep cut-off wall is believed to have been constructed by the US Army Corps of Engineers, to reduce seepage through a levee embankment on the Mississippi River (Kramer, 1946). The trench was supported by clay slurry during excavation and was backfilled with low-permeability puddle clay. The construction of structural diaphragm walls was first conceived by the Austrian engineer Christian Veder slightly earlier in 1938 (Glossop, 1963; Xanthakos, 1979). Veder (1953) described a method of constructing deep cast-in-situ piles placed side by side using bentonite fluids, and gave two early projects in Italy as examples. Veder (1963) reviewed some of the early work, and noted that the first diaphragm walls were constructed in 1950 'using circular concrete elements partly embedded one in another'; that is, secant pile walls as we know them today. Veder also mentioned that the first applications were 'impermeable cut-offs below earth dams in deposits of sands and gravels with large boulders' and for 'making a reservoir watertight'. These walls were about 35–40 m deep and would have been backfilled with concrete.

It can be seen from these early records that by the late 1950s foundation engineers had already mastered the basic technique of supporting deep excavations with a fluid. Of course, further developments of the technique did not stop and have continued until the present day. In the 1960s and 1970s, the slurry excavation technique remained a popular topic both in research and industry; this led to much improved excavation tool design, and better understanding of the required fluid properties and their impact on the stability of the excavation. For example, by the early 1960s it was already possible to construct rectilinear elements one after the other, giving a continuous trench 5–10 m long and reaching a depth of 40 m (Veder, 1963), and different types of wall joints were already in use (Sadleir and Dominioni, 1963). The required properties of bentonite slurry were also much better understood, and the Federation of Piling Specialists published the first of a series of bentonite slurry specifications in 1975 and a revision in 1977 (FPS, 1975, 1977). Specifications and guidance were also proposed by practitioners including Hutchinson *et al.* (1975), Hodgson *et al.* (1977) and Holden (1984).

At about the same time, the UK Construction Industry Research and Information Association (CIRIA) published two reports on the use of bentonite in construction (Fleming and Sliwinski, 1977; Harris and Walker, 1967). In the year 2000, a document entitled *Bentonite Support Fluids in Civil Engineering* was published by the FPS, and was later reprinted in 2006 (FPS, 2000, 2006). These documents have been widely used by foundation engineers, and are referenced in industry specifications such as the *ICE Specification of Piling and Embedded Retaining Walls* (ICE, 1996, 2007, 2017). In the US, the Department of Transportation of each state has its own support fluid specification. At the national level, the American Association of State Highway and Transportation Officials and the Federal Highway Administration have published their own drilled shaft (bored pile) specifications, including bentonite fluid requirements (AASHTO, 2010; Brown *et al.*, 2010; O'Neill and Reese, 1999).

Over time, in response to the evolving requirements of the civil engineering industry, bentonite producers have developed a range of products to suit different needs. For civil engineering works, sodium-activated bentonite is most commonly and indeed almost universally used. Some civil engineering bentonites may contain polymer additives to enhance their performance.

1.3.3 Polymer

Although by the late 1970s the use of bentonite fluids in civil engineering was well established and a mature technology, due to concerns about reserves of good-quality Wyoming bentonite and the costs of liquid waste (used slurry) disposal, engineers started to look for alternatives. A proprietary polymer product known as Telmarch is reported to have been used in Japan for diaphragm walling in the 1970s, but the application of this product seems to have been short lived (Ikuta, 1974; Xanthakos, 1979). From that period and until the early 1990s, natural biodegradable polymers such as carboxymethyl cellulose, xanthan gum and guar gum were tried at intervals, but none of them managed to make any real impact on industry practice. It is thought that these polymers may have suffered from problems such as incompatibility with concrete, high fluid loss and cleaning difficulties, but perhaps most significantly there was insufficient understanding of the behaviour of polymer fluids in the field. Possibly due to their limited market penetration, there are only a few publications that describe this early phase of polymer fluid development (however, see, for example, Beresford *et al.*, 1989; Tallard, 1992; Zapico, 1991). Although these natural polymers and many others are still available on the market, they are now mainly used for drilling site investigation boreholes (BDA, 1992) and for the construction of permeable reactive barriers and deep trench drains (e.g. Day and Ryan, 1992; Day *et al.*, 1999).

For the construction of structural foundations formed under polymer fluids, a major breakthrough came in the early 1990s when a specific type of synthetic polymer known as partially hydrolysed polyacrylamide was promoted by a small number of manufacturers, and at the same time was investigated by researchers in the USA (e.g. Kheng *et al.*, 1991; O'Neill, 1991). Since then, the use of synthetic polymer fluids has gradually increased and the practice has spread to other parts of the world. Synthetic polymer technology has also developed to include a range of additional chemicals to fulfil specific functions, such as fluid loss control, accelerated particle sedimentation and soil encapsulation. These chemicals may be added as additives to the base polymer during use on site or pre-blended into a single product by the manufacturer. As a result, polymers are now often sold as a whole package of materials and described as a 'polymer system'. Due to these developments, the former clear distinction between synthetic and natural polymers has become blurred, as some polymer systems may include synthetic and natural products (which may be chemically modified), depending on the properties required on site (e.g. Goodhue and Holmes, 1995).

1.4. How support fluids work

All support fluids work by exerting a hydrostatic fluid pressure on the base and the side walls of an excavation. In the presence of groundwater, the stabilising pressure at all depths needs to be greater than the free-field pore water pressure in order to stop flow of groundwater into the excavation. To illustrate the mechanism of stabilisation, Figure 1.1 shows a schematic diagram of a fluid-supported excavation. It can be seen that the pressure inside the excavation is greater than the free-field pore water pressure in the soil due to the head difference between the support fluid and the groundwater. A net stabilising pressure is exerted on the excavation base and side walls, and this leads to an outward fluid flow into the surrounding soil. Therefore, there is a tendency for the fluid within the excavation to be continually lost to the ground, and thus the support fluid will need to be regularly topped up if the head difference is to be maintained (Figure 1.2).

The rate of fluid loss will principally depend on the type and properties of the support fluid, the fluid head difference and, of course, the permeability of the in situ soil. In practice, fluid loss is only an issue in coarse soil (sand and gravel), and is usually negligible in fine soil (silt and clay), as the low permeability of the latter does not permit fast flow. If a bentonite fluid is used to stabilise an excavation in permeable soils, a low-permeability filter cake will develop on the soil surface, and this will reduce the rate of fluid loss into the ground. However, for coarse soils a rheological blocking process will first have to be established in the pore space before the filtration process can take place on the soil surface.

The stabilising mechanism for polymer fluids is, however, rather different, and in fact goes against the conventional understanding that only a thixotropic slurry with a density greater than that of water is capable of supporting an excavation. As polymer fluids are prepared at low concentrations of polymer (typically <2 kg/m^3), they are of essentially the same density as water (1.0 g/cm^3) and contain no suspended solids when freshly prepared. Therefore, instead of forming an impermeable filter cake on the cutting

Figure 1.1 A schematic diagram of a fluid-supported excavation

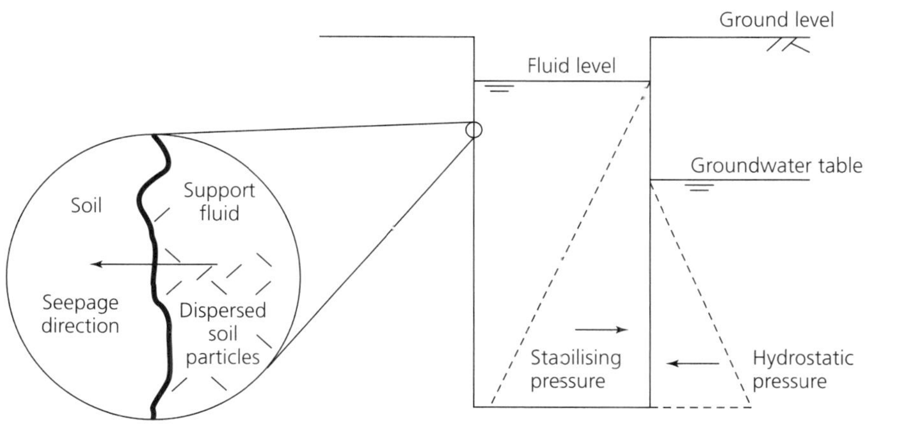

Figure 1.2 Pumping support fluids into a pile bore

surface, fresh polymer fluids will flow continuously into the soil due to the head difference. However, as polymer fluids are designed to be of high viscosity, up to 100 000 times that of water at low shear rates, the flow rate is actually very slow compared with the flow of water under the same conditions. Also a thin filter cake may form from excavated soil suspended in the fluid; thin because PHPA polymer fluids inhibit soil dispersion. In coarse grounds, flow rates can still be high, and it may be necessary to use additives to control loss rates if the clogging action of soil particles suspended in the support fluid is not sufficient. For these reasons, polymer fluids can also maintain the stability of an excavation even in coarse ground. Some polymers may be capable of forming a fluid film at the soil surface comparable to a filter cake, but further research is needed to elucidate mechanisms. It is worth noting that water is only suitable for use in excavation in fine soils due to its low viscosity (1 mPa s). Experience has shown that in coarse ground the rate of fluid loss is usually so high that an immediate loss of stability is inevitable.

Regardless of the type of support fluid used, special procedures needs to be adopted when working in areas where the groundwater table is close to the ground surface. This is because when an excavation tool is lowered into a pile bore or trench, some support fluids will be displaced and lost to the surrounding ground, and on retrieval of the tool the fluid level will drop below the original level, thus leading to possible instability of the excavation. In such situations, a temporary support fluid reservoir may have to be built around the excavation, so that a sufficiently high support fluid level can be maintained at all times.

1.5. Critical functions of support fluids

In addition to its main function of keeping an excavation open, a support fluid is also expected to perform many other functions. In general, the following characteristics are required of a support fluid during the course of an excavation project:

- fully functional at all times
- reliable and simple to use
- controllable, with properties easily monitored and maintained
- good health and safety and environmental performance
- low running cost
- simple and low-cost disposal on project completion
- meets the designer's and contractor's other objectives.

Some of the above characteristics are subjective, and interpretation of requirements will differ between projects and between different contractors and consultants. Table 1.1 lists the functions of an ideal polymer support fluid for four stages of construction of a foundation element: excavation, fluid handling, concreting and foundation loading. Each of these functions is discussed briefly below.

1.5.1 The excavation stage

1.5.1.1 Keep excavation open with little loss of support fluid

This requirement is the most important of all, because significant fluid loss into the ground will lead to a rapid drop in the stabilising pressure, a rise in local pore water pressure and, therefore, instability of the excavation. As discussed in the previous section, to minimise fluid loss in coarse soils, the support fluid either has to form a low-permeability filter cake at the excavation face (as bentonite slurry does) or possess a high solution viscosity to minimise the flow rate (in the case of a polymer solution). The filtration and rheological properties of the support fluid are therefore important considerations. Several other factors can also influence the amount of fluid loss, such as the groundwater level, the chemistry of the pore water, the choice of fluid additives and the design of the excavation tools.

Fluid loss through spillage onto the ground surface as the excavation tool is removed from the excavation, although not in itself an issue for the excavation stability, can still be expensive, and should be avoided by using free-draining augers, grabs or buckets.

1.5.1.2 Minimise soil softening at the walls and base of the excavation

The ideal support fluid should also minimise the degree of soil softening at the walls and base of the excavation. When a cutting is made in clay, the horizontal total stress on the vertical section reduces from the 'at rest' pressure to zero, and the pore water pressure becomes negative. When support fluid under its own hydrostatic pressure acts against the cutting face, the two pressures tend to equalise; that is, the clay takes up water and loses strength accordingly (Nash and Jones, 1963). Burland (1963) presented a case study showing an additional increase in the moisture content of London Clay near the surface of a diaphragm wall due to the use of bentonite slurry. Similar results were also observed by Chang and Zhu (2004). To prevent the softening and disintegration of soil, the ideal

Table 1.1 Functions of an ideal support fluid and the required properties

Construction stage	Required functions	Related fluid properties	Underlying mechanism	Other considerations
Excavation	Keep excavation open with little loss of support fluid	Rheological properties	Transient and steady state seepage	Support fluid and groundwater levels; soil strength; salinity of pore water; choice of fluid additives; design of excavation tools
		Density	Hydrostatic pressure	
	Minimise softening of soil at the side wall and base	Residual polymer concentration	Amount of active polymer in the system	Soil type; pile bore or trench open time
		Polymer chemical reactivity	Polymer–soil sorption	
		Polymer chain length	Soil particle encapsulation	
	Keep excavated materials dry by preventing dispersion	Polymer chain length	Soil particle encapsulation	Excavation tool design
	Can be cleaned in situ to reduce the amount of suspended particles before concreting	Density and sand content	Amount of suspended soil particles in the fluid	Waiting time after in-hole treatment
		Polymer chemical reactivity	Reaction with cleaning chemicals	
		Rheological properties	Particle-holding ability	

Fluid handling	Fluid performance can be maintained during the course of the project	Density	Amount of suspended soil particles in the fluid	Fluid tank volume; fluid maintenance procedures
		Residual polymer concentration	Amount of active polymer in the system	
	Enable solid concrete without inclusions	Rheological properties Sand content	Viscosity and gel strength Amount of suspended coarse particles in the fluid	Rheological properties of wet concrete; stability of the side wall (no collapsing soil)
	Does not react with fresh concrete	Polymer chemical reactivity	Possible reaction with concrete	Concrete (superplasticiser) chemical properties
Foundation loading	Does not reduce concrete bonding to reinforcement	Polymer chemical reactivity	Affinity with steel or rust	Rheological properties of wet concrete
		Fluid rheological properties	Flow around reinforcing bars during displacement by wet concrete	
	Give a clean concrete–soil interface at the pile shaft	Fluid density	Physical blocking of soil pores on the side wall surface by suspended soil particles	Soil type; stability of excavation; degree of soil softening; design of excavation tools
		Filter cake thickness	Seepage and filtration	
	Give a clean pile base	Fluid rheological properties	Particle-holding ability controlling the settling rate of soil particles	Base cleaning and checking procedures; discharge rate of the first load of concrete
		Filter cake thickness	Seepage and filtration	

support fluid should be able to bind to and encapsulate the exposed soil surface, to preserve the suction in the soil and hence the soil strength. When polymer fluid is used, the relevant properties are the length of the polymer chains and their chemical activity. Both of these also affect the viscosity of the fluid. The amount of active polymer in the fluid can also affect the degree of soil softening.

1.5.1.3 Keep excavated materials dry by preventing dispersion

The ideal support fluid should be able to keep the excavated materials dry by avoiding them becoming slurrified and useless as construction materials. Excavated materials maintained at their natural moisture content are more likely to be reusable as construction materials either on the same construction site or elsewhere. Since bentonite slurry is simply a clay–water suspension, it does not inhibit dispersion of soils; indeed, the sodium-rich environment of most bentonite slurries will promote soil dispersion. A bentonite fluid, therefore, will not inhibit the breaking up of clods of excavated soil into individual soil particles. On the other hand, some polymers can inhibit soil dispersion due to their ability to bind with the soil. This ability is influenced by the length of the polymer chains and their chemical activity. The amount of active polymer in the fluid is again a controlling factor.

1.5.1.4 Can be cleaned in situ to reduce the amount of suspended particles before concreting

Due to the action of the digging tool, some of the native soil particles will inevitably become suspended in the fluid during the excavation process. If these suspended particles are not removed before concreting, they will settle to the base of the excavation to form a soft base for the completed foundation element. In the industry, a soft base is commonly referred to as a ‘soft toe’. Suspended soil particles remaining in the fluid may also settle on the rising wet concrete during casting and become trapped as soil inclusions (e.g. Deese, 2004). To prevent the occurrence of soft toes and soil inclusions in concrete, the fluid must be cleaned prior to concreting. The ideal support fluid should allow in situ cleaning. In other words, it must be possible to clean the fluid, to an acceptable solids content, in the excavation without the requirement for external tanks and plant.

Since bentonite slurry cannot be cleaned in situ in an excavation, used slurry has to be exchanged for fresh slurry if the fine and coarse soil contents which are typically assessed by measurement of the fluid density and sand content, was greater than the allowable upper limits. In the UK, the upper limit for the sand content prior to concreting was commonly taken as 4% according to the Federation of Piling Specialists (FPS, 2006), although lower values may be required for some projects. In the ICE Specification for Piles and Embedded Retaining Walls (ICE, 2017), the sand content limit for bentonite slurries has been reduced to 2% with 4% permitted only when ‘cleanliness of the base and end bearing performance are not critical’. For deep excavations, the fluid exchange process can take several hours, as a long tremie pipe is required to remove the slurry from the bottom of the excavation and fluid volumes may be large. Polymer fluids, on the other hand, can be chemically treated in hole to promote the sedimentation of the suspended soil particles. Factors that can influence the efficiency of in-hole cleaning include the initial density and sand content of the fluid, the particle-holding ability of the fluid, the sizes of the suspended particles, the type and dosage of the cleaning chemicals, and the waiting time for particle settlement allowed or available after the introduction of the chemicals.

1.5.2 Fluid handling

1.5.2.1 Fluid performance can be maintained during the course of a project

During the course of a piling or diaphragm walling project, the support fluid recovered from the completed bores or trenches typically will be reused as many times as possible so as to minimise the amount of fresh fluid that has to be made up and the volume of used fluid that has to be disposed of. It is therefore important that the fluid is suitable for multiple uses. However, for both bentonite and polymer fluids there can be a problem of loss of active material. For bentonite fluids, bentonite particles will be lost to the filter cake with the fines in the fluid–solid separation (cleaning) plant. For polymer fluids, the amount of active polymers in the fluid will decline with use due to sorption onto soils. Additionally, support fluids also will be diluted by pore water from excavated soil dispersed into the fluid during digging.

Both types of fluid can also pick up fines from the ground that can be very difficult to remove. There are many case studies where the fluids became too heavy due to accumulation of fines and had to be disposed of completely. Therefore, the chosen support fluid must be designed for maintenance (reconditioning) during the course of the project, otherwise the quality of the completed foundation elements may suffer.

1.5.2.2 Enable solid concrete without inclusions

As discussed above, inclusions of excavated soil can occur if the support fluid is not of sufficiently low solids content prior to concreting. However, inclusions also can occur as a result of the intermixing of wet concrete and support fluid and/or failure of the concrete to displace the support fluid. To avoid intermixing of concrete and the support fluid, concrete is always cast via a tremie pipe; a single tremie will be sufficient for piles, but two to three pipes typically will be required for diaphragm wall panels. As concrete is placed in an excavation, the support fluid is withdrawn from the top, reconditioned for reuse and then returned to the storage tanks. During the concreting process, the support fluid in the excavation should not interfere with the flow of the wet concrete or reduce its quality. However, if the viscosity of the support fluid becomes too high due to accumulation of native soil particles, the flow of the wet concrete may be impeded. Problems with the fluid becoming too thick can occur with both bentonite and polymer-based fluids. The ideal support fluid should be able to be displaced by the rising concrete with as little intermixing and contamination as possible.

1.5.2.3 Does not react with fresh concrete

Previous research has shown that, despite the use of tremie pipes for placing concrete, a certain degree of intermixing between the wet concrete and the support fluid is still possible and is, in fact, almost unavoidable (Lam *et al.*, 2010). For this reason, the support fluid should not chemically react with the fresh concrete to reduce its quality, although if intermixing does occur there will be an inevitable increase in the water/cement ratio and thus some degradation of the concrete quality. The compatibility between the two materials will obviously depend on their respective chemical compositions, and hence on the type of support fluid and the choice of concrete admixtures such as superplasticisers. It is well recognised that chemical interaction between wet concrete and bentonite support fluid can significantly alter the rheological properties of the fluid, including viscosity, yield stress and thixotropy (Lubach, 2010). As a result, it is common industry practice to discard the support fluid from the fluid–concrete interface zone. Furthermore, to minimise the effect of the support fluid on the concrete, it is also common practice to over-pour the concrete so that the concrete in the interface region is either displaced from the trench or raised above the final trim level of the concrete. It should be noted that over-pouring deals only with the potential for near-surface defects. Reduced quality can still be present at greater depths if proper fluid management and tremie concreting procedures are not observed. Polymer fluids typically have a rather modest impact on concrete other than that resulting from any increase in the water/cement ratio should there be significant intermixing. Calcium and polyvalent cations from cement can precipitate many polymer products (e.g. Ata and O'Neill, 2000), so reducing the active polymer concentration in polymer fluids. However, the effect is typically modest, typically limited to the contact zone, and can be beneficial, as it can destroy any polymer film at the wall of an excavation – a film that might otherwise reduce interface friction.

1.5.3 Foundation loading

1.5.3.1 Does not reduce concrete bonding to reinforcement

Support fluids should not reduce the bonding between reinforcing steel bars (rebars) and concrete. A reduction in bond strength can occur for several reasons. Firstly, some chemical additives that have been used in polymer support fluids are known to show an affinity with the rust on the rebars and to form a film around them, and hence the use of these chemicals must be carefully controlled. Secondly, as rebar bond strength is related to the tensile strength of the concrete, the support fluid should not intermix or interact with concrete and so reduce its quality. Thirdly, the support fluid should be free of settling solids. Such solids can settle and remain on horizontal bars or horizontal ribs of deformed vertical bars. Lastly, as full rebar bond strength requires complete displacement of the support fluid by concrete during casting, the flow behaviour of the support fluid should not be such that it will impede the flow of wet concrete around the rebars, especially when they are ribbed.

1.5.3.2 Ensure a clean concrete–soil interface at the pile shaft

The ideal support fluid should not reduce the interface shear resistance between the foundation element and the adjacent soil. The interface resistance will depend on the ground conditions, the stability of the pile bore during excavation, the degree of soil softening at the walls, the side wall profiles and the thickness of the filter cake, if any. The thickness of a filter cake, in turn, depends on factors such the type and properties of the support fluid, the permeability of the soil, the excavation open time, and the choice of digging and hole-cleaning tools. In relation to hole cleaning, there have been cases where it was possible to improve shaft resistance by scraping the walls (e.g. Randolph, 2003) but on other projects scraping has achieved no improvement. Although concrete–soil interface resistance is a complex matter, the choice and correct use of support fluids has been found to be extremely important, as many of the potential issues are fluid related.

1.5.3.3 Ensure a clean pile base

Whether a support fluid can give a clean pile base will largely depend on the type of support fluid and, equally importantly, the base-cleaning procedures adopted on site. For fluid-supported excavations, base cleaning is necessary to prevent the occurrence of soft toes and to minimise the risk of settled solids becoming intermixed with the concrete. As discussed before, bentonite fluids with a sand content exceeding the specified limit have to be exchanged with fresh fluids, and polymer fluids can be cleaned in situ by using chemical additives. If done properly, both types of fluid should be able to ensure a clean pile base. Currently, the problem of 'soft toes' can be more an issue with polymer fluids, as their use requires special attention from contractors and a good understanding of how polymer systems work by the engineers. Unfortunately, this is currently not always the case, and hence it is the purpose of this book to discuss the best practice and pitfalls in the use of polymer support fluids.

REFERENCES

AASHTO (American Association of State Highway and Transportation Officials) (2010) *LRFD Bridge Construction Specifications*, 3rd edn. AASHTO, Washington DC, USA.

Ata A and O'Neill M (2000) The physicochemical interaction between PHPA polymer slurry and cement mortar. *Geotechnical Testing Journal* **23(2)**: 225–235.

BDA (British Drilling Association) (1992) *Drilling Technology*. British Drilling Association, Brentwood, UK.

Beart R (1845) *Apparatus for Boring in the Earth and in Stone*. English Patent 10 258.

Beresford JJ, Cashman PM and Hollamby RG (1989) The merits of polymeric fluids as support slurries. *Proceedings of the 3rd International Conference on Deep Foundations, London* (Burland J and Mitchell J (eds)). Balkema, Rotterdam, the Netherlands, vol. 1, pp. 3–10.

Brantly JE (1971) *History of Oil Well Drilling*. Gulf Publishing, Houston, TX, USA.

Brown DA, Turner JP and Castelli RJ (2010) *Drilled Shafts Construction Procedures and LRFD Design Methods*. US Federal Highway Administration, Department of Transportation, Washington, DC, USA, Report FHWA NHI-10-016.

Burland JB (1963) Contribution to discussion. *Grouts and Drilling Muds in Engineering Practice: Symposium*. Butterworth, London, pp. 223–225.

Chang MF and Zhu H (2004) Construction effect on load transfer along bored piles. *Journal of Geotechnical and Geoenvironmental Engineering* **130(4)**: 426–437.

Darley HCH and Gray GR (1988) *Composition and Properties of Drilling and Completion Fluids*, 5th edn. Gulf Publishing, Houston, TX, USA.

Day SR and Ryan CR (1999) State of the art in bio-polymer drain construction. In *Slurry Walls: Design, Construction, and Quality Control* (Paul DB, Davidson RR and Cavalli NJ (eds)). American Society for Testing and Materials, Philadelphia, PA, USA, pp. 333–343.

Day SR, O'Hannesin SF and Marsden L (1999) Geotechnical techniques for the construction of reactive barriers. *Journal of Hazardous Materials* **67(3)**: 285–297.

Deese GG (2004) *Slurry Sand Content and Concrete Interaction in Drilled Shaft Construction*. Master's thesis, University of South Florida, Tampa, FL, USA.

Fauvelle M (1846) On a new method of boring for artesian springs. *Journal of the Franklin Institute* **42(6)**: 369–372.

Fleming WK and Sliwinski ZJ (1977) *The Use and Influence of Bentonite in Bored Pile Construction*. Construction Industry Research and Information Association, London, UK, DOE and CIRIA Piling and Development Group Report PG3.

FPS (Federation of Piling Specialists) (1975) Specification for cast-in-place piles formed under bentonite suspension. *Ground Engineering* **8(2)**.

FPS (1977) Modifications to specification for cast-in-place concrete diaphragm walling (1973) and specification for cast-in-place piles formed under bentonite suspension (1975) *Ground Engineering* **10(8)**.

FPS (2000) *Bentonite Support Fluids in Civil Engineering*, 1st edn. FPS, Beckenham, UK.

FPS (2006) *Bentonite Support Fluids in Civil Engineering*, 2nd edn. FPS, Beckenham, UK.

Glossop M (1963) Opening address. *Grouts and Drilling Muds in Engineering Practice: Symposium*. Butterworth, London, UK, pp. 1–3.

Goodhue Jr KG and Holmes M (1995) *Earth Support Fluid Composition and Method for its Use*. US Patent 5 407 909.

Harris DP and Walker WR (1967) *Effect of Bentonite on the Bond between Steel Reinforcement and Concrete*. Construction Industry Research and Information Association, London, UK, CIRIA Interim Research Report 9.

Hodgson FT (1977) Design and control of bentonite/clay suspensions and concrete in diaphragm wall construction. *A Review of Diaphragm Wall: Seminar*. ICE Publishing, London, UK, pp. 79–92.

Holden JC (1984) The construction of bored piles in weathered sedimentary rock. *Proceedings of the 4th Australia–New Zealand Conference on Geomechanics, Perth*. Institution of Engineers Australia, Barton, Australia, pp. 378–384.

Hutchinson MT, Daw GP, Shotton PG and James AN (1975) The properties of bentonite slurries used in diaphragm walling and their control. *Diaphragm Walls and Anchorages: Conference*. Institution of Civil Engineers, London, UK, pp. 33–39.

ICE (Institution of Civil Engineers) (1996) *ICE Specification for Piling and Embedded Retaining Walls*. Thomas Telford, London, UK.

ICE (2007) *ICE Specification for Piling and Embedded Retaining Walls*, 2nd edn. Thomas Telford, London, UK.

ICE (2017) *ICE Specification for Piling and Embedded Retaining Walls*, 3rd edn. ICE Publishing, London, UK.

Ikuta Y (1974) Diaphragm walling in Japan. *Ground Engineering* **7(5)**: 39–44.

Kheng H, Bloomquist D, Townsend FC, Thompson PY and Dobson E (1991) *Slurry Performance in Drilled Shaft Construction*. Department of Civil Engineering, University of Florida, Gainesville, FL, USA, Final Report. (Available from Association of Drilled Shaft Contractors, Irving, TX, USA.)

Kramer H (1946) Deep cutoff trench of puddle clay for earth dam and levee protection. *Engineering News-Record* **136(26)**: 76–80.

Lam C, Troughton V, Jefferis S and Suckling T (2010) Effect of support fluids on pile performance. *Ground Engineering* **43(10)**: 28–31.

Lubach, AJ (2010) *Bentonite Cavities in Diaphragm Walls: Case Studies, Process Decomposition, Scenario Analysis and Laboratory Experiments*. MSc dissertation, Delft University of Technology, Delft, the Netherlands.

Nash JKTL and Jones GK (1963) The support of trenches using fluid mud. *Grouts and Drilling Muds in Engineering Practice: Symposium*. Butterworth, London, pp. 177–180.

O'Neill MW (1991) Effects of mineral and polymer slurries on drilled shafts. *Proceedings of the 16th DFI Annual Members' Conference*. Deep Foundations Institute, Hawthorne, NJ, USA.

O'Neill MW and Reese LC (1999) *Drilled Shafts: Construction Procedures and Design Methods*. Federal Highway Administration, US Department of Transportation, Washington, DC, USA, Report FHWA-IF-99-025.

Randolph, MF (2003) Science and empiricism in pile foundation design. *Géotechnique* **53(10)**: 847–875.

Sadleir NA and Dominioni GC (1963) Underground structural concrete walls. *Grouts and Drilling Muds in Engineering Practice: Symposium*. Butterworth, London, UK, pp. 189–195.

Tallard GR (1992) New trenching method using synthetic bio-polymers. In *Slurry Walls: Design, Construction, and Quality Control* (Paul DB, Davidson RR and Cavalli NJ (eds)). American Society for Testing and Materials, Philadelphia, PA, USA, pp. 86–102.

Veder C (1953) Method for the construction of impermeable diaphragms at great depth by means of thixotropic muds. *Proceedings of the 3rd International Conference of Soil Mechanics and Foundation Engineering, Zurich, Switzerland*, pp. 91–94. (In French.)

Veder C (1963) Excavation of trenches in the presence of bentonite suspensions for the construction of impermeable and load-bearing diaphragms. *Grouts and Drilling Muds in Engineering Practice: Symposium*. Butterworth, London, UK, pp. 181–188.

Xanthakos P (1979) *Slurry Walls*. McGraw-Hill, New York, NY, USA.

Zapico J (1991) *Diaphragm Wall Construction Gelatin Composition*. US Patent 4 990 608.

Polymer Support Fluids in Civil Engineering
ISBN 978-0-7277-5786-9

http://dx.doi.org/10.1680/psfce.57869.017

Chapter 2
Constraints on the use of bentonite support fluids

2.1. Introduction

Clay slurries such as bentonite fluids have been used for civil engineering works since the 1940s. As a result of this long history of use, contractors around the world had become very familiar with this construction material as early as the end of the 1970s. However, as they continue to compete with one another for projects and with ever more stringent technical, operational and environmental constraints, it has become apparent that bentonite fluids are no longer able to satisfy all the constraints of current foundation work and that more functionally designed systems, such as synthetic polymers, need to be considered.

In this chapter, the major limitations of bentonite fluids will be discussed. These include: (a) declining reserves and a perceived reduction in performance of some bentonites; (b) the constraints imposed by the requirement for vigorous mixing and significant bentonite hydration time; (c) the need for surface soil/slurry separation plant; and (d) disposal costs and possible environmental impact of used slurry. How polymers can be used to solve these problems will also be discussed.

2.2. The steady decline in the reserves and the perceived reduction in performance of some bentonites

Bentonite is a depleting natural material composed predominantly of clay minerals of the smectite (montmorillonite) group. In its natural state, bentonite is found mainly in calcium or sodium forms – the former being the most commonly occurring, but the latter more industrially important due to its higher swelling and absorptive properties. Today, sodium bentonite is used in many different industries, including cosmetics, foundry moulding, construction and waste disposal. Bentonite has therefore earned a name as the 'clay of 1000 uses'. Unsurprisingly, since its first commercial production in the late 19th century in the USA, production around the world has risen significantly, to a current level of between 11 million and 14 million tonnes a year, with the USA being the largest producer – supplying 30–40% of the world's total demand (British Geological Survey, 2011; US Geological Survey, 2016).

In terms of reserves, there are currently no reliable estimates, and different reports give very different figures. What is certain is that the amount of high-quality natural sodium bentonite, which is almost only mined in the State of Wyoming in the USA, has been in steady decline for many years (Hosterman and Patterson, 1992). The best reserves of sodium bentonite in the USA are said to be in the Bighorn Basin in Wyoming (Tannehill, 2015). Currently, most bentonites used for oil well drilling and construction are sodium-activated calcium bentonites produced by an ion exchange process. Some of these bentonites may also be treated with an extender such as carboxymethyl cellulose to increase slurry viscosity and reduce fluid loss. Today, many bentonites are significantly polymer modified at the time of production, although the user may not be aware of this. Polymer modification can enable bentonites of moderate activity to develop sufficient properties for use in excavation support and/or to develop systems with the enhanced rheological properties that may be required in horizontal directional drilling.

Despite their chemical treatment, the performance of sodium-activated and/or polymer-modified bentonites is often perceived as not matching that of their natural counterparts, which of course may themselves be modified during processing. In reality, bentonites are natural products, and the behaviour of materials from different sources must be expected to be different both in terms of the proportion of smectite clay and the performance of this clay (Bol, 1986). Even a single mine may produce clays of several grades. It is therefore important for the producer to blend and process bentonites to produce a consistent product.

Those using bentonite to form excavation support fluids should be aware of these differences and recognise that specifications tend to be local rather than international. Thus, specifications tend to be written around locally available bentonites in different parts of the world. Specifications from one area of the world should not be used in another unless the clay source is also respected.

An issue that is sometimes raised is that sodium-activated calcium bentonites will contain a significant amount (typically several per cent) of calcium carbonate. This is a by-product of the ion exchange process in which sodium carbonate is used to precipitate the calcium from the clay as calcium carbonate, thus allowing the sodium to enter the exchange sites of the clay. Calcium carbonate is of very low solubility, and there is little potential for the reaction to reverse in a bentonite slurry.

Concerns about declining reserves and the performance of available bentonites are one of the reasons why deep foundation contractors are seeking alternative support fluid systems, although the principal reason is more stringent performance requirements. Synthetic polymers are the ideal alternatives in this respect, as they are synthetic materials with, in theory, fully controllable properties. Parameters such as molecular weight, charge density and form (granular or emulsion) can be varied by the producer, although the user will not have any control of these parameters and indeed may have no knowledge of them. A downside of polymers is that the synthetic ones are made from petroleum oil – which is also a finite resource. However, there are polymers that are derived from natural renewable materials.

2.3. Mixing and hydration requirements of bentonite fluids

Because bentonite is a smectite clay, microscopically it consists of numerous platelets stacked one on another, much like a stack of playing cards. To create a support fluid with

Figure 2.1 A high-speed vortex mixer and storage tanks for small- to medium-scale projects using bentonite fluid

useful rheological and filter loss properties, it is necessary to disperse the platelets as much as possible during mixing. Jefferis (1982) showed that a poorly mixed slurry never quite develops as good rheological properties as better-mixed slurries. Therefore, to maximise the performance of a bentonite slurry, high-shear mixing should always be employed on site to promote dispersion of the platelets. Nowadays, most foundation contractors are aware of this requirement and commonly employ a high-speed mixer to prepare their slurries, but unfortunately it is not always possible or financially feasible for contractors working on smaller projects to do the same. In these cases, a small paddle mixer is often used, but lumps of non-dispersed bentonite can often be found at the base due to the lack of fluid agitation in this type of mixer.

Figure 2.1 shows a vortex mixer and slurry hydration tanks used for a medium-sized piling project in the UK. The bentonite powder was added manually to the swirling water inside the mixer, as shown in Figure 2.2. For projects requiring a large amount of slurry (>100 m^3), it is common to employ an automated mixing system and dry bentonite silos, as shown in Figure 2.3. The major disadvantage of a bentonite mixing system is that it requires a considerable amount of space on site, both for the actual mixing operation and for material storage. As will be shown later (see Chapter 5), polymer mixing is much simpler and involves much less material, as the dosage is typically 1–2 kg/m^3 for polymers compared with 40–60 kg/m^3 for bentonite, and the plant can be more basic.

Figure 2.2 Adding bentonite powder to a vortex mixer

Figure 2.3 An automated bentonite mixing system and silos for large-scale projects

In addition to the requirement of vigorous mixing, bentonite also requires considerable hydration time to develop its rheological properties. The Federation of Piling Specialists notes that after mixing the slurry is normally stored for at least 12 hours prior to use (FPS, 2006). However, recent experience has shown that up to 24 hours' hydration may be beneficial, to minimise flocculation of the bentonite particles in contaminated ground (Suckling *et al.*, 2011). The requirement for a hydration period plus the initial set-up of the bentonite mixing plant often means that the slurry has to be prepared well in advance of the start of an excavation. This can absorb valuable programme time for projects of short duration.

Polymer fluids, on the other hand, can be used immediately after mixing, as effectively no hydration period is required. This shorter time frame is a major reason why contractors increasingly favour polymer fluids over bentonite for projects with tight schedules or on small or congested sites.

2.4. Requirement of surface treatment plant

When bentonite fluids are used in an excavation, soil particles will accumulate in the slurry due to its relatively high viscosity and gelling (thixotropic) behaviour, although the gel strength can be reduced by shearing but will still rebuild on standing. If the suspended soil particles are not removed from the used bentonite fluid, they can

- thicken the slurry, so that it is not readily displaced by concrete
- settle from the slurry during concreting, so leading to inclusions in the concrete
- form a thick layer of filter cake on the excavation walls, so potentially reducing the cover to reinforcement
- build up to form soft sediment at the base of an excavation.

All of these effects are highly undesirable for the performance of foundation elements. To eliminate them, it has become common practice to use separate digging and concreting slurries unless, at the end of the excavation, the digging fluid can meet the relevant specifications (Chapter 10). The digging slurry will then be treated by a separation plant to remove the soil particles prior to reuse. Typically, a separation plant will have a relatively coarse scalping screen and then desanding and desilting hydrocyclones discharging onto a dewatering screen. A decanting centrifuge may also be necessary if the used fluid contains a significant amount of fine soil particles. Figure 2.4 shows the separation plant and storage tanks for a substantial diaphragm walling project in the UK.

It follows from the above analysis that, for many urban sites with limited space and restricted access, the size of the mixing and separation plant makes the use of bentonite challenging. In such cases, options for pile construction become the use of full-length casing or polymer fluids. Because the full-length casing option is more time consuming and may still require a considerable amount of on-site storage space, polymer fluids have become the preferred solution for many urban sites as they do not require any treatment plant. For example, Lennon *et al.* (2006) note that, for developments in central Glasgow, the size and cost of the temporary casings or the treatment plant required for bentonite fluids make both options relatively uneconomic, and polymer fluids are a practical

Figure 2.4 The bentonite cleaning plant employed for a diaphragm wall project in the UK. In the foreground from left to right: desilter, desander, desander (a different make) and decanting centrifuge. In the background: bentonite slurry storage tanks

alternative. In fact, for many urban redevelopment sites requiring bored piles, using polymer fluids was the only viable option, as the treatment plant required for bentonite fluids or the steel tubes required for the cased option would simply require too much working space.

The construction of the East Kent Ashford to Cheriton section of the Channel Tunnel Rail Link (Contract 440) is another example of polymer fluid use due to space restrictions (Anon., 2001). Some 32 m-deep bored piles were required for this project. Because the soils were water bearing, excavation support was required during piling to keep the bores open, but full-length temporary casings were not a feasible option due to the length of the piles. Bentonite fluids were considered, but the space restrictions on some of the sites led to the conclusion that setting up a slurry separation plant would be impracticable. Polymer fluid was therefore chosen for this project because it required none of the bulky and elaborate equipment that bentonite would need. The saving of time for site set-up was another associated advantage of using polymer fluid. Because support fluid was needed at 19 different locations – many of which required only a small number of piles – the compact nature of the polymer fluid set-up meant moving from site to site was relatively easy, whereas mobilising a full bentonite set-up for each location would have taken a considerable amount of the programme time.

2.5. Disposal of used fluids: environmental regulations and on-site treatment

The introduction of more stringent environmental regulations in recent years is making the disposal of used bentonite slurry more difficult and expensive around the world.

For example in the UK, the EU Landfill Directive (EC, 1999) was brought into force by The Landfill (England and Wales) Regulations 2002 (HMG, 2002). Since the end of the transitional period in 2007, liquid waste disposal has been banned at all UK landfill sites. According to a guidance document published by the UK Department for Environment, Food and Rural Affairs, liquid waste is defined as any waste that near instantaneously flows into an indentation void made in the surface of the waste, or any waste (load) containing free-draining liquid substance in excess of 250 litres or 10% of the load volume – whichever represents the lesser amount (DEFRA 2009). The first definition is used to distinguish between liquids and sludges, and the second is used to assess cases where liquid forms part of a load of solid waste such as milk cartons. By these definitions, bentonite slurries, even if heavily soil laden, will be classed as liquid wastes, but dewatered bentonite sludge following some form of on-site treatment need not be. Three possible methods can be used to treat used bentonite slurry according to the Federation of Piling Specialists (FPS, 2006), including (a) continuous belt filtration units, which produce a clay product with a solids content of approximately 65%; (b) decanting centrifuge units, producing a similar product to the preceding method and; (c) flocculation of the bentonite particles, followed by the addition of cement, to form a clay-like product, generally with a solids content of approximately 25%. Decanting centrifuges are the most commonly used on piling and diaphragm walling sites though cake solids contents as high as 65% may not be achieved with bentonite rich feed slurries. Belt filtration is not appropriate for short-term use on site; it is more appropriate to fixed set-ups or long-term use such as on long slurry tunnelling projects. The pH increase on cement addition can make disposal of the solidified material difficult, thus leaving the decanting centrifuge as the most practical option for deep foundation projects.

Figure 2.5 also shows a dewatered bentonite sludge collected from a diaphragm walling site in London. The sample was produced by a decanting centrifuge operating at a high speed to separate the water from the solids. It can be seen that the sludge did not flow into the indentation at the surface that was made with a palette knife, and thus is not a liquid as regards disposal to landfill in the UK. This is because the water content of this sample although about 100% was less than the liquid limit of sodium bentonite which is typically over 300% (although the soil–bentonite sludge will have a slightly lower liquid limit). The treated bentonite sludge did not release any free-flowing material, and thus it could be disposed of as a solid waste in the UK. However, the required on-site treatment tends to be economical only for large-scale projects involving several thousand cubic metres of slurry due to the relatively high set-up and rental costs of the treatment plant.

Despite the possibility of on-site treatment, tankering away used bentonite slurry by a licensed carrier is currently still the most cost-effective and sometimes the only possible option for many small- to medium-scale excavation projects. The cost of this option, however, has risen dramatically over the last decade in the UK (Wheeler, 2003), and is mainly due to the increasingly more stringent environmental regulations including the Environmental Protection Act 1990 and the EU Landfill Directive. Before the introduction of these regulations, the cost of disposing bentonite was low in the UK because

Figure 2.5 Bentonite sludge from a decanting centrifuge

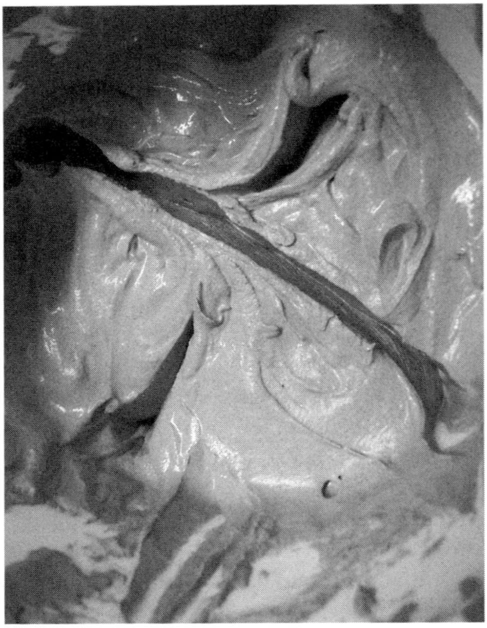

many farmers across the country would accept bentonite slurry and use it to improve the quality of their soil. In fact, the benefits of using bentonite in agriculture have been confirmed by various reports, and it is a retrograde step that by regulation it is now a waste. It has also been shown that bentonite can be used in arid regions to improve the water-holding capacity of degraded or sandy soils to improve crop yields (IWMI, 2009); however, if the soil is over-treated, then the high level of salts from the slurry can have the opposite effect and prevent plant growth (Moseley, 1983). Nevertheless, as in the UK, disposing of bentonite slurry even on private lands is now generally prohibited by law in many countries. However, depending on the jurisdiction and the nature of the project, it can be possible to obtain a bespoke environmental permit from the relevant authority to spread used slurry direct to land without treatment (Environment Agency, 2012). For example, a bespoke permit was obtained for the London Gateway project. This permit allowed the disposal of used bentonite slurry from diaphragm wall construction by mixing it with previously placed fill material (Environment Agency, 2011).

Contractors outside Europe are also facing similar restrictions. For example, in Hong Kong it is common practice to mix used bentonite slurry with dry soil on site before disposal in a public filling area or, as the last resort, at a marine spoil ground, but this will require a marine dumping licence from the Environmental Protection Department (EPD, 1994). In the USA, Oliver (2015) noted that many states still allow spreading unwanted drill spoils over wide areas on lands (a practice known as landfarming), as long as approval from regulating agencies and the landowner is obtained. Oliver (2015) also mentioned that, over the years, used drilling fluids have been left at the drill site to solidify on their own, and, sometimes, drill cuttings, soil, sawdust, concrete or lime have been added to speed up this process. However, Oliver commented that the use of these materials does not always prevent leaching of drilling fluids into surrounding soils. With the new tougher regulations introduced at federal and state levels, these practices are not as acceptable as they once were in the USA. In Brazil, Paschoalin Filho *et al.* (2013) noted how the increasingly stringent restrictions imposed by Brazilian environmental agencies on the use of bentonite slurry have created a demand for new technologies for the stabilisation of excavations.

In summary, the rising cost of bentonite disposal due to the introduction of more stringent environmental regulations has been an important driver for contractors to seek alternatives to reduce costs. Polymer fluids are ideal in this respect, as the polymer molecules can be easily broken down on site using simple chemicals, and the treated fluids can be disposed of to the sewer system with the water undertaker's consent. Details about polymer fluid disposal will be discussed in Chapter 5.

2.6. Environmental impact

Although clean, unused bentonite slurry would be classed as non-hazardous waste according to the current regulations in the UK if it were to be disposed of, it can still be highly polluting if it is released into the aquatic environment. In the USA, O'Neill (1991) noted that polymer fluids can be permitted for excavations which penetrate aquifers, whereas at many locations clay-based slurries are not permitted. For projects near watercourses, bentonite fluids are also deemed unsuitable because they can pose a danger to fish by building up on their gills and causing suffocation. For example, Schünmann (2004) reported a case study in Essex, UK, where bored piles were required next to a river for a new crossing and any pollutants that reached the river would then flow on to a site of special scientific interest. Polymer fluid was favoured over bentonite since if spilled into the river the former is less likely to cause harm to the environment. Using bentonite fluid would also have increased the construction cost because a physical barrier would have been mandated to surround the bentonite operations in order to minimise the environmental risk. Similar case studies on the environmental impact of bentonite can be found in other countries. For example, in the USA, Holmes (1993) reported the construction of fluid-supported bored piles for a new bridge across the Ohio River near Owensboro, Kentucky. Firstly, there were concerns about the environmental impact of the support fluids on the mussel beds immediately downstream from the bridge site. It was also felt that discharges of conventional mineral slurry or of fluids made from oil-bearing emulsion polymers might adversely impact on the river ecosystem. Secondly, the costs of off-site disposal to a landfill or similar facility would have been significant. A support fluid system prepared using dry granular polymers, which are free of oil and are non-toxic, was therefore chosen for the bridge project, as it presented an environmentally sound alternative. In Japan, the increase in waste bentonite slurry has become a serious problem due to its negative environmental impact and the lack of disposal sites (Uchiyama and Horiuchi, 1994). In Bangkok, Thasnanipan *et al.* (2003) noted that in most cases the primary reason for switching to polymers was to

Figure 2.6 A stockpile of excavated material from a polymer-supported pile bore

minimise the environmental issues associated with bentonite. In Portugal, Caputo (2009) also expressed concerns about the environmental impact of bentonite fluids for the bored piles for a new bridge across the Tagus River.

2.7. Soil dispersion

Another limitation of bentonite fluids is that they do not inhibit the dispersion of soils; that is, they do not inhibit the breaking up of clods of excavated soil into individual soil particles. Therefore, in practice there is usually little prospect for the arisings to be reused because of their high moisture content. On the other hand, as some polymers have the ability to encapsulate excavated materials and retain their natural moisture contents (Zumsteg *et al.*, 2016), the arisings can be reused and need not be disposed of as wet spoil to landfill. Even in sands, the excavated material may appear no wetter than originally in the ground; the soil behaves as if wrapped in a plastic bag, with little penetration of the polymer fluid into the soil. Figure 2.6 shows a stockpile of gravelly clay (Lambeth Group) removed from a polymer-supported pile bore at a construction site in London. It can be seen that the soil remained relatively dry, and this property allowed it to be reused elsewhere on the same site. Paschoalin Filho *et al.* (2013) made a similar comment on the arisings from diaphragm wall trenches stabilised by polymer fluid at a site in São Paulo, Brazil. They also noted that arisings from bentonite-supported trenches had a higher level of moisture, and this made it difficult to transport and dispose of the excavated materials.

REFERENCES

Anon. (2001) Way out of a tight spot. *Ground Engineering* **34(5)**: 16–17.

Bol GM (1986) Bentonite quality and quality-evaluation methods. *SPE Drilling Engineering* **1(4)**: 288–296.

British Geological Survey (2011) *World Mineral Production 2005-09*. British Geological Survey, Keyworth, UK.

Caputo A (2009) Instrumented large diameter bored piles. *Proceedings of the 5th International Symposium on Deep Foundations on Bored and Auger Piles*. CRC Press, Leiden, the Netherlands, pp. 213–227.

DEFRA (Department for Environment, Food and Rural Affairs) (2009) *Environmental Permitting Guidance – The Landfill Directive*. DEFRA, London UK.

EC (European Community) (1999) Directive 1999/31/EC of the European Parliament and of the Council of 26 April 1999 on the landfill of waste. *Official Journal of the European Communities* **L182/1**.

Environment Agency (2011) *Regulatory Position Statement – The Use of Waste Bentonite for Engineering Purposes as Part of the Creation of a Dock*. Environment Agency, Bristol, UK.

Environment Agency (2012) *Position Statement – Regulating the Recovery of Waste Drilling Muds*. Environment Agency, Bristol. UK.

EPD (Environmental Protection Department) (1994) *Practice Note for Professional Persons – Construction Site Drainage, ProPECC PN 1/94*. EPD, Hong Kong Government, Hong Kong.

FPS (Federation of Piling Specialists) (2006) *Bentonite Support Fluids in Civil Engineering*, 2nd edn. FPS, Beckenham, UK.

HMG (Her Majesty's Government) (2002) Environmental Protection, England and Wales. The Landfill (England and Wales) Regulations 2002. The Stationery Office, London, UK.

Holmes MM (1993) Drilled shafts and new slurry technology used instead of driven piles: sets record on the Ohio River. *Foundation Drilling* **32(5)**: 13–15.

Hosterman JW and Patterson SH (1992) *Bentonite and Fuller's Earth Resources of the United States*. US Geological Survey, Denver, CO, USA.

IWMI (International Water Management Institute) (2009) *Economic Gains of Improving Soil Fertility and Water Holding Capacity with Clay Application: The Impact of Soil Remediation Research in Northeast Thailand*. IWMI, Colombo, Sri Lanka, Research Report 130.

Jefferis SA (1982) Effects of mixing on bentonite slurries and grouts. *Proceedings of the Conference on Grouting in Geotechnical Engineering*. American Society of Civil Engineers, New York, NY, USA, pp. 62–76.

Lennon DJ, Ritchie D, Parry GO and Suckling TP (2006) Piling projects constructed with vinyl polymer in Glasgow, Scotland. *Proceedings of the 10th International Conference on Piling and Deep Foundations, Amsterdam*. Deep Foundations Institute, Hawthorne, NJ, USA, pp. 499–506.

Moseley Jr HR (1983) Summary of API onshore drilling mud and produced water environmental studies. *Proceedings of the IADC/SPE 1983 Drilling Conference*. Society of Petroleum Engineers, Houston, TX, USA, pp. 321–325.

Oliver B (2015) Solid answers to questions about drilling fluids disposal. *National Driller* **36(2)**: 28.

O'Neill MW (1991) Effects of mineral and polymer slurries on drilled shafts. *Proceedings of the 16th DFI Annual Members' Conference, Chicago*. Deep Foundations Institute, Hawthorne, NJ, USA, pp. 141–154.

Paschoalin Filho JA, Guerner Dias AJ, Brum de Lima Duarte E (2013) The use of polymer slurry as an economic and sustainable way to build diaphragm walls – a case study of a construction work in São Paulo/Brazil. *Exacta* **11(3)**: 299–306.

Schünmann D (2004) Fisherman's friend. *Ground Engineering* **37(12)**: 17.

Suckling TP, Lam C, Jefferis SA and Pantling CJ (2011) Evaluation of causes of bleeding of free water from a bentonite slurry. *Proceedings of the 36th Annual Conference on Deep Foundations*. Deep Foundations Institute, Hawthorne, NJ, USA, pp. 401–407.

Tannehill T (2015) What happens when we run out of bentonite? *National Driller* **36(5)**: 32.

Thasnanipan N, Aye ZZ and Boonyarak T (2003) Behaviour of polymer-based slurry for deep-seated bored piles in multi-layered soil of Bangkok. *Proceedings of the 4th International Geotechnical Seminar on Deep Foundations on Bored and Auger Piles (BAP IV), Ghent*. Millpress, Rotterdam, the Netherlands, pp. 267–274.

Uchiyama N and Horiuchi S (1994) In-situ utilization of waste bentonite slurry. *Environmental Aspects of Construction with Waste Materials: Proceedings of the International Conference on Environmental Implications of Construction Materials and Technology Developments*. Elsevier, Amsterdam, the Netherlands, pp. 217–226.

US Geological Survey (2016) USGS National Minerals Information Center. http://minerals.usgs.gov/minerals/ (accessed 15/08/2016).

Wheeler P (2003) Clear as mud. *New Civil Engineer*, 17 Apr.: pp. 22–24.

Zumsteg R, Puzrin AM and Anagnostou G (2016) Effects of slurry on stickiness of excavated clays and clogging of equipment in fluid supported excavations. *Tunnelling and Underground Space Technology* **28**: 197–208.

Polymer Support Fluids in Civil Engineering
ISBN 978-0-7277-5786-9

http://dx.doi.org/10.1680/psfce.57869.029

Chapter 3
Introduction to polymers and polymer fluids

3.1. Introduction
This chapter introduces basic concepts of polymers and their solutions (polymer fluids), which are essential to the understanding of the following chapters. No prior knowledge of polymer science is required, although it is assumed that the reader will have received some basic training in chemistry to secondary school level. Where appropriate, everyday examples are used to help introduce the subject.

3.2. What is a polymer?
3.2.1 Polymers in everyday life
Polymers are chemical compounds made up of large molecules with repeating units called monomers. Some polymers are soluble in water, but many are not. Most synthetic polymers (e.g. polyethylene, used for plastic bags) cannot be readily broken down, but natural polymers such as cellulose in green plants are often biodegradable; that is, they can be broken down to smaller, simpler molecules by microbes. The ultimate degradation products will include carbon dioxide and water. Many well-known materials are polymers, for example PVC (polyvinyl chloride), Teflon (PTFE, polytetrafluoroethylene) and DNA (deoxyribonucleic acid).

Polymer fluids, which are aqueous solutions of polymeric compounds, are literally part of our everyday life because the human body produces polymer fluids to help us function. For example, saliva is a natural polymer fluid, as it contains carbohydrate molecules known as mucopolysaccharides, which can also be found in other parts of the body such as in the synovial fluid inside the cavities of joints. In any solution the molecules forming the solution will be in constant movement, and this is equally true for the macromolecules of polymers. For large polymer molecules, this movement leads to hydrodynamic and molecular interactions that can give rise to some special fluid behaviours such as the stringiness and sliminess that we commonly observe in bodily fluids. In the food industry, xanthan gum, which is a polysaccharide, is commonly used as a thickening agent (food additive code E415) in the production of low-fat mayonnaise, ice cream and salad dressings. It is the presence of xanthan gum that prevents the formation of ice crystals in ice cream, giving it a 'smooth' texture. Natural polymers such as xanthan gum and carboxymethyl cellulose (CMC) are also widely used in household products such as toothpaste, lotions, creams and cleaning fluids – to name just a few. It is therefore no exaggeration to say that polymer fluids surround us in our daily life.

3.2.2 Polymers in civil engineering

In civil engineering, polymer fluids have found their ways into many different applications. For example, polymer fluids based on the chemistry of lignosulfonates, sulfonated melamine formaldehyde (SMF) and sulfonated naphthalene-formaldehyde (SNF) are all commonly used as water reducers or superplasticisers to make high-workability and self-consolidating concrete. To lubricate and prime concrete pumps and hoses on construction sites, liquid polymer primers are used to coat the internal surfaces of these components in order to reduce friction and prevent blockages.

Of course, polymer fluids are also commonly used in geotechnical construction, and it is the purpose of this book to discuss this last category of use. At present, a major issue in the use of polymer support fluids is that no differentiation is made between the different types of polymer products that are available in the marketplace. Unlike bentonite, which is basically sodium or calcium montmorillonite clay, polymers are much more complex, both in their chemical composition and in their variety. As chemistry is not a core subject in many civil engineering curricula, few engineers in the foundation industry currently understand the difference between the various polymer systems and their relative strengths and weaknesses. Many engineers are scared off the subject when they hear chemical terms such as 'polyacrylamide' or 'carboxymethyl cellulose', even though products containing these polymers are commonly used in civil engineering. As chemical information is seldom provided, and when provided, rarely understood, most engineers currently recognise only two types of support fluid: bentonite and polymers. This lack of awareness of the variety of polymers is undoubtedly a major obstacle to the further advancement of the technology, since without an understanding of what polymers are and how they behave, it is impossible to learn from past experience and move forward – if there is a problem with one type of polymer, it is assumed that all polymers will show the same defect. It is therefore the aim of this chapter to demystify polymers by introducing readers to the various types of polymers commonly used in civil engineering, especially in geotechnical construction.

3.3. Basic polymer science

3.3.1 Nomenclature

For many civil engineers, the names of polymers can be very confusing because they often contain unfamiliar chemical terms linked together to make a long name such as 'partially hydrolysed polyacrylamides'. However, there are some simple rules that can help the understanding. After many years of standardisation within the chemical industry, polymers are now commonly named according to the rules developed by the International Union of Pure and Applied Chemistry (IUPAC, 2009). For many simple polymers, the name of the polymer is just the name of the monomer prefixed by the word 'poly'; that is, the name of the polymer taking the name of its source. For example, polyethylene consists of ethylene molecules ($CH_2{=}CH_2$, the monomer) linked together, and polyvinyl chloride (PVC) is formed by linking vinyl chloride molecules ($CH_2{=}CHCl$). To avoid confusion, proper naming practice also requires brackets to be put around the name of the monomer if it contains two or more separate words. Therefore, 'polyvinyl chloride' strictly should be written as 'poly(vinyl chloride)', although both usages can be found in the literature. However, no brackets are required for names such as 'polytetrafluoroethylene' (Teflon) because 'tetrafluoroethylene' (C_2F_4) is a single word.

The chemical term 'vinyl' has become a household name due to widespread use of products such as vinyl records and vinyl flooring (both are made of PVC). The term 'vinyl' is used for a family of monomers, all which have the chemical formula CH=CHR, where R may represent any pendant or side group. Thus, in vinyl chloride, the R represents chlorine (Cl). It follows that any polymer containing the vinyl linkage ($CH_2{=}CH^-$) can be described as a vinyl polymer. The term 'vinyl' is highlighted here because polymers with a vinyl linkage are widely used in polymer support fluids.

Although the IUPAC rules recommend the prefix 'poly', there are a significant number of polymers whose names do not start with this prefix. For example, carboxymethyl cellulose (CMC), which is a natural polymer commonly used as an additive in bentonite slurry or on its own as a support fluid, gets its name from the carboxymethyl side groups (CH_2COO^-) attached to a cellulose backbone. Lignosulfonates (sulfonated lignins) are water-soluble anionic polymers commonly used as a superplasticiser in making concrete, and acquire their name from the sulfonic group ($-SO_3H^-$) and lignin, which is a chemical compound found in wood.

A polymer can also be named or classified according to its chemical structure. For example, polymers that contain repeating units of a single type of monomer are called 'homopolymers', and those that contain two or more types of repeating units are called 'copolymers', as shown in the following examples for monomers A and B:

Homopolymer:	—A—A—A—A—A—A—A—
Copolymer (alternating):	—A—B—A—B—A—B—A—
Copolymer (random):	—A—B—B—A—B—A—A—

As can be seen, copolymers can be further classified as 'alternating' or 'random', depending on the occurring pattern of the monomers. Other types of copolymers exist, but further discussion is not required for the present purposes. Interested readers are referred to polymer science textbooks such as that by Sperling (2006) for further information.

The difference between natural and synthetic polymers must be made clear here, as these two terms are often used to describe the polymers used in civil engineering. In essence, synthetic polymers are artificial polymeric compounds derived from petroleum oil. Nylon, Teflon and PVC are all examples of synthetic polymers. Natural polymers, on the other hand, are those that occur in nature. DNA, cellulose and starch are all natural polymers. Polymers that are derived or extracted from natural materials are also classified as natural in this context, and one such example is CMC, which is derived from cellulose by reactions with chemicals. Although natural polymers tend to be biodegradable, the terms 'natural' and 'biodegradable' are not synonymous. Some synthetic polymers are biodegradable, and indeed may be designed to be. Synthetic polylactides (PLAs) are a good example of this category of polymer. Finally, there are some polymers that can be natural or synthetic, such as latex, from which rubbers are manufactured.

Polymer macromolecules can carry a charge, and very often do. Polymers where the macromolecule carries a negative charge are described as anionic, and those with a positive charge as cationic. The charge on the polymer must be balanced by ions of opposite charge. Very often the charge on anionic polymers will be balanced by sodium, and that on cationic polymers by chlorine. The polymer charge can have a major impact on the behaviour of polymers in solution.

3.3.2 A simple example: ethylene and polyethylene

At the start of this chapter, polyethylene was cited as the polymer used for making plastic bags. This polymer is used here to demonstrate some of the basic elements of polymer nomenclature as described above. Polyethylene has the chemical formula $(C_2H_4)_n$, and is created by joining ethylene (strictly, the IUPAC name is now ethene) molecules (C_2H_4) together. The ethylene molecules, which are the building blocks of the polymer, are called the 'monomer', and the process of joining them together is called 'polymerisation'. The symbol n in the chemical formula $(C_2H_4)_n$ represents the number of monomer units in the polymer, which may range from as few as two to many millions. The words 'monomer' and 'polymer' are derived by combining the relevant Greek words: *μόνος* (*mono*), meaning 'single'; *πολύς* (*polus*), meaning 'much' or 'many'; and *μέρος* (*meros*), meaning 'parts'. It is worth noting that for a chemical to function as a monomer, it must have at least two reactive sites to which other monomers can be joined. For example, the carbon–carbon double bond (C=C) of ethylene, if broken to form a single bond, releases two reactive sites, thus allowing the monomer unit to be joined with others as shown in the equation

$$n(CH_2{=}CH_2) \rightarrow (-CH_2-)_n \tag{3.1}$$

The value of n controls the molecular weight of the polymer, and so the degree of polymerisation will have a significant effect on the physical properties of the polymer. However, rather than specifying the numeric value of n, it is common to specify the molecular weight or the density of the product, as this can be measured more easily. In the case of polyethylene, if the molecular weight of the polymer is high, the product will have a higher density and material strength. High-density polyethylene (HDPE), which has an n value of between 10 000 and 100 000 and a molecular weight of between 2×10^5 and 3×10^6, is the material used for making plastic pipes and geomembranes for construction. In low-density polyethylene (LDPE), the molecules are more than 100 times smaller than their high-density counterparts. LDPE is commonly used for making plastic bags and wrapping films. Both HDPE and LDPE are recyclable, and their plastic identification codes are 2 and 4, respectively (SPI, 1988).

3.3.3 Carbohydrates

Carbohydrates are the most abundant organic compounds in living organisms, and are the building blocks of all natural polymers, including those used for civil engineering works. It is therefore necessary to discuss their structures here to aid the understanding of other sections in this book. In essence, carbohydrates are compounds consisting of carbon, hydrogen and oxygen; the hydrogen:oxygen ratio is usually 2, and so the chemical formula of carbohydrates can be written as $C_m(H_2O)_n$, where m can be

Figure 3.1 Chemical structures of three monosaccharides found in natural polymers

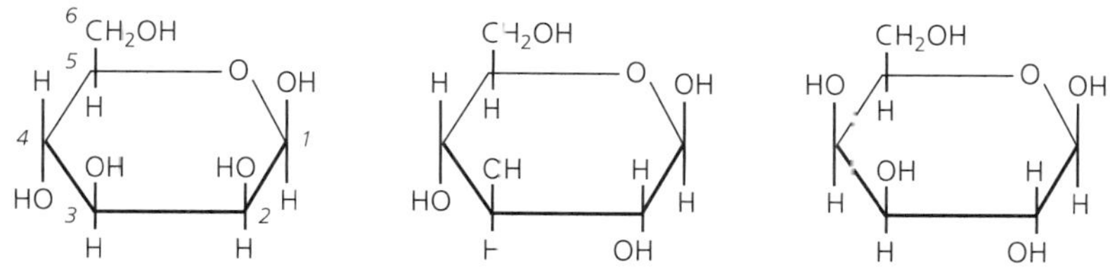

different from n. Carbohydrates are also called saccharides, which comes from the Greek word σάκχαρο (*sákcharo*), meaning 'sugar'. Many simple carbohydrates are sugars. In terms of classification, the most basic saccharides are known as monosaccharides, which are the building blocks of more complicated compounds. The combination of two monosaccharides is a disaccharide, and long molecules consisting of a large number of monosaccharides are known as polysaccharides. Many natural polymers are polysaccharides.

Figure 3.1 shows three monosaccharides commonly found in natural polymers: mannose, glucose and galactose. Note their names all end with '-ose', as this suffix is used in chemistry to form the names of sugars. Other examples include cellulose, lactose and sucrose – the last of these is table sugar. In the ring form representation shown in Figure 3.1, there is a carbon (not shown) at each vertex of the ring except for the one occupied by oxygen. There are a total of six carbon atoms in these sugar molecules, and they are called hexoses. As shown in the structure for mannose, the carbon atom at the far right vertex is labelled as the first C-1 carbon and the next carbon atom clockwise around the ring is labelled C-2, and so on. Although these molecules all have a common chemical formula ($C_6H_{12}O_6$), as can be seen in the figure the orientation of the hydroxyl (—OH) groups is different. More specifically, mannose differs from glucose in respect of the orientation of the hydroxyl group on C-2, and glucose differs from galactose with respect to the orientation of the hydroxyl group on C-4. While this level of chemical detail may seem unnecessary, it is actually essential for the understanding of the different types of natural polymers discussed later.

Lastly, a homely example, although not a civil engineering one: it may be noted that the sugar level in the human body is determined by the amount of glucose in the blood, and that lactose ($C_{12}H_{22}O_{11}$), which makes up around 2–8% of milk, is a disaccharide derived from galactose and glucose. Hopefully, these facts will make the study of polymer chemistry more interesting to civil engineers.

3.3.4 Dry powder and liquid emulsion forms

Similarly to how simple chemicals such as sugar and salt are supplied as solids, most water-soluble polymers are supplied in dry granular or powder form. This allows for easy storage and long shelf life with the added benefit of having the product in a highly concentrated form. However, some polymers are also supplied in liquid emulsion form for various reasons such as easy and immediate dissolution in water – dry powders can be difficult to disperse in water, and may lump and form 'fish eyes' that do not easily dissolve. The term 'emulsion' requires further clarification as it is not a term commonly used in civil engineering. Essentially, an emulsion is a mixture of two or more liquids that are immiscible (e.g. oil and water) but one liquid is dispersed in the other. Milk is a natural oil-in-water emulsion containing about 87% water and 13% fat. In fact, the word 'emulsion' actually comes from the Latin word *ēmulsiōn*, meaning 'to milk'. Milk is stable because it also contains proteins, which are natural emulsifiers and coat the fat droplets to allow them to stay dispersed in the water phase of the milk. These proteins can act as emulsifiers, as one end of the protein molecule is hydrophilic (water liking) and the other is lipophilic (fat liking). This enables the molecule to coat the surface of oil droplets and effectively insulate them from the water, thereby stopping them from coalescing and separating from the water. Natural rubber latexes, which have been harvested from Pará rubber trees for hundreds of years, are another example of natural emulsions. These latexes consist of colloidal isoprene polymer droplets dispersed in water, hence the milky colour of the liquid.

The examples discussed above are oil-in-water emulsions; but the emulsion polymers commonly used in civil engineering are water-in-oil emulsions known as inverse or invert emulsions. In these emulsions the polymer-containing water droplets are dispersed in a continuous oil phase. Emulsion polymers are manufactured by first developing a homogeneous network of water droplets surrounded by surfactants (emulsifiers) in order to keep the droplets stable in the continuous oil phase. Monomers and catalysts are then introduced into the system, initiating millions of individual reactions within each water droplet. At the end of the polymerisation process, the system will consist of numerous dispersed water droplets, containing dissolved polymer molecules, in a continuous oil phase. Importantly, the viscosity of an emulsion polymer is primarily controlled by the viscosity of the continuous oil phase because the water droplets are isolated one from another. To allow easy handling of the emulsion, low-viscosity oil is usually used for the continuous phase. For use, the emulsion is mixed with the required quantity of water which allows the long-chain polymer molecules to uncoil and swell, thereby developing the required fluid behaviour. An important feature of emulsion polymers is that the polymer molecules are already dissolved in water, albeit a very small amount of water – the droplets. Therefore, there is no potential for lumping when more water is added, as could occur with solid polymer granules.

3.4. Basic classification of polymers of interest for use as support fluids

Over the short history of polymer support fluids, many different types of polymers have been tried with varying degrees of success. This section discusses the basic types of polymer that are commonly used in geotechnical construction. Four types of polymer will be considered, namely cellulose and its derivatives, guar gum, xanthan gum and polyacrylamides. Figure 3.2 shows photographs of these products, as supplied, in granular or powder form. As can be seen, it is rather difficult to differentiate them with the naked eye. Although it is possible to distinguish them by microscopy, such an exercise will still give no information on the polymer type, chemistry, etc. Microscopic

Figure 3.2 Basic types of polymer commonly used in geotechnical construction

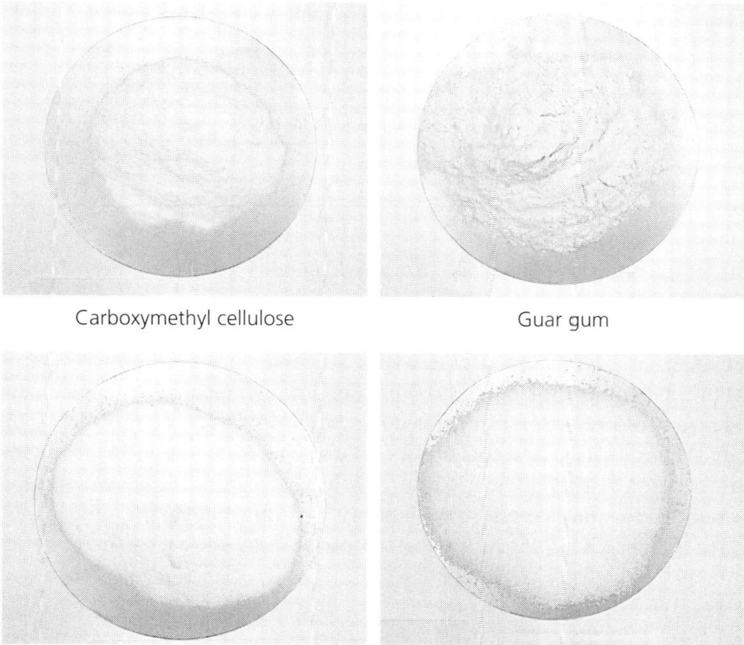

Carboxymethyl cellulose — Guar gum — Xanthan gum — Partially hydrolysed polyacrylamide

investigation of solid particles, however, can sometimes be useful in identifying whether a material as supplied contains more than one type of particle and thus potentially more than one system component.

Polymer fluids, both natural and synthetic, have been used for oil well drilling and reservoir flooding since the early 1950s, and a vast amount of information about polymers is available in the oil and gas literature (e.g. Caenn *et al.*, 1989; Chatterji and Borchardt, 1981; Darley and Gray, 1988; Littmann, 1988; Sorbie, 1991). The purpose of this section is not to replicate what has already been published elsewhere but to distil the information down to a level that is useful to a civil engineer using polymer fluids. The applications of the various polymer types in oilfields also will be briefly discussed, as this prior experience, although gained in a different industry, will help illustrate the strengths and weaknesses of the different polymers.

3.4.1 Cellulose and its derivatives

Cellulose is an insoluble organic compound that exists naturally in the primary cell wall of plants. It is the most abundant organic polymer in nature and accounts for at least one-third of all vegetable matter on the Earth. For example, cotton fibres are 90% cellulose, and wood is around 50% cellulose. Cellulose has the chemical formula $(C_6H_{10}O_5)_n$ and the molecular structure shown in Figure 3.3.

Figure 3.3 The chemical structure of cellulose consisting of repeating anhydroglucose units

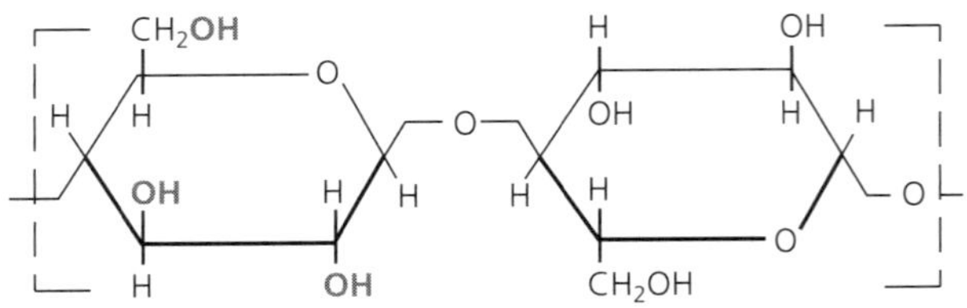

The fundamental repeating block is an anhydroglucose unit (i.e. dehydrated glucose), which contains three hydroxyl (OH^-) groups. There are many hydroxyl groups in the cellulose molecule and they are capable of forming hydrogen bonds, which occur when a hydrogen atom in a molecule interacts with a nearby atom such as oxygen. Hydrogen bonds within a cellulose molecule and with neighbouring cellulose molecules make for very strong intermolecular and intramolecular bonding. Water cannot break these bonds to dissociate the molecules, and this makes cellulose insoluble in water. This is the reason why wood does not dissolve when it rains. However, water-soluble cellulose derivatives can be formed by reacting cellulose with alkali (sodium hydroxide) and sodium monochloroacetate ($ClCH_2COONa$) in a two-step process, as shown in the equations

$$ROH + NaOH \rightarrow RONa + H_2O \tag{3.2}$$

$$RONa + ClCH_2OCOONa \rightarrow ROCH_2COONa + NaCl \tag{3.3}$$

where R represents $C_6H_9O_4$, which is the part of the polymer molecule that does not participate in the chemical reaction. It can be seen from Equations 3.2 and 3.3 that, as a result of the reaction, the hydrogens of the hydroxyls are substituted by the sodium carboxymethyl groups ($CH_2COO^-Na^+$) to form sodium CMC, an anionic polymer. Although in theory all three hydrogen atoms of the hydroxyl groups (OH^-) on the anhydroglucose unit can be replaced by sodium in the reaction, in practice the reaction is controlled in such a way that only a partial substitution occurs. To express it numerically, the average number of hydroxyl groups subject to substitution per anhydroglucose unit is referred to as the degree of substitution (DS). As the DS required to achieve optimal solubility and other desirable properties is much less than 3, for commercial use the DS value of CMC typically ranges from 0.7 to 1.2 because the most desirable properties are obtained in this range. For example, a CMC with a DS value of 0.7 means an average of seven carboxymethyl groups per ten anhydroglucose units. Figure 3.4 shows the structure of a CMC with a DS value of 1, and thus there is a carboxymethyl group in every anhydroglucose unit. Uniformity of substitution along the cellulose molecules is also important in the development of desirable polymer properties.

When selecting a CMC product, one needs to be aware that higher degrees of substitution generally give improved compatibility with other soluble species, such as salts. Just as for bentonite, polymer systems can be damaged or their performance reduced by polyvalent ions such as those of calcium (Ca^{2+}), so salt tolerance is a key

Figure 3.4 The chemical structure of CMC with a DS of 1

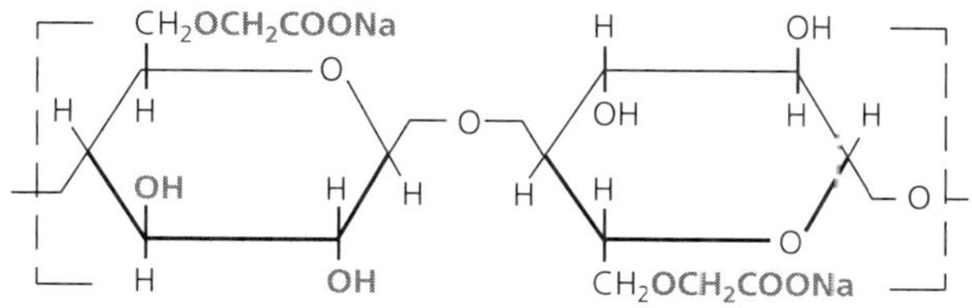

consideration when selecting polymers. Higher DS values can also reduce biodegradability (Wirick, 1968). In addition to the DS, fluid properties such as viscosity are also dependent on the molecular weight of the polymer molecules. At a given value of the DS, the higher the molecular weight of the polymer, the higher the corresponding solution viscosity. For example, a high-viscosity product typically has a molecular weight of about 700 000 g/mol and a DS of between 0.7 and 0.8. High-viscosity CMC is prepared from cotton linters, and medium-viscosity products are produced from wood pulp, whereas low-viscosity products are prepared by ageing the shredded alkali cellulose and by using chemical oxidants. As there is a wide range of possible physical and chemical properties, CMC manufacturers commonly specify their products according to the DS (typically between 0.7 and 1.2), the viscosity grade (low, medium and high), industry grade (standard (for industrial use), food and pharmaceutical) and sometimes particle size. CMC is an approved food additive, and its E number is E466. High-purity premium CMC with less residual NaCl and a higher degree of carboxymethyl substitution is often referred to as polyanionic cellulose (PAC) in the drilling industry.

Another industrially useful cellulose derivative is hydroxyethyl cellulose (HEC), which is formed by substituting hydroxyethyl groups ($-CH_2CH_2OH$) for some of the hydrogens in the hydroxyl groups, as shown in Figure 3.5. As for CMC, the production of HEC is a two-step process. The cellulose is first treated with sodium hydroxide to produce alkali cellulose, which is more reactive. The treated cellulose is then reacted with ethylene oxide (CH_2CH_2O), to form HEC. Once a hydroxyethyl group is attached to the molecule, it can further react with additional groups in an end-to-end formation, hence the two different types of side chain configurations shown in Figure 3.5. Therefore, for HEC, it is also important to count the number of moles of chemical reagent (ethylene oxide) that become attached to each anhydroglucose unit, and the result is expressed as moles of substitution (MS). The example shown in Figure 3.5 has an MS of 1.5 (three ethylene

Figure 3.5 The chemical structure of HEC

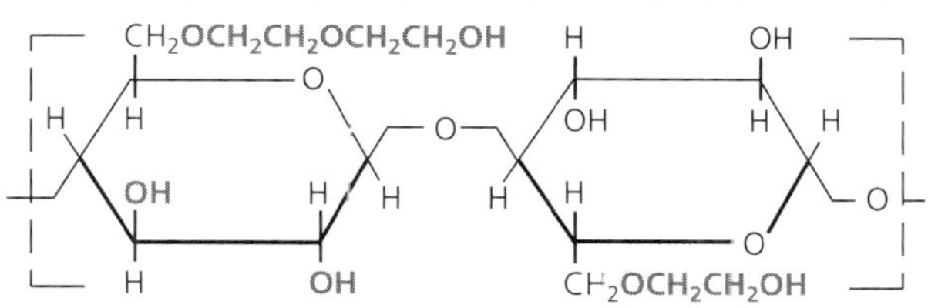

oxide groups per two units) and a DS of 1.0 (two hydroxyl groups substituted per two units). It is worth noting that it is the hydroxyethyl groups that confer water solubility to the polymer molecule. Commercial water-soluble HECs typically have MS values in the range 1.3–3.4 and DS values of between 0.9 and 1.4. Optimal solubility in water is usually obtained at an MS value of around 2.5. Unlike CMC, HEC is non-ionic when dissolved in water, and is therefore stable in polyvalent brines, such as calcium chloride ($CaCl_2$) and magnesium chloride ($MgCl_2$) solutions. As a consequence of this property, HEC is often used to provide the viscosity of the solids-free brines that are commonly used as completion fluids in oil well drilling. As with CMC, HEC is manufactured to different viscosity and industrial grades. The molecular weight of high-viscosity HEC can be up to about 1.3 million grams per mole. In terms of chemical classification, both CMC and HEC are cellulose ethers, as they contain the ether group, which has the general formula $R{-}O{-}R'$.

There are many other types of cellulose derivatives that are available for a wide range of industrial applications. For example, sodium carboxymethyl hydroxyethyl cellulose (CMHEC) can be formed by treating cellulose with ethylene oxide and then sodium monochloroacetate. The end product has attributes of both CMC and HEC due to the presence of both non-ionic and anionic side groups. This gives the polymer higher salt tolerance than pure CMC but allows it to retain some of its anionic characteristics, which can be useful for flocculation, etc. For many years, CMHEC has been successfully used as a fluid-loss control additive and as a retarder in cement slurries for sealing oil wells.

3.4.2 Guar gum

Guar gum is a water-soluble polysaccharide produced from the beans of the guar plant. Guar (scientific name: *Cyamopsis tetragonoloba*) is a plant grown mainly in India, which accounts for about 80% of world production. In the production process, the guar seeds are dehusked, milled and then screened to obtain the product in powder form. The word 'gum' in the context of guar gum means a group of natural water-soluble polymers. In the food industry, gums are also commonly known as hydrocolloids. Chemically, guar gum is a type of galactomannan with a galactose on every other mannose unit, as shown in Figure 3.6. The mannose-to-galactose ratio is therefore 2 : 1. Guar gum is completely non-ionic, and is therefore compatible with cations from salts such as sodium, magnesium and calcium chlorides. This allows guar to be used to viscosify a wide variety of brines, and this is a particularly useful feature in situations where fresh water is in short supply or not available. The aqueous solutions of guar gum can be cross-linked by reagents such as sodium borate (borax) to form a gel. In oil and gas applications, guar gum is commonly used to provide the necessary high viscosity of hydraulic fracturing (fracking) fluids used for shale gas production and to provide the viscosity of 'slugging' fluids for flushing boreholes in oilfields.

3.4.3 Xanthan gum

Xanthan gum is an approved food additive with the E number E415. It can be found in all kinds of food products, ranging from bakery and dairy products to sauces and gravies. Originally named polysaccharide B-1459, xanthan gum was first developed by

Figure 3.6 The chemical structure of guar gum

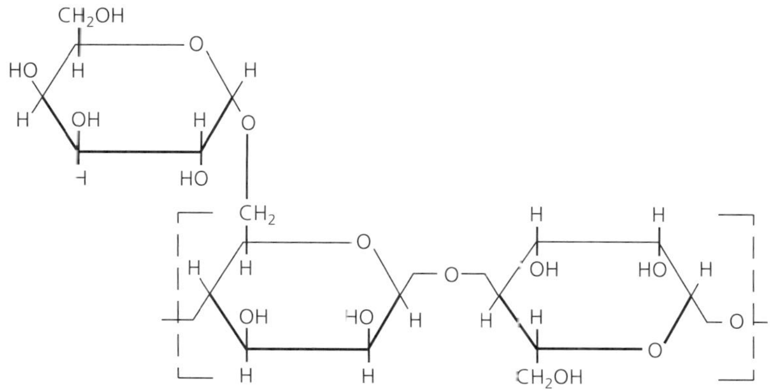

Jeanes *et al.* (1961) to overcome the limitations in the usefulness of plant-derived polymers. Xanthan gum is a natural polysaccharide that can be found on the leaves of brassica vegetables such as cabbages. Industrially, xanthan gum is produced by culturing the micro-organism *Xanthomonas campestris* in a well-aerated medium containing commercial glucose, a nitrogen source and mineral salts. Throughout the fermentation process, the pH, aeration, temperature and agitation are carefully monitored and controlled. At the end of the process, xanthan gum is recovered from the culture fluids by precipitation in alcohol. The precipitate is separated, rinsed, dried and milled to produce the final product in powder form.

Xanthan has the chemical formula $(C_{35}H_{49}O_{29})_n$. The structure of xanthan, which is given elsewhere (Philips *et al.*, 1985), is considerably more complicated than that of other polymers, but this should not be a concern to users. Essentially, the xanthan molecule consists of a cellulose-like backbone with a long trisaccharide side chain attached to every other glucose unit in the backbone. The side chain consists of three different saccharide units, namely a mannose acetate (the one close to the backbone), a glucuronic acid salt (the middle one) and a terminal mannose unit that is linked to a pyruvate group at the end. The glucuronic acid and the pyruvic acid groups give xanthan an anionic charge and its water solubility.

The solution conformation of the xanthan molecule is such that the side chains lie along and shield the backbone, giving the molecule a rod-like rigid structure. This grants the polymer solution some unique properties, including relative insensitivity to temperature, pH and the ionic strength of the solution which may be due to hardness or salinity. Also because of its rigid molecule structure, xanthan is resistant to physical damage caused by high-speed mixing or pumping. These characteristics make xanthan a very useful polymer in difficult ground conditions such as pH or high salinity, or when the fluid is subjected to a high shear stress. For oil and gas explorations, xanthan gum is commonly used to increase the solids-carrying capacity of the drilling fluid due to its high viscosity at low shear rates. As a consequence of its excellent rheological properties and

Figure 3.7 The chemical structures of (a) polyacrylamide and (b) its partially hydrolysed form

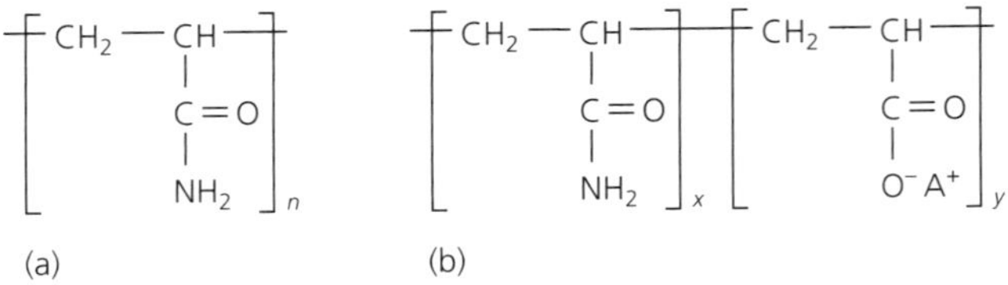

resistance to high temperatures (up to 135°C), solutions of xanthan gum are widely used for flooding oil reservoirs in enhanced oil recovery operations (Sorbie, 1991).

Synergistic effects can be obtained by using xanthan gum together with certain types of polymers. For example, the viscosity of a fluid containing both xanthan and guar gums is much higher than that of a solution of either one of the polymers alone. The synergistic effect with locust bean gum (a vegetable gum extracted from the seeds of the carob tree) and konjac (a plant native to the Far East) is even greater, and together they produce strong thermo-reversible gels (Williams, 2016).

3.4.4 Polyacrylamides

Polyacrylamide is a synthetic homopolymer consisting of repeating acrylamide monomer units. It has the chemical formula $(C_3H_5NO)_n$, and its structure is shown in Figure 3.7(a). As the molecule does not carry any positive or negative charges, it is non-ionic and has limited functionality. Therefore, for industrial applications it is often used in its partially hydrolysed form, with some of the acrylamide units substituted by acrylates, making it an anionic copolymer. The chemical formula of the partially hydrolysed form is $(C_3H_5NO)_x\text{–}(C_3H_3O_2A)_y$, where A is a positive ion (typically sodium), and the values of the variables x and y will depend on the degree of hydrolysis. Figure 3.7(b) shows its chemical structure. It should be noted that, in the literature, partially hydrolysed polyacrylamide is commonly abbreviated to PHPA or HPAM. The former is commonly used in the petroleum and civil engineering industries, and in this book. Despite its name, PHPA may be prepared by direct copolymerisation of acrylate with acrylamide monomers as well as by partial hydrolysis of polyacrylamides. If the polymer is fully hydrolysed so that it has only acrylate units, then it becomes a polyacrylate – which acts a waterlock due to its ability to absorb as much as 200–300 times its solid mass in water. The British standard BS EN 1407 contains some further information on polyacrylamides and their use as flocculants in water treatment (BSI, 2008a). Both cationic and anionic polyacrylamides are available, but the cationic versions are mainly used for water treatment and are not discussed further. Information about cationic polyacrylamides can be found in another BS standard (BSI, 2008b).

To produce PHPA in dry granular form such as previously shown in Figure 3.2(d), in the factory, after the polymerisation stage, the polymer gel is taken through several drying and pulverising processes. The end product is not completely dry and may contain up to 10–15% water. Further information about the production process of PHPA granules can be found in patents (e.g. Kudomi and Sato, 1984; Landolt and Allen, 1979). PHPA can also be obtained in liquid emulsion form, which is favoured over the dry powder form in some circumstances, for example when rapid dissolution is required. The production process of emulsion polymer has been briefly discussed in Section 3.3.4, and further information can be found in patents (e.g. Reekmans and Cornet, 2004; Vanderhoff and Wiley, 1966).

The acrylate units in PHPA make it anionic, with the result that there can be a significant increase in the solution viscosity over a range of alkaline pH values. The effect of solution pH on the properties of PHPA will be discussed in detail later. Martin and Sherwood (1975) note that, if mixed with fresh water, polyacrylamides with a 25–35% degree of hydrolysis produce the most viscous solutions, but in high-calcium brines a 15% hydrolysed polyacrylamide can produce greater solution viscosity. Caenn *et al.* (1989) reported that a 35% degree of hydrolysis gives the maximum solution viscosity. As the degree of hydrolysis increases, water solubility increases but the viscosity of the polymer solution at a given concentration decreases. As PHPA is anionic, it is sensitive to the ionic strength of the solvent; that is, the concentration and charge of dissolved ions. The quality of mix water is therefore very important for this class of polymer. When dissolved in water, PHPA has a flexible molecular structure, which can lead to a random coil configuration. It is known that PHPA interacts with polyvalent ions such as iron, aluminium and chromium to form three-dimensional gel networks.

Prior to its introduction to civil engineering, PHPA had been used successfully in oilfields for a considerable time. For example, 30–40% hydrolysed PHPA is often used in a potassium-based (KCl) fluid system to drill in water-sensitive shale because it can help to protect the walls of the drill holes from disintegration. This protective mechanism is thought to be due to the anionic molecules attaching themselves to the positive sites on the surface of the shale, thus inhibiting its swelling and disintegration. Like xanthan, PHPA is commonly used for enhanced oil recovery because of its rheological properties (Caenn *et al.*, 1989; Sorbie, 1991).

3.5. Dissociation of polymers in solution

3.5.1 Fundamentals

In addition to recognising the different polymer types, appreciating the polymer dissociation process is also necessary for the understanding of polymer fluid properties. This section briefly describes how a polymer (solute) may dissolve in water (solvent) to form a polymer solution. In essence, for a polymer to be water soluble, it has to be able to form hydrogen bonds with the water molecules. Polyacrylamide, although having an overall non-ionic character, can still dissolve in water because the molecule is polar and can form hydrogen bonds with the water molecules. However, even when a chemical is soluble in water it does not necessarily mean that it will dissociate completely in the solvent. Water solubility and the degree of dissociation (ionisation) of a chemical are two different phenomena.

To understand how a chemical dissociates or ionises in water, it is best to start the discussion with a simple non-polymeric chemical. For example, sulfuric acid (H_2SO_4) dissociates effectively completely into positive (H^+) and negative (SO_4^-) ions in dilute aqueous solutions because it is a strong electrolyte (acid) with a pK_a value of -3. pK_a is the negative log of the acid dissociation constant (K_a),

$$pK_a = -\log K_a \tag{3.4}$$

and K_a is defined as

$$K_a = \frac{[H^+][A^-]}{[HA]} \tag{3.5}$$

where $[A^-]$ and [HA] denote the concentrations (strictly the activities) of the ionised and non-ionised forms of the acid, respectively. Generally, an acid having a negative pK_a value is considered to be strong, and one with a positive value to be weak. Acetic acid (CH_3COOH) is an example of a weak acid, and has a relatively high pK_a value of 4.76. In water, acetic acid dissociates only to a limited extent, with the majority of the molecules existing in the unionised form.

Polymers containing charged functional electrolyte groups are known as polyelectrolytes. Similar to their non-polymerised counterparts, they can also be divided into 'strong' or 'weak', depending on the dissociation properties of the electrolyte groups. PHPA, because it contains an acrylate unit and the pK_a value of acrylic acid is 4.25, is a weak polyelectrolyte. Therefore, when dissolved in water a significant proportion of the molecules may remain unionised. The degree of dissociation is also controlled by the pH of the solution, and this can be modelled via the Henderson–Hasselbalch equation as

$$pH = pK_a + \log \frac{[A^-]}{[HA]} \tag{3.6}$$

Equation 3.6 can be used to estimate the degree of dissociation if the pK_a of the acid and the pH of the final solution are both known. It should be noted that in an unbuffered system the final pH of the solution and the degree of dissociation will depend on the amount of acid added. For example, a 1 M (1 molar, 60 g/litre) acetic acid solution will give a final pH of 2.4, and hence the degree of dissociation will only be 0.4%, as shown in the following calculations.

$$pK_a = 4.76 \text{ (acetic acid)} \tag{3.7}$$

Recall that

$$pK_a = -\log K_a \tag{3.8}$$

Therefore,

$$K_a = 1.7378 \times 10^{-5} \tag{3.9}$$

For weak acids when there is no other source of hydrogen ions, $[H^+]$ equals $[A^-]$, and so Equation 3.5 for K_a can be written as

$$K_a = \frac{[H^+]^2}{[HA]} \tag{3.10}$$

Now, we know the acid in question is very weak, and for a 1 M solution $[HA] \approx 1$, so the concentration of protons is

$$[H^+] = \sqrt{K_a[HA]} = \sqrt{1.7378 \times 10^{-5} \times 1} = 4.1687 \times 10^{-3} \tag{3.11}$$

Therefore,

$$pH = -\log[H^+] = 2.38 \tag{3.12}$$

Substituting this pH value into the rearranged form of the Henderson–Hasselbalch equation (Equation 3.6) gives

$$\frac{[CH_3COO^-]}{[CH_3COOH]} = \frac{[A^-]}{[HA]} = 10^{(2.38-4.76)} = 0.4\% \tag{3.13}$$

Thus, the predicted degree of dissociation is indeed small. Further information about the dissociation (ionisation) behaviour of chemicals in solutions can be found in textbooks such as that by Pandit and Soltis (2011).

3.5.2 A synthetic polymer example: PHPA

Most PHPA supplied for civil engineering use is in the sodium form; that is, sodium balances the negative charge of the acrylate unit, and this gives the PHPA solution an alkaline pH. In a good solvent such as water and in the absence of other cations such as those due to salinity or hardness, the negatively charged acrylate units are unshielded and repel one another. As a result, the polymer chains become stretched out (otherwise they would take up a random coiled configuration), causing an increase in the solution viscosity and many other interesting non-Newtonian properties. This is known as the electroviscous effect.

Since the degree of dissociation will have a significant effect on the solution viscosity of a water-soluble polymer and it is dependent on the pH of the solution, to maximise the viscosifying power of PHPA at a given concentration one would therefore need to raise the pH of the solvent to increase the degree of dissociation. This is why PHPA manufacturers typically recommend contractors to raise the pH of mix water from the natural pH value of around 7 to 10–11 by using chemicals such as sodium carbonate, as the degree of dissociation will in theory be increased by 1000–10 000 times, according to the Henderson–Hasselbalch equation – note the logarithmic relationship between pH and $[A^-]/[HA]$ (see Equation 3.6). However, if the calculation result is examined more closely, then it can be seen that at pH 7 the $[A^-]/[HA]$ ratio of a PHPA is already in excess of 250 if calculated using a pK_a value of 4.6 as estimated by Sun (2011). When expressed as a percentage, this means that the polymer is already 99.6% ionised at this pH. It can

Figure 3.8 The carboxylate base: hydrogen bonding and ion–dipole attraction

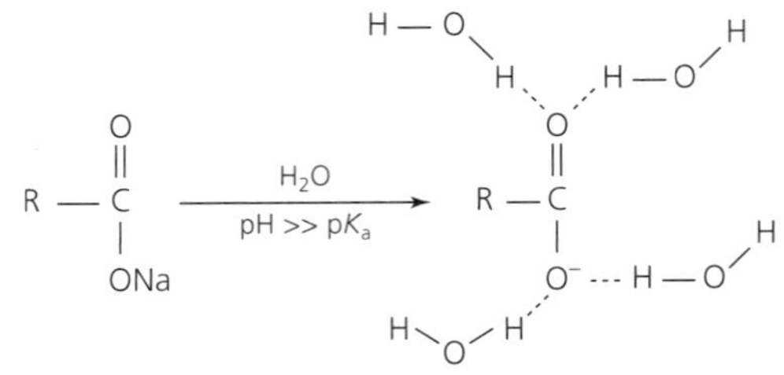

therefore be said that a high pH is not in itself a prerequisite for achieving a high degree of dissociation for high solution viscosity, but that it only serves to keep the pH at a 'healthy' alkaline level in case of contamination and/or dilution by acidic groundwater. This explains why some earlier successful polymer projects reported reasonable Mash funnel times of up to 65 s at relatively low pH values of between 6 and 7 (Holmes, 1993). Another major reason for pH adjustment is to remove water hardness by reducing the amount of free calcium ions, which if left in the water can significantly reduce the viscosity of the fluid by removing the repulsion between the anionic polymers, and therefore reducing their effective hydrodynamic volumes. The hydrodynamic volume of a polymer molecule in solution depends on how well it interacts with the solvent and also other molecules. Despite the advantages of pH adjustment, an unnecessarily high pH should also be avoided. Nasr-El-Din and Taylor (1996) reported that the viscosity of PHPA solutions actually reduces with increasing sodium hydroxide concentration. They attributed this to the sodium ions shielding the negative charges on the polymer chains.

Lastly, it is worth noting that because the dissociated $-COO^-$ groups of a PHPA are highly polar, they attract a cloud of water molecules because of hydrogen bonding and ion–dipole attractions, as depicted in Figure 3.8. In addition, the hydrogen atoms in the amide groups ($-CONH_2$) are also sufficiently positive to form hydrogen bonds with similar molecules and with water, as shown in Figure 3.9. These intermolecular forces,

Figure 3.9 The amide group: hydrogen bonding (a) with like molecules and (b) with water

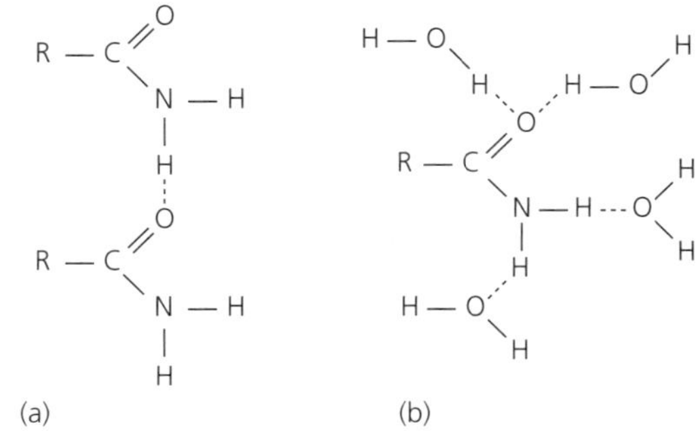

among other physicochemical effects, are thought to play a role in reducing the swelling of water-sensitive shales because the water molecules are 'tied up' by the dissolved polymers.

3.5.3 A natural polymer example: CMC

The chemical nature of CMC makes it an anionic polyelectrolyte when dissolved in water. As a weak base, the degree of dissociation of CMC can be modelled with the Henderson–Hasselbalch equation (see Equation 3.6), and is dependent on the pH of the solvent. For good solubility, the pH of the solvent needs to be above 7. In general, CMC solutions exhibit their maximum level of viscosity at a pH range of between 7 and 9. If the pH is less than 6, the dried CMC powder does not dissolve well, and the solutions will contain some gel particles and become hazy. If the pH is below 4, CMC does not dissolve in water.

3.6. Key properties of polyacrylamides

3.6.1 Introduction

PHPAs are currently by far the most common type of polymer used in foundation construction. This is because they are good viscosifiers and are non-biodegradable, so their properties can be maintained for a considerable period of time if used properly. However, the fundamental properties of PHPAs can vary considerably between brands. For PHPAs, the molecular weight and the degree of hydrolysis are undoubtedly the two most important parameters because together they dictate the physicochemical properties of the polymer solution. The purpose of this section is to discuss the test methods that can be used to measure these properties, so that the end users can verify the properties of a PHPA product if required. Authorities might also specify the fundamental properties of PHPAs to be used based on the results of these tests.

3.6.2 Molecular weight

Many methods have been developed by polymer scientists to determine the molecular weights of polymers. These include gel permeation chromatography (also called size exclusion chromatography), ultracentrifugation, light scattering, intrinsic viscosity and hydrodynamic chromatography (Langhorst *et al.*, 1986). However, many of them require expensive laboratory equipment. For PHPAs, the lack of calibrants available for molecular weights higher than 10×10^6 g/mol and the presence of undissolved microgels in the solutions can confound some test procedures. Among the possible methods, that based on the measurement of intrinsic viscosity is probably the most suitable for non-specialists because it is the easiest to perform and does not require any expensive equipment. This method also has a long history of use in polymer science, so there is an abundance of relevant literature, including standardised test procedures (ASTM, 2012; BSI, 2009; CSBTS, 1989a). The theoretical basis for this test can be found in many standard polymer science textbooks, so only the basic operating procedure is described here.

In essence, the intrinsic viscosity method involves measuring the efflux times of dilute polymer solutions using a simple device known as the Ubbelohde viscometer, which is named after the German chemist Leo Ubbelohde. Figure 3.10 shows an Ubbelohde

Figure 3.10 The Ubbelohde viscometer: the annotations are described in the text

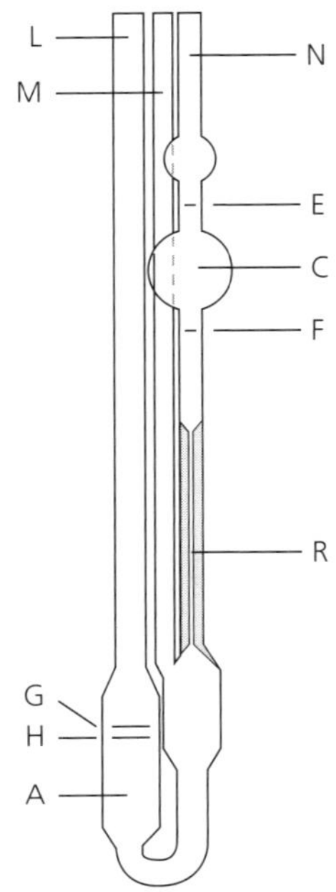

viscometer. As can be seen, this viscometer has a U shape and three tube sections. It has an overall length of 283 mm. In one of the tubes there is a capillary. During the test the liquid is suspended in the capillary, which it fills completely, so Ubbelohde viscometers are a type of suspended level viscometer. There are many viscometer (capillary) sizes available for fluids of different viscosities: a list of them can be found in standards (ASTM, 2012; BSI, 2009). For the purpose of viscometer selection, viscosity is expressed as kinematic viscosity in centistokes (cSt), which is numerically the same as mm^2/s. For the testing of high-molecular-weight PHPA polymer solutions, viscometer size No. 1 has been found suitable by the authors. This viscometer size has a capillary diameter of 0.58 mm, and is designed to measure fluids with a kinematic viscosity range of between 2 and 10 cSt.

To run a test, a water bath and a stopwatch are needed to control the fluid temperature (typically 25 or 30°C) and to measure the efflux time, respectively. The following procedure may be followed (all designations refer to Figure 3.10):

- Prepare the test fluid at the required concentration.
- Introduce the fluid into reservoir A through tube L until the fluid level is between marks G and H.
- Mount the viscometer in a holder in a water bath set at the required temperature. Ensure that tube N is vertical.
- Charge the measuring bulb C by closing tube M and applying suction to tube N until the liquid level is about 5 mm above mark E.

- Hold the liquid at this level by closing tube N.
- Open tube M so that the liquid drops away from the lower end of capillary tube R.
- Open tube N and measure the time taken to the nearest 0.2 s for the fluid level to drop from mark E to F.
- Without emptying or recharging, repeat the measurement of the efflux time immediately until two successive times agree to within 0.25%. Take the mean of these two values as the efflux time.

The above procedure should be carried out for the basic solvent and polymer solutions prepared at three or more different concentrations (c_p). The solvent should be a salt solution such as 1.0 M sodium chloride. The use of a salt solution as the solvent ensures that the repulsive charges between molecules and hence the electroviscous effect are eliminated, so that only the viscosifying effect of the molecular chain length (molecular weight) is measured. To interpret the results of a test, the measured efflux times are used to calculate the relative viscosity (η_{rel}) and the specific viscosity (η_{sp}) of the fluid, expressed as

$$\eta_{rel} = \frac{\eta}{\eta_{sol}} \approx \frac{t}{t_{sol}} \tag{3.14}$$

$$\eta_{sp} = \eta_{rel} - 1 \tag{3.15}$$

where η and t are the viscosity and the efflux time of the polymer solution, and η_{sol} and t_{sol} are the viscosity and the efflux time of the solvent, respectively. The parameters η_{rel} and η_{sp} are then used to calculate two other viscosity numbers known as the reduced viscosity (η_{sp}/c_p) and the inherent viscosity ($\ln \eta_{rel}/c_p$). By plotting them against c_p, the intrinsic viscosity, $[\eta]$, is obtained from the y intercept of the two curves, which, in theory, should have the same value as schematically shown in Figure 3.11.

Once the intrinsic viscosity has been obtained, the molecular weight of the polymer (M) can be determined using the Mark–Houwink equation

$$[\eta] = KM^a \tag{3.16}$$

where the parameters K and a depend on the particular polymer–solvent system and the solution temperature. For polyacrylamides, many studies have been carried out to establish their values; a summary of them is provided by Munk *et al.* (1980). Among them, the one reported by the American Cyanamid Company (1955) is probably the most commonly cited in the literature:

$$[\eta] = 3.73 \times 10^{-2} M^{0.66} \tag{3.17}$$

Equation 3.17 was originally proposed for polyacrylamide in 1.0 M sodium nitrate at 30°C, but the replacement of sodium nitrate with sodium chloride does not appear to make it less useful. It is interesting to note that the value of the exponent a (0.66) lies well within the typical range of 0.5–0.8 for polymers behaving as random coils, whereas

Figure 3.11 Determination of the intrinsic viscosity from the reduced and inherent viscosities

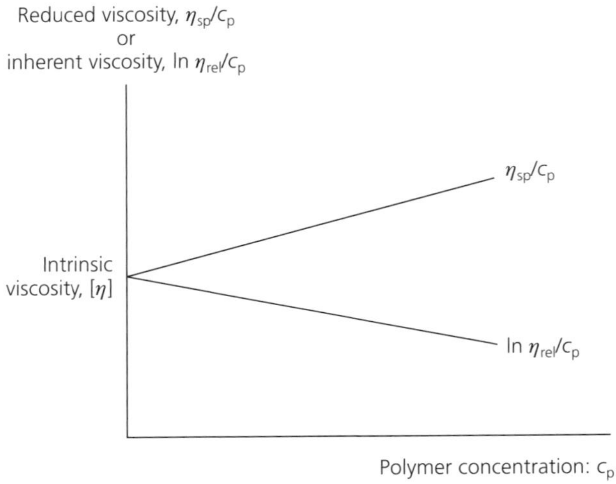

for polymer chains behaving like stiff rods, the value of this parameter could be as high as 2 (Sperling, 2006).

To give an example of the calculation procedures, a non-ionised polyacrylamide with a claimed molecular weight of 18×10^6 g/mol was tested. The reduced and inherent viscosities (η_{sp}/c_p and ln η_{rel}/c_p) were calculated using the measured efflux times and plotted against polymer concentration (c_p), as shown in Figure 3.12. It can be seen from the figure that the *y* intercepts of the two fitted lines are reasonably close to each other, together giving an average intrinsic viscosity of 2147 cm^3/g. This corresponds to a

Figure 3.12 Determination of the intrinsic viscosity of a polyacrylamide

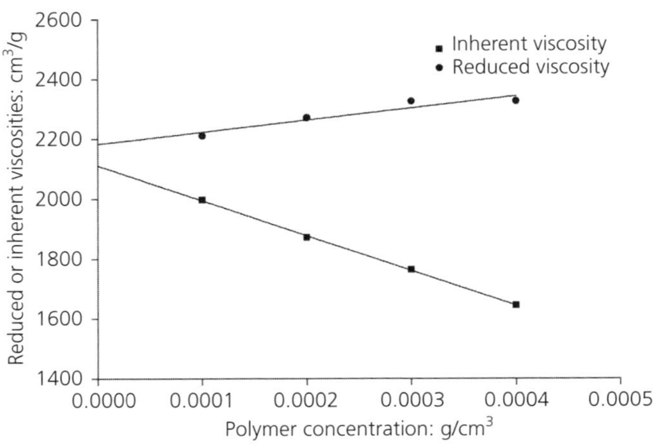

molecular weight of 16.3×10^6 g/mol if calculated using the Mark–Houwink parameters proposed by the American Cyanamid Company. The result is 9.4% lower than the expected value of 18×10^6 g/mol, but this difference is considered acceptable given the possible errors resulting from the indirect nature of the measurement. Nevertheless, as shown in this example, the intrinsic viscosity method offers non-specialists a relatively easy way to confirm the molecular weight of polyacrylamides.

3.6.3 Degree of hydrolysis

The degree of hydrolysis of a PHPA is the percentage of acrylamide units in a polyacrylamide chain that have been transformed into acrylic acid or acrylate units. If an acrylamide unit is denoted as x and an acrylate unit as y (see Figure 3.7), then the degree of hydrolysis can be expressed as $y/(x + y) \times 100\%$, and it will range from 0 for non-ionic polyacrylamides to 100% for polyacrylic acids. Although it is now common practice to produce PHPAs by copolymerising chosen proportions of acrylamide and acrylate directly, the term 'degree of hydrolysis' is still used to refer to the proportion of acrylate units present in the materials. Anionicity, which is a better term in this regard, is also used in some literature to refer to this same material property (Arinaitwe and Pawlik, 2009).

The degree of hydrolysis of PHPAs can be determined by at least eight different methods, ranging from the classical acid–base titration to more sophisticated methods such as carbon-13 nuclear magnetic resonance (Taylor and Nasr-El-Din, 1994). For non-polymer scientists, the titration method is probably the best option because it does not require any specialist skills or expensive instruments. Titration is a common laboratory method that involves adding one solution of known concentration and volume (the titrant) to another solution of unknown concentration (the analyte) until the reaction reaches neutralisation. For this test, the titrant can be a 0.1 M hydrochloric acid (HCl) solution, and the analyte is the polymer fluid mixed using deionised water. The Chinese standard GB 12005.6-89 (CSBTS, 1989b) specifies the use of 100 ml polymer solution as the analyte, but the authors have found that the accuracy of the test can be improved by increasing the sample volume to 200 ml or more. During the test, the acid is added to the polymer solution, drop by drop from a burette until the end point is detected.

During the titration, the $-COO^-$ groups of the polymer will become $-COOH$, and sodium chloride will form, as shown in Figure 3.13. The equivalence point of the reaction is reached when all $-COO^-$ groups have become $-COOH$ groups. With the titration volume at the equivalence point known, the degree of hydrolysis of the polymer can be calculated using the following equation given by Chen (1985):

$$\mathrm{DH} = \frac{c_{\mathrm{HCl}} V_{\mathrm{HCl}}[94(\mathrm{DH}) + 71(1 - \mathrm{DH})]}{W} \times 100\% \tag{3.18}$$

where DH is the degree of hydrolysis (in %), c_{HCl} is the concentration of the HCl solution (in mol/l), V_{HCl} is the titration volume (in ml) at the titration end point, and W is the weight of the polymer in the test specimen (in mg). The number 94 is the

Figure 3.13 The chemical reaction during the titration of PHPA with HCl

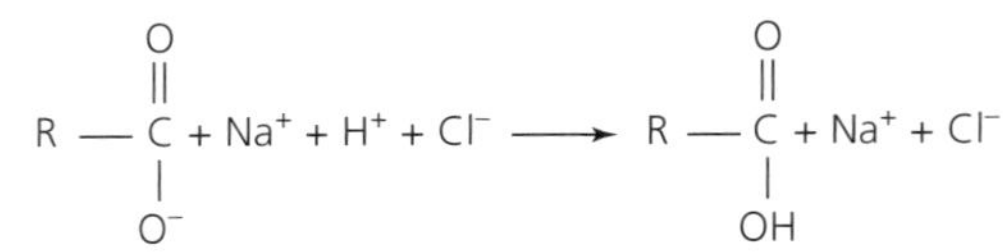

molecular mass of a sodium acrylate unit in g/mol, and the number 71 is the molecular mass of an acrylamide unit, also in g/mol. The square bracket term in Equation 3.18 takes into account the effect of the degree of hydrolysis on the molecular weight of the reactive monomer units.

Equation 3.18 can be rearranged to express the degree of hydrolysis as follows:

$$\mathrm{DH} = \frac{71c_{\mathrm{HCl}}V_{\mathrm{HCl}}}{1000m_{\mathrm{p}}s_{\mathrm{p}} - 23c_{\mathrm{HCl}}V_{\mathrm{HCl}}} \times 100\% \qquad (3.19)$$

In this rearranged form, the parameter W is replaced by $1000m_{\mathrm{p}}s_{\mathrm{p}}$, where m_{p} is the mass of the polymer as supplied (in g) and s_{p} is the solids content of the polymer granules. This latter parameter takes into account the fact that granular polymers are not totally dry and contain some moisture. The value of the solid content can be determined by weighing some polymer granules before and after oven-drying overnight.

The titration end point of this test can be determined by several methods. For example, CSBTS (1989b) specifies the use of a mixed colour indicator consisting of one drop of 0.1% methyl orange solution and one drop of 0.25% indigo carmine solution. This indicator changes its colour sharply from green to grey at a pH of about 3.6–3.7 and to violet if the pH continues to decrease. If this indicator is added to the polymer fluid, then the end point of the titration can be determined when the colour of the indicator becomes grey. This method is slightly user dependent due to the fact that people vary to a degree in how they see colour. A second way to determine the titration end point is by measuring the electrical conductivity (EC) of the mixture solution (see Section 6.2 for the measurement techniques). This method is based on the fact that, before the equivalence point is reached during the titration, the $-\mathrm{COO}^-$ groups will receive protons (H^+) from the acid to give $-\mathrm{COOH}$, so that the amount of free Cl^- ions in the solution will increase linearly with the amount of HCl added. However, after the equivalence point, there will be no more available $-\mathrm{COO}^-$ groups, and hence the number of $\mathrm{H}^+$and Cl^- ions will increase together. Since the EC is a measure of free ions in the solution, it will have a bi-linear profile if plotted against the titration volume; the turning point of the curve will be the equivalence point. The use of the EC curve as a replacement for the indicator method has been proposed by Zhao (2006) due to its higher repeatability and accuracy. The titration end point can also be determined by measuring the changing pH value of the solution, but the authors have found it less useful than the EC method.

To give an example of this test, the degree of hydrolysis of a PHPA standard with a claimed degree of hydrolysis of 70% has been determined using the indicator and the

Figure 3.14 The titration curve of a PHPA measured by the electrical conductivity method

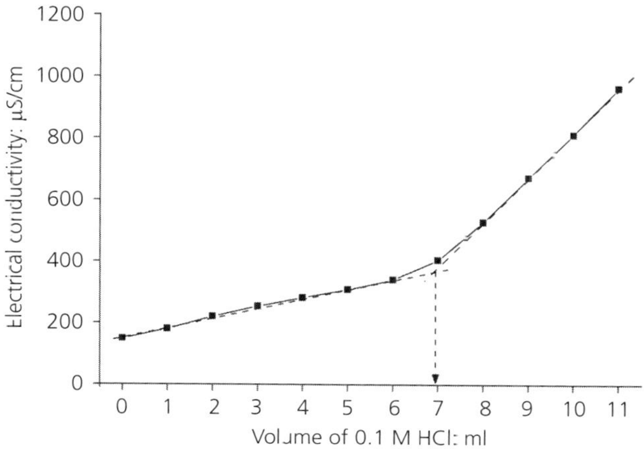

EC methods. The test conditions were $m_p = 0.1$ g, $s_p = 0.87$ and $c_{HCl} = 0.1$ mol/l. During the test, the colour of the indicator changed from green to grey after the seventh milliliter of acid had been added to the polymer solution. In other words, the titration end point was reached when V_{HCl} equalled 7 ml, but the true equivalence point could have been reached anywhere between 6 and 7 ml. Reducing the drop volume would have led to higher accuracy, but it would have taken longer to complete a test. Nonetheless, substituting $V_{HCl} = 7$ ml and values of the other parameters into Equation 3.19 gave a degree of hydrolysis of 70.1%, which is the expected value for this polymer. Figure 3.14 also shows the EC curve for the experiment. As can be seen, the curve has a bi-linear profile, and the turning point occurs when the titration volume reaches 7 ml.

It is worth mentioning that this test can also be carried out on PHPAs supplied in emulsion form, but the mineral oil first needs to be removed from the actual polymers. This can be easily achieved by precipitating the polymeric materials using a suitable chemical such as acetone. The precipitates can then be used to prepare the polymer solution for the test.

3.6.4 Microstructure in solution

The microstructure of a polyacrylamide solution refers to the configuration and orientation of the individual polymer chains and their interaction. Although the molecular weight and the degree of hydrolysis of polyacrylamides have been discussed, it is interesting to directly observe the polymer chains, to see how these basic properties translate into fluid structures. An understanding of the microstructure is particularly useful for the interpretation of the rheological behaviour of polymer fluids, which will be discussed in Chapter 9. The following describes such an investigation of PHPA by the authors.

Since polymer fluids are transparent liquids such as water, the microstructure of the fluid is invisible to the naked eye. An environmental scanning electron microscope (ESEM)

was therefore used by the authors to obtain electron micrograph images of the PHPA fluid after a treatment procedure involving freeze-drying and sublimation. The ESEM was a Carl Zeiss Evo LS15, and the images were captured using a variable-pressure secondary electron detector. An ESEM was used because it does not require the specimen to be coated with a layer of conductive material as a conventional scanning electron microscope does, so that the microstructure of the fluid is therefore more likely to be preserved. The sample preparation generally followed the procedure employed by Feng *et al.* (2002), which is summarised as follows: a small droplet of the polymer solution was freeze-dried on a specimen holder to $-15°C$ using a Peltier cooling system before being moved into the ESEM chamber. To reveal the polymer microstructure, the frozen ice needed to be removed, and this was achieved by gradually reducing the chamber pressure to about 133 Pa (1 torr) absolute. This allowed the ice within the sample to sublime without damaging the microstructure of the polymer. Micrographs of the sample were then acquired under the secondary electron mode, and processed by a computer.

Figure 3.15 shows the ESEM micrograph of a PHPA solution prepared at a concentration of 2 kg/m^3. Deionised water was used for the sample preparation in order to maximise the electroviscous effect. An acceleration voltage of 15 kV and a working distance of 7.5 mm were used to produce this image. The PHPA polymer had a molecular weight of about 20×10^6 g/mol and a degree of hydrolysis of approximately 40%. It

Figure 3.15 An ESEM micrograph of a high-molecular-weight PHPA fluid

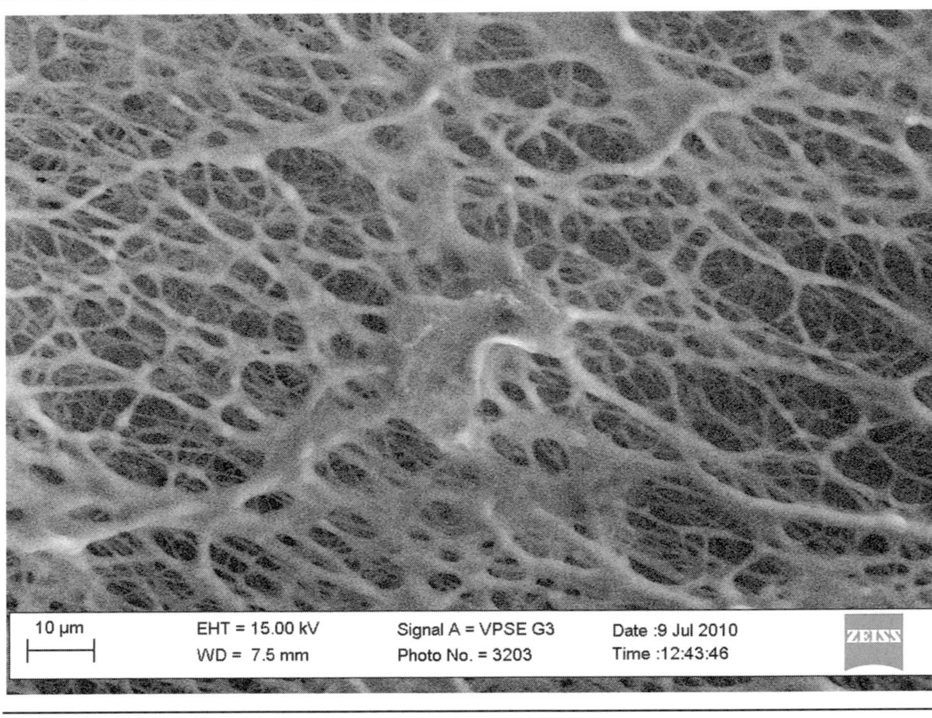

can be seen that the polymer fluid had a three-dimensional network structure that spread throughout the specimen and slightly resembles the appearance of bone tissue. This micrograph shows that the polymer chains, instead of existing individually within their hydrodynamic volumes, were actually entangled with one another in the solution. It is also interesting to note that the polymer chains, although entangled with one another, also appeared to be quite rigid. This was likely to be the result of the electrostatic repulsion between the anionic charged sites distributed along the chains. If the electrostatic charges were shielded by, say, dissolved cations in the fluids, then the chains would be expected to be much more flexible and to behave like random coils. Lastly, it is worth adding that if the polymer chains were not frozen, then according to Barnes (2000) the movement of the chains, under the incessant action of Brownian motion, would be like 'snakes in a snake pit', continuously sliding over each other, forming and disengaging from individual entanglements as they move.

REFERENCES

American Cyanamid Company (1955) *Polyacrylamide. New Product Bulletin No. 34*. American Cyanamid Company, New York, NY, USA.

Arinaitwe E and Pawlik M (2009) A method for measuring the degree of anionicity of polyacrylamide-based flocculants. *International Journal of Mineral Processing* **91(1–2)**: 50–54.

ASTM (American Society for Testing and Materials) (2012) D446-12. Standard specifications and operating instructions for glass capillary kinematic viscometers. ASTM International, West Conshohocken, PA, USA.

Barnes HA (2000) *A Handbook of Elementary Rheology*. Institute of Non-Newtonian Fluid Mechanics, University of Wales, Aberystwyth, UK.

BSI (British Standards Institution) (2008a) BS EN 1407:2008. Chemicals used for treatment of water intended for human consumption. Anionic and non-ionic polyacrylamides. BSI, London, UK.

BSI (2008b) BS EN 1410:2008. Chemicals used for treatment of water intended for human consumption. Cationic polyacrylamides. BSI, London, UK.

BSI (2009) BS EN ISO 1628-1:2009. Plastics. Determination of the viscosity of polymers in dilute solution using capillary viscometers. Part 1: General principles. BSI, London, UK.

Caenn R, Burnet DB, and Chilingarian GV (1989) Polymer Flooding. In: EC Donaldson, GV Chilingarian, and TF Yen (eds) *Enhanced Oil Recovery, II: Processes and Operations*. Elsevier, Amsterdam, the Netherlands.

Chatterji J and Borchardt JK (1981) Applications of water-soluble polymers in the oil field. *Journal of Petroleum Technology* **33(11)**: 2042–2056.

Chen J (1985) A new formula for calculating degree of hydrolysis of partially hydrolysed polyacrylamides (PHP). *Oilfield Chemistry* **2(1)**: 165–166. (In Chinese.)

CSBTS (China State Bureau of Technical Supervision) (1989a) GB 12005.1-89. Determination for limiting viscosity number of polyacrylamide. CSBTS, Beijing, China (in Chinese).

CSBTS (1989b) GB 12005.6-89. Determination for hydrolysis degree of partly hydrolysis polyacrylamide. CSBTS, Beijing, China. (In Chinese.)

Darley HCH and Gray GR (1988) *Composition and Properties of Drilling and Completion Fluids*, 5th edn. Gulf Professional Publishing, Houston, TX, USA.

Feng Y, Luo P, Luo C and Yan Q (2002) Direct visualization of microstructures in hydrophobically modified polyacrylamide aqueous solution by environmental scanning electron microscopy. *Polymer International* **51(10)**: 931–938.

Holmes MM (1993) Drilled shafts and new slurry technology used instead of driven piles: sets record on the Ohio River. *Foundation Drilling* **32(5)**: 13–15.

IUPAC (Internal Union of Pure and Applied Chemistry) (2009) *Compendium of Polymer Terminology and Nomenclature: IUPAC Recommendations 28*. RSC Publishing, London, UK.

Jeanes A, Pittsley JE and Senti FR (1961) Polysaccharide B-1459: a new hydrocolloid polyelectrolyte produced from glucose and bacterial fermentation. *Journal of Applied Polymer Science* **5(17)**: 519–526.

Kudomi H and Sato S (1984) *Process for Preparing Water-soluble Partially-hydrolyzed, Solid Acrylamide Polymer*. US Patent 4 482 682.

Landolt PH and Allen LN (1979) *Process for Water-soluble Synthetic Polymer in Powder Form*. US Patent 4 138 539.

Langhorst MA, Stanley Jr FW, Cutié SS, Sugarman JH, Wilson LR, Hoagland DA and Prud'homme RK (1986) Determination of nonionic and partially hydrolyzed polyacrylamide molecular weight distributions using hydrodynamic chromatography. *Analytical Chemistry* **58(11)**: 2242–2247.

Littmann W (1988) *Polymer Flooding*. Elsevier, Amsterdam, the Netherlands.

Martin FD and Sherwood NS (1975) The effect of hydrolysis of polyacrylamide on solution viscosity, polymer retention and flow resistance properties. *Proceedings of SPE Rocky Mountain Regional Meeting*. Society of Petroleum Engineers, Richardson, TX, USA.

Munk P, Aminabhavi TM, Williams P, Hoffman DE and Chmelir M (1980) Some solution properties of polyacrylamide. *Macromolecules* **13(4)**: 871–876.

Nasr-El-Din HA and Taylor KC (1996) Rheology of water-soluble polymers used for improved oil recovery. In *Advances in engineering fluid mechanics: Multiphase reactor and polymerization system hydrodynamics* (Cheremisinoff NP (ed.)). Gulf Publishing, Houston, TX, USA, pp. 615–668.

Pandit NK and Soltis RP (2011) *Introduction to pharmaceutical sciences: an integrated approach*, 2nd edn. Lippincott Williams and Wilkins, Philadelphia, PA, USA.

Philips JC, Miller JW, Wernau WC, Tate BE and Auerbach MH (1985) A high-pyruvate xanthan for EOR. *Society of Petroleum Engineers Journal* **25(4)**: 594–602.

Reekmans SIJ and Cornet P (2004) *Surfactant Composition for Inverse Emulsion Polymerization of Polyacrylamide and Process of Using the Same*. US Patent 6 686 417 B1.

Sorbie KS (1991) *Polymer-improved Oil Recovery*. Blackie, Glasgow, UK.

Sperling LH (2006) *Introduction to Physical Polymer Science*, 4th edn. Wiley, New York, NY, USA.

SPI (Society of the Plastics Industry) (1988) *Resin Identification Code*. SPI, Washington, DC, USA.

Sun QJ (2011) Theoretical analysis and correction for the determination of the hydrolysing degree of polyacrylamide. *Guangzhou Chemical Industry* **39(3)**: 113–116. (In Chinese.)

Taylor KC and Nasr-El-Din A (1994) Acrylamide copolymers: a review of methods for the determination of concentration and degree of hydrolysis. *Journal of Petroleum Science and Engineering* **12(1)**: 9–23.

Vanderhoff JW and Wiley RM (1966) *Water-in-oil Emulsion Polymerization Process for Polymerizing Water-soluble Monomers*. US Patent 3 284 393.

Williams PA (2016) Gums: properties and uses. In *Encyclopedia of Food and Health* (Caballero B, Finglas PM and Toldrá F (eds)). Academic Press, Oxford, UK, pp. 283–289.

Wirick MG (1968) A study of the enzymic degradation of CMC and other cellulose ethers. *Journal of Polymer Science Part A: Polymer Chemistry* **6(7)**: 1965–1974.

Zhao JY (2006) Determining hydrolysis degree of HPAM polymers by conductometric titration. *Oilfield Chemistry* **23(2)**: 170–172. (In Chinese.)

Polymer Support Fluids in Civil Engineering
ISBN 978-0-7277-5786-9

http://dx.doi.org/10.1680/psfce.57869.057

Chapter 4
Commercial polymer systems

4.1. Introduction

Over the past two decades, the types of polymer available for foundation drilling have expanded from the early days of a few biodegradable and partially hydrolysed polyacrylamide (PHPA) polymers to a much wider range of proprietary chemical systems designed to suit a wide range of situations. While this reflects the growing maturity of the technology, allowing polymers to be used even in difficult ground conditions, it has created two major problems. Firstly, those not familiar with the technology tend to regard all polymers as if they are a single product, and, therefore, problems with a polymer-supported bore or trench at one site may be interpreted as a failure of the polymer technology as a whole. This is why, after more than two decades of use, polymers are still rejected for some projects and a significant amount of deep excavation work has to be carried out under bentonite slurries, even if they are not optimal for the work. Secondly, as polymer products are usually given trade names rather than chemical identifiers, it is difficult for practising engineers to identify the differences between various products. In fact, knowing the specific chemical names alone is not sufficient for polymers, where the chain length and the arrangement of pendant groups from the backbone may be key to performance and also manufacturer specific. Although the use of trade names is normal business practice, it does create an obstacle for those who wish to gain a better understanding of the polymer technology and to select the most appropriate polymer for a project.

A further problem is that, unlike bentonite for which the basic properties (water content, particle size and methylene blue index) are usually available upon request to the manufacturer, the properties of the polymer materials (type, molecular weight and charge density) are usually not provided. This makes it difficult for professional organisations and clients to specify the required properties of polymers and to standardise their use. As a result, for most current and potential users the polymer technology remains a 'black box', as neither the polymer type nor its key properties are known; typically, all that is available is a trade name and an application sheet. As a first step towards better understanding of the various commercial products, this chapter discusses their development since the 1970s when polymers were first used. Figure 4.1 summarises the key events in the development of polymer fluid technology in the form of a timeline. For those products that are not discussed in detail in the following chapters, evidence of actual use such as case studies will also be presented where possible. It should be emphasised that, as shown in Figure 4.1, the development of polymer products for the civil engineering industry has been very rapid over the last three decades, and this is expected to continue,

Figure 4.1 The timeline of the development of polymer fluid technology for geotechnical applications

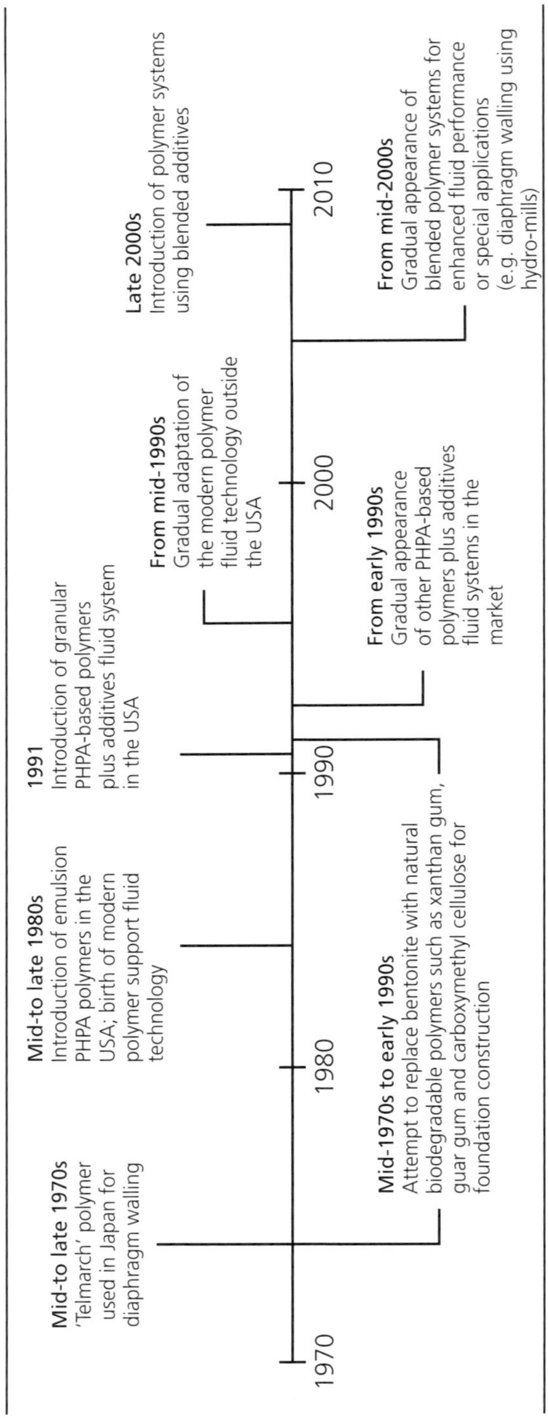

so it is suggested that users consult the polymer manufacturers for the latest advice on the applications of their polymer products. In addition, due to the ever expanding range of polymer products on the market, it is not possible to include every product in this chapter. For this reason, polymers that have evidence of actual field use (e.g. documented case studies) are given more coverage. Also, some of the historical products described in this chapter are known to have been discontinued or may be no longer available in the market. However, their descriptions in this chapter and elsewhere in the book should still serve as a point of reference for the development of this evolving construction technology.

4.2. Natural biodegradable polymers

4.2.1 Telmarch

A polymer known as Telmarch and described as a polymer stabiliser was introduced to the diaphragm walling market in Japan in the 1970s. According to Xanthakos (1979), there were two types of Telmarch – one was available as an organic polymer for general use and the other as an inorganic silicate salt for mixing with sea water. Unfortunately, based on these descriptions alone it is not possible to deduce the actual chemistry of the polymer. Ikuta (1974) reported that 30 tons of Telmarch were produced every month at that time, most of which appears to have been used in Japan. Figure 4.2, which has been plotted from data given by Xanthakos (1979), shows the Marsh funnel time curve for polymer concentrations ranging from 0.8% to 1.6%, which correspond to 8–16 kg of polymer per cubic metre of mix water. At these concentrations, the density of the fluid is very low and should be less than 1.01 g/cm^3. Note that the Marsh funnel time shown in Figure 4.2 is determined according to the original method proposed by Marsh (1931), which specifies pouring 500 cm^3 of the test fluid into the funnel and measuring the efflux time for the same volume. For a 500 cm^3 sample volume, the efflux time of clean water

Figure 4.2 Marsh funnel times of Telmarch polymer fluids at different concentrations. (From Lam and Jefferis, 2015)

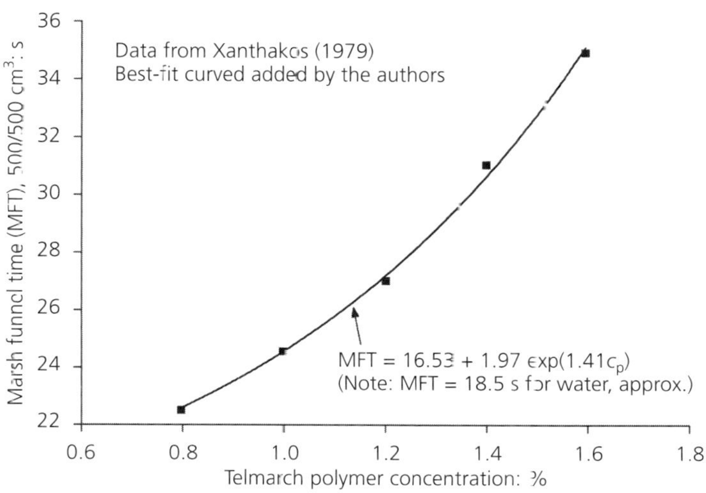

Figure 4.3 The effect of cement contamination on Telmarch and bentonite fluids. (Modified after Ikuta, 1974)

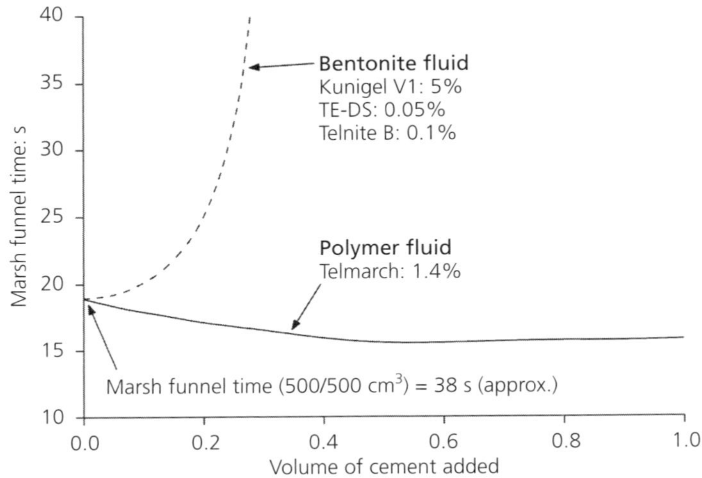

should be about 18.5 s; further details of the Marsh funnel test can be found in Section 6.5. It can be seen from the figure that a significant viscosity was achieved at these polymer concentrations, although by the current standard the dosages were very high even when compared with today's natural polymers. Typical current-day polymers require a concentration of less than 3 kg/m^3 (0.3%).

Figure 4.3 shows the apparent viscosity of the Telmarch fluid over a range of cement contamination levels. A bentonite fluid was also used for comparison; Kunigel V1 is the commercial name of the bentonite material, and both TE-DS and Telnite B are probably additives. Unfortunately, no units were given in the original paper for either the concentration level (possibly in %) or the apparent viscosity (possibly in mPa s), nor was the shear rate used for the viscosity tests specified. Nevertheless, it can be seen that the addition of cement only slightly reduced the viscosity of the Telmarch fluid, whereas a significant increase in viscosity can be seen for the bentonite fluid. Note that there is a small discrepancy between Figures 4.2 and 4.3 – the former shows a Marsh funnel time of 33 s at a polymer concentration of 1.4% while the latter shows 38 s. The reason for this is unknown. Ikuta (1974) also noted that the specific gravity (density) of the Telmarch polymer fluid changed very little during use, and this facilitated recycling of the fluid.

In addition to the effect of cement, Xanthakos (1979) noted that during excavation the viscosity of the Telmarch fluid tended to remain constant or underwent only minor changes even when the suspended solids concentration reached 15–20%. While the effect will depend on the nature of the solids suspended, this finding is queried because recent research has shown that soil particles can have a significant effect on the rheological properties of polymer fluids (Lam and Jefferis, 2015). Xanthakos also noted that during concrete placement – which usually causes an increase in fluid pH – the viscosity of the Telmarch fluid tended to decrease whereas the viscosity of the bentonite fluid would increase due to flocculation of the bentonite particles. In terms of the effect of the polymer on exposed soil surfaces, Xanthakos reported that the long-chain polymers could penetrate 'vigorously' into the soil pores to form a 'filter cake' of unusually high shear strength, and this gave the face of the side walls a remarkable stability even in regions of overhanging boundaries. For the same reason, soil cuttings did not disperse easily, and this resulted in a lower rise of the solids content in the fluid during excavation. Despite the various advantages of the Telmarch fluid, no mention of this product can be found elsewhere. A recent enquiry to the Japanese company Telnite has revealed further information about the history of the product. According to Kodama (personal communication, 2013), Telnite produced Telmarch for use in construction in Japan in the 1970s. Due to the ease of mixing and application, Japanese companies also used this material in some overseas projects. The major problem was that the Telmarch fluid biodegraded in deep excavations, which required a longer time to complete. The use of Telmarch therefore gradually declined, and Telnite eventually stopped producing the material. From this information, it can be concluded that Telmarch was a natural biodegradable polymer, although the exact type remains confidential to the manufacturer.

4.2.2 OP-4 and OP-7

OP-4 is the trade name of another polymer product that was used in Japan in the 1970s and 1980s. 'OP' probably denotes 'organic polymer'. In a discussion of the development of diaphragm walling techniques, Hosoi and Matsushita (2016) and Hosoi (personal communication, 2016) mentioned that OP-4 was a carboxymethyl cellulose (CMC) and the degree of substitution (DS) was originally less than 1.0 at around 0.6–0.7. However, CMC at this level of DS was sensitive to biodegradation, so antiseptic (biocide) solution was also needed. The CMC was subsequently reformulated to increase the DS to between 1.3 and 1.5, and this made the polymer more tolerant to salt and biodegradation – this description agrees with the generic properties of CMC as described in Section 3.4.1. Takeshi and Shinya also noted that CMC polymer fluid was specified by the Japan National Railroad for the construction of the Nagoya Subway Line 6 in 1983. After that, CMC fluids became widely used for diaphragm wall construction in Japan. Some properties of the OP-4 fluid are given by Kita and Kawachi (1979). According to Hosoi (personal communication, 2016), CMC fluids are still used in Japan. OP-7 is the trade name of a guar gum product that was also used, although there is relatively little information about its history of use. Kawachi (1987) discusses the physicochemical properties of both OP-4 and OP-7 and their use for excavation works in Japan.

4.2.3 VisKeau

VisKeau is a polymer product that is mentioned several times in the civil engineering literature but is no longer available. One of the early reports on this polymer is by White Jefferis and Associates (1989), which categorised VisKeau as a xanthan gum. More recently, Smith *et al.* (2005) described VisKeau as a naturally occurring polymer manufactured on an industrial scale by a fermentation process. It is also described as a non-toxic complex sugar (polysaccharide) that is fully biodegradable and commonly

Figure 4.4 The apparent viscosity curve of the VisKeau polymer fluid

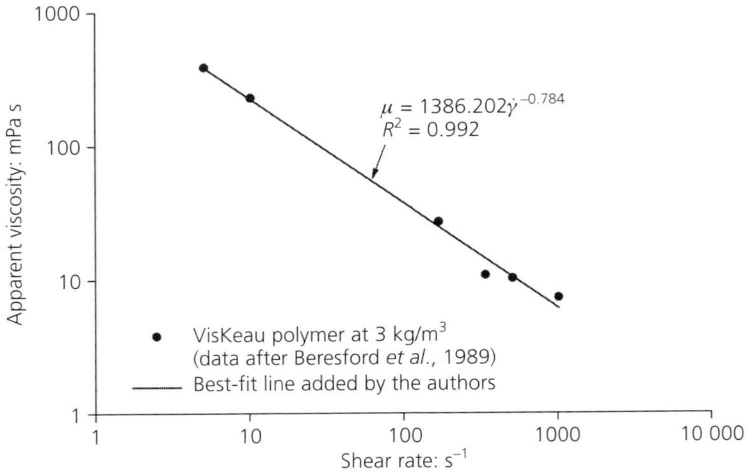

used as a food additive. These descriptions effectively confirm that it was a xanthan gum, as xanthan is the only commonly used natural polymer that is produced by a fermentation process. Since the xanthan gum is biodegradable, a biocide is usually required (Smith *et al.*, 2005).

For on-site fluid preparation, Beresford *et al.* (1989) reported the use of a high-speed mixer with sodium hydroxide solution added to the mix water to raise its pH to between 8.4 and 8.6. Beresford *et al.* (1989) presented viscosity data of a VisKeau fluid mixed at a concentration of 3 kg/m^3 in fresh water. The viscosity was measured using a direct-indicating viscometer. Figure 4.4 shows the viscosity curve.

It can be seen from Figure 4.4 that the VisKeau fluid shows decreasing apparent viscosity with increasing shear rate. This behaviour is known as shear thinning, and is a rheological property common to most polymer fluids. Smith *et al.* (2005) also noted that as a consequence of this shear-thinning property, during pumping the frictional resistance will be lower than that of traditional bentonite-based slurries. As would be expected for a xanthan gum (see Section 3.4.3), Smith *et al.* (2005) also reported that the VisKeau fluid was resistant to high shear forces and tolerant of a wide range of salts in water.

4.2.4 PREMIUM

PREMIUM is a biodegradable low-molecular-weight polymer based on cellulosic derivatives supplied by Liquid Earth Support in the USA. This product is also referred to as GS TP Premium in the manufacturer's literature. It is supplied as a creamy white powder, and is compatible with bentonite slurry. According to the material information sheet, this polymer is a 'semi-synthetic polymer formulation based on cellulosic derivatives'. It is thus possibly a CMC derivative. Also according to the manufacturer, this polymer has been used for several structural diaphragm wall and bio-polymer drain projects in the USA. For example, it was used for the structural diaphragm (slurry) walls of the International Cultural and Trade Center in the Federal Triangle, Washington, DC. For this project, the PREMIUM polymer was used together with bentonite at concentrations of 0.15% and 1.5%, respectively. When used in this way, the manufacturer states that the resulting fluid will behave as a mineral slurry with respect to filter cake formation and as a polymer fluid with respect to its inability to suspend sand and ability to self-clean. The reported filtrate volume of the resulting fluid was 20–40 cm^3, measured using the conventional American Petroleum Institute fluid loss cell, and the Marsh funnel time was 32–38 s.

The manufacturer also recommends the product for drilling potable water wells because of its very low biochemical oxygen demand (BOD) at 5–15 (unit not given, probably mg/l). The BOD is a measure of the quantity of oxygen used by micro-organisms in the oxidation of organic matter, and is therefore used as an indicator of potential impacts on water quality. It is worth mentioning that the use of certain types of biodegradable polymer additive (e.g. guar gum) for drilling water wells caused controversy in the past due to the residual polymers contributing to bacterial growth (Richard 1979a, 1979b). Howsam *et al.* (1995) reported a trend away from the use of biodegradable polymers for water wells, but bacterial growth promotion can be addressed by, for example, flushing and polymer oxidation. The manufacturer of the material notes that the biodegradable PREMIUM polymer is now recommended for the construction of deep drainage (leachate collection) trenches where biodegradability and cleanliness are essential for the drain performance.

4.2.5 AquaVIS

The AquaVIS product line from Ashland comprises a range of CMC polymers marketed and sold for civil engineering works, including micro-tunnelling, piling, diaphragm walling and horizontal direction drilling (Ashland, 2013). The available products include AquaVIS 633, which is supplied in powder form, and AquaVIS SC 50, which is supplied as a liquid emulsion. The benefit of the latter is quicker mixing, but the active concentration in the liquid is only 45%. The manufacturer also has an ETD ('easy to disperse') powder product range that has improved dissolution performance compared with conventional CMC products. All the products are designed to be used as extenders (additives) to bentonite fluids, although the manufacturer also states that for clayey formations the polymers can be used alone. The benefits of using CMC in bentonite fluids include a reduction in filter loss, increased tolerance of contamination, and development of rheological properties leading to higher fluid viscosity. The increase in viscosity also increases the solids-carrying capacity of the fluid. This feature is particularly useful for applications such as horizontal directional drilling that require the fluid to suspend solid cuttings for transport to the surface plant. For most applications, the manufacturer recommends a typical bentonite dosage of between 20 and 60 kg/m^3 and a polymer dosage of between 0.5 and 3.0 kg/m^3. Compared with other types of bentonite extenders, CMC is noted to have good salt tolerance and excellent fluid loss control properties. It is worth adding that guar gum, which also has good salt tolerance, can have fermentation issues and is sensitive to microbial degradation.

4.2.6 SC GUM, SC XGUM and SC VIS

Süd-Chemie, now part of the Clariant group, offers several biodegradable polymers for civil engineering use. These include SC GUM, which is a guar gum, SC XGUM, which is a xanthan gum, and SC VIS, which is a cellulosic ether. The first two, SC GUM and SC XGUM, are recommended for applications such as drainage works, water well drilling and horizontal directional drilling. The last, SC VIS, although described by the manufacturer as a 'cellulosic ether', which may mean one of the many cellulosic polymers, is categorised by Lesemann and Vogt (2012) as a CMC. According to the manufacturer, SC VIS has a molecular weight of less than 1 million g/mol and is available in both low- and high-viscosity versions (LV and HV), as is typical for CMCs. The product can be supplied in either powder or liquid form, with the letter 'P' or 'L' as appropriate added at the end of the product name. In terms of application, both LV and HV versions (the latter is also and confusingly known as SC GSTP) can be used as a fluid-loss control additive for bentonite fluids, especially the manufacturer's own brand BENTONIL GTC4. According to the manufacturer, a combination of BENTONIL GTC4, SC GSTP and a thinner was used for the construction of some diaphragm walls and deep bored piles for the Dubai Metro project. The high-viscosity CMC (SC VIS HV P/GSTP) at a concentration of between 0.1% and 0.4% (1–4 kg/m^3) can be used on its own as a pure polymer fluid. Both SC VIS HV P and SC XGUM were used by Lesemann and Vogt (2012) for field and laboratory trials (discussed in Chapter 7).

4.3. Polyacrylamide-based systems

4.3.1 GS 550

GS 550 is a polymer that has been primarily used for drilling small-diameter site investigation boreholes. According to Campanella *et al.* (1990), GS 550 is a granular copolymer of sodium acrylamides and acrylates; that is, a PHPA (see Figure 3.7). Scarrow and Gosling (1986) stated that the molecular weight of this polymer is about 25 million and the preferred pH range for site use is 7–9. It is believed that this polymer is a high-molecular-weight polyacrylamide. Figure 4.5 shows Marsh funnel viscosities for solutions of the GS 550 polymer fluid over a range of concentrations, c_p.

As can be seen from Figure 4.5, the viscosity of the fluid increases rapidly with increasing polymer concentration. Probably due to the marked shear-thinning property, Scarrow and Gosling (1986) noted that this fluid can be pumped (circulated) at relatively low pressures in boreholes, and that this is an especially useful feature with wire-line drilling equipment, which has a restricted annular return space. Scarrow and Gosling also mentioned that GS 500 is biodegradable but that the use of bactericide is not required. This most probably stems from a misunderstanding by them that all organic polymers are biodegradable. Actually, PHPA is not biodegradable, as it cannot be used as a food source by micro-organisms. However, its properties can degrade with time due to other reasons such as ageing. Furthermore, biodegradable polymers will always require a biocide to inhibit fermentation.

Some brief case studies of use of this polymer are presented here, as it will not be discussed further in later chapters. The first case study identified is that provided by

Figure 4.5 Marsh funnel times of GS 550 fluids at different concentrations

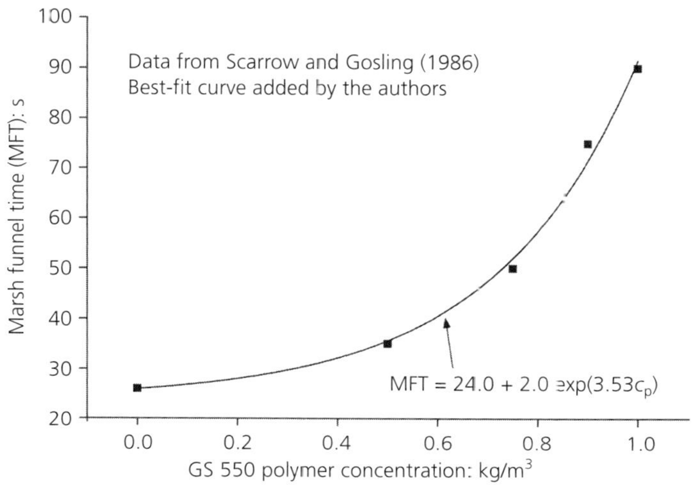

Jannaire (1981), who described the use of the GS 550 polymer fluid for the undisturbed sampling of soils during rotary coring. Jannaire stated that the success of the technique depended on the long polymer chains depositing themselves onto the cuttings and linked polymer chains forming a stabilising web or thin skin on the borehole wall and on the outer surface of the cores. Jannaire also noted that the drilling method and the drilling fluid needed to be complementary to ensure complete recovery of intact core.

Scarrow and Gosling (1986) described another rotary coring project, in the Middle East. This case study is also discussed by Clayton *et al.* (1995). Ground conditions at the site consisted of very stiff clay overlying silty and slightly gravelly sand. Water was used to drill the upper clay layer in each borehole, and the GS 550 fluid was used to drill the sand layers at depth. Bentonite fluid was also tried in a few boreholes in the sand layers but was found to lead to unacceptably low core recovery. As regards fluid mixing and maintenance, the polymer fluid was prepared in a pre-dug pit, with polymer granules introduced into the mix water through a small funnel. Based on actual consumption data, the polymer concentration was around 0.7 kg/m^3. After mixing, an initial Marsh funnel time of about 60 s was achieved. Note that this viscosity value is different from that shown in Figure 4.5, although not by a large amount; the difference was probably caused by site factors such as mix water quality. During drilling, the viscosity of the fluid was maintained by adding extra polymer material to compensate for the polymer molecules deposited on the borehole walls and taken up in encapsulating the cuttings. No problems were experienced with the cuttings not settling out in the sedimentation pit, and the polymer fluid was described as self-cleaning. Core recovery was at or very close to 100% in both the clay and sand layers. From the recovered cores, it was found that, as mentioned by Jannaire (1981), the GS 550 fluid formed a thin skin on the surface of the cores and did not permeate into the body of the core. Therefore, the natural density and moisture content of the cores could be determined, so enabling direct assessments of the relative density of the sand.

The use of the GS 550 drilling fluid in rotary coring has also been reported in several other case studies, including those of Beckwith *et al.* (1995), Johnson *et al.* (1995), and Pickles and Everton (1995). Readers can refer to these publications for further information on the history of use of this polymer product.

4.3.2 Geomud-15

Compared with other polymer products currently available in the marketplace, relatively little information has been found about the Geomud-15 polymer, except for the description by Bustamante *et al.* (1998) that it is a high-molecular-weight polymer. According to the data sheets provided by the manufacturer (Petrind), Geomud-15 is a polymer designed for civil engineering use, including piling, diaphragm walling and the drilling of water wells. On the data sheets it is listed as a 'copolymer P(AA-AM)'; that is, a copolymer of acrylic acid (AA) and acrylamide (AM), or in other words a PHPA. It should be noted that, strictly speaking, describing the monomer as an 'acrylic acid' ($CH_2{=}CHCOOH$) is incorrect, as it should be the salt (typically sodium) of the acid's conjugate base ($CH_2{=}CHCOONa$), as shown in Figure 3.7. Geomud-15 is supplied as white granules, and its natural pH is between 7 and 9.

Figure 4.6 shows the Marsh funnel time plot provided by the manufacturer. In this instance, the Marsh funnel time is that determined for the efflux time of 1 litre from a funnel filled with 1.5 litres of fluid. The efflux time for water should be 28 ± 0.5 s for 1 litre. It can be seen that, similar to the viscosity curves of Telmarch and GS 550 (see Figures 4.2 and 4.5), the Marsh funnel time of the Geomud-15 fluid increases rapidly with increasing polymer concentration. In terms of fluid specifications, the manufacturer

Figure 4.6 Marsh funnel times of Geomud-15 polymer fluids at different concentrations

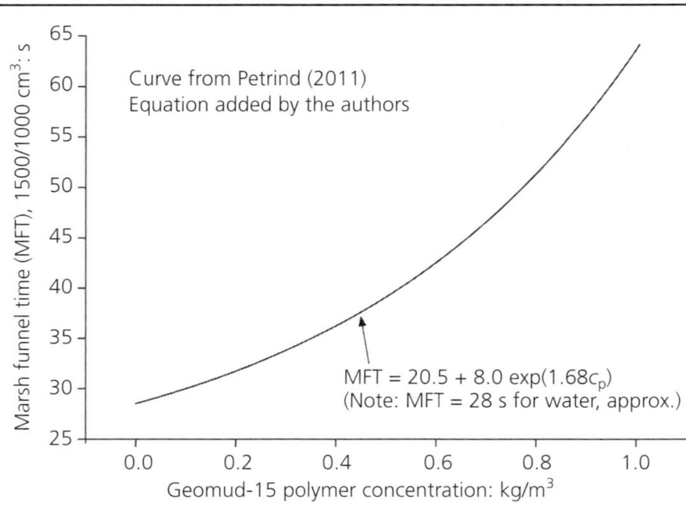

suggests that the fluid density should not be higher than 1.08 g/cm^3 and that the Marsh funnel time should be kept within 38–55 s. This polymer has been used in various countries, including Portugal, Italy and Benin, and these case studies are discussed in Chapter 7.

4.3.3 SC MUD

SC MUD is a synthetic polymer supplied by Süd-Chemie (now part of the Clariant group) for geotechnical applications such as soil sampling, bored piling, tunnelling and pipe jacking (Süd-Chemie, 2013). This polymer is described by the manufacturer and also Lesemann and Vogt (2012) as an anionic polyacrylamide; that is, a PHPA. The product is supplied in both powder (designated as 'P') and liquid emulsion forms (designated as 'L'). According to the manufacturer, the benefits of this polymer include high fluid viscosity at small dosages, quick hydration time, reduced clay stickiness and non-fermenting. As for some other PHPAs, it is recommended that soda ash is used to raise the pH of the mix water to between 8 and 10. The manufacturer recommends that the electrical conductivity of the mix water, an indication of the concentration of dissolved ionic species in the water, should be less than 1000 μS/cm. Measuring the electrical conductivity of the mix water is a quick way of checking the amount of dissolved salts of the water, although at present it is not commonly done in construction practice.

Figure 4.7 shows the manufacturer's Marsh funnel time data for the SC MUD (P and L) fluids. It can be seen that a range of Marsh funnel times is shown for each polymer type, perhaps reflecting the range of results that can be expected in typical waters and with typical mixing equipment.

Perhaps due to a potential issue of high fluid loss in coarse permeable grounds, the manufacturer recommends adding some bentonite to the fluid. The recommended mix

Figure 4.7 Marsh funnel times of SC MUD P and SC MUD L

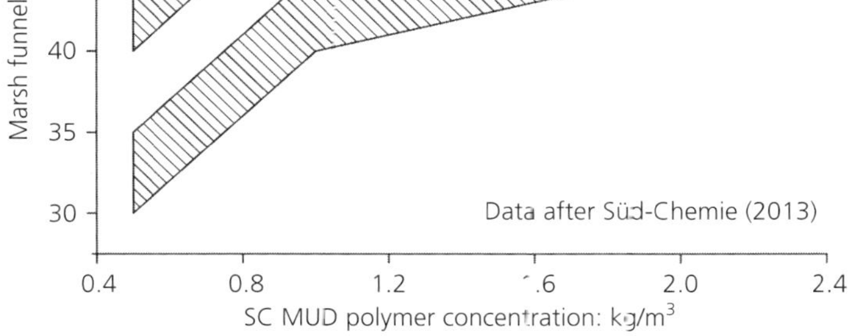

procedure is to add bentonite at a concentration of between 5 and 10 kg/m^3 and then the SC MUD (P or L) to obtain a target Marsh funnel time of between 40 and 60 s. The manufacturer notes that the polymer fluid will have rather limited ‘wall-building’ (filter cake formation) properties without the bentonite. As will be discussed in Chapter 7, the use of polymer–bentonite mixtures seems to have become common practice in some countries but can have some drawbacks.

4.3.4 Super Mud

Super Mud is a liquid emulsion polymer that has been available in the market for over two decades. It is an anionic polyacrylamide (i.e. a PHPA) supplied in a water-in-oil emulsion form. There is also Super Mud Dry, which is again a PHPA, but supplied in dry granular form. According to the manufacturer (PDS, 2013), the advantages of synthetic polymers include the following: they (*a*) are easy to mix and require little or no mixing equipment, (*b*) can be mixed in either fresh or salt water, (*c*) increase the speed of production, (*d*) reduce wear on tools and equipment, (*e*) control fluid loss, (*f*) are recyclable and reusable, (*g*) lower disposal costs and (*h*) are environmentally friendly. These are realistic claims for a PHPA, although their realisation always requires good site practice (see Chapter 5). In terms of mixing, the manufacturer recommends treating the mix water with sodium carbonate, which it markets under the trade name Water Treat, to raise the pH to between 8 and 10. Raising the pH by the addition of sodium carbonate will precipitate calcium hardness in the water as calcium carbonate. The recommended dosage for water treatment is 0.5 kg for either 800 litres of fresh water or 400 litres of brackish or salt water. The manufacturer states that polymer mixing can be achieved in a variety of ways. The polymer (liquid or dry powder form) can be added via a Venturi-type mixer or slowly and directly into a rapid, turbulent stream of water filling a mixing tank. For in-hole or at-hole mixing, the emulsion polymer can be added directly into a stream of fluid at the excavation. An auger or drilling bucket should then be slowly moved up and down in the fluid column to distribute the polymer and homogenise the fluid. Slow rotation of the tool can also be applied. Because the properties of the fluid can be more readily controlled in a surface mixing tank, this mixing option is recommended by the manufacturer, especially for large-scale projects.

For the emulsion polymer, the manufacturer recommends a typical mixing ratio of 1 : 800 (0.125%) for fresh water and 1 : 600 (0.17%) for brackish water. Mixing ratios for specific soil types for both granular and emulsion forms are also given in Table 4.1. It should be noted that emulsion polymers are sensitive to freezing and have a shorter shelf life than their dry granular counterparts. By way of example, Figure 4.8 shows a drum of emulsion PHPA polymer that appears to have passed its shelf life: the emulsion has become a semi-solid whereas it was originally a viscous liquid.

In addition to the base polymer, PDS also provides three additives. The first is an additive called Aquasorb, which, according to the safety data sheet, is also a PHPA, but the molecules are cross-linked. Due to the cross-linking, the polymer can absorb many hundreds of times its own weight of water, and swell to form a durable crystalline gel. In coarse, highly permeable grounds, the gel particles can help to clog the pores of the formation, and so reduce bulk fluid loss. To achieve maximum results, the manufacturer

Table 4.1 Recommended dosages and Marsh funnel times for Super Mud and Super Mud Dry

Formation type	Super Mud		Super Mud Dry	
	Dosage: v/v	Marsh funnel time: s	Dosage: kg/m^3	Marsh funnel time: s
Clay and shale	1/800	35–45	0.2–0.5	40–50
Silt and fine to medium sand	1/600	45–60	0.5–1.0	50–60
Coarse sand to pea gravel	1/400	>60	1.0–1.5	60–80
Gravel to cobbles	—	—	1.5–2.0	>80

PDS (2013).

recommends pre-hydrating 1 lb (0.45 kg) of Aquasorb with 3 US gallons (11.36 litres) of water for 10–15 min before introduction to the excavation. It may be noted that cross-linked polyacrylamides and polyacrylates, which are key ingredients of superabsorbent polymers, are commonly used to make disposable hygiene products such as sanitary towels and baby nappies. Superabsorbent polymers have also been used to absorb free water in unwanted fluids at the end of drilling projects (Oliver, 2015).

The second additive, which is also a fluid-loss control agent, is a granular bentonite in pellet form with the trade name Granular Seal. The manufacturer describes it as a high-swelling, chemically unaltered sodium bentonite. Bentonite pellets will swell in

Figure 4.8 Emulsion PHPA polymer past its shelf life. (Courtesy of KB International)

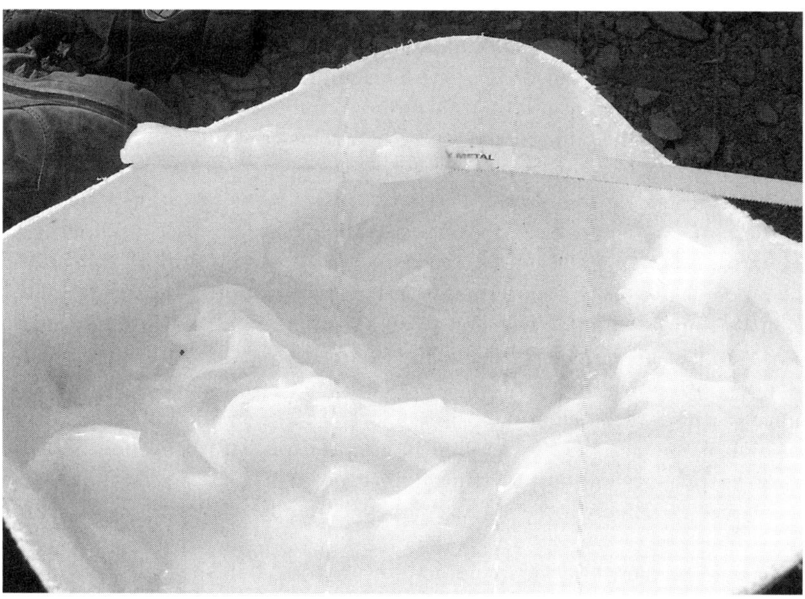

water, and the swollen mass can help to clog formation voids, so reducing fluid loss. Bentonite pellets are commonly used to form a seal in site investigation and piezometer boreholes. The difference between the two additives – Aquasorb and Granular Seal – in terms of performance and situations where they should be used is not stated by the manufacturer. As for Aquasorb, the manufacturer recommends pre-hydrating Granular Seal in water before introduction to the excavation. This is because, if added directly to an excavation, the polymer fluid will retard the swelling of the bentonite due to sorption (Inyang *et al.*, 2007) – this is the same reason why polymer fluids can retard the swelling of excavation side walls.

The third additive, known as Quick Floc, is a non-ionic polyacrylamide supplied in liquid emulsion form. The manufacturer states that this product can rapidly agglomerate silt and other micrometre-sized particles that are suspended in the fluid and thus reduce the time required for their settlement to the bottom of the excavation. Due to the non-ionic nature of the product, it is salt tolerant. For spoil settlement, the manufacturer recommends a dose rate of 1–2 parts of Quick Floc to 15 000 parts of support fluid. The product can be premixed with the base polymer in a surface tank, or added directly into a rapidly moving stream of support fluid being pumped to an excavation. According to the manufacturer, settling of the suspended particles should occur within 30–60 min, depending on the amount of fines to be settled and the settling depth. The manufacturer also notes that the use of centrifugal pumps should be avoided, as they tend to break down the molecular chains and thus reduce the effectiveness of the product – this point will be discussed further in Section 5.3. Published case studies documenting the use of Super Mud include those by Kheng *et al.* (1991) and Hardy (2009).

4.3.5 POLY-BORE

POLY-BORE is a dry granular polymer product marketed by Baroid Industrial Drilling Products for bored piling, horizontal directional drilling, trench excavation and reverse circulation rotary drilling. According to the manufacturer, POLY-BORE is a PHPA with a high molecular weight of about 20 million g/mol and an anionic charge of about 38% (Baroid Industrial Drilling Products, 2013). This combination allows the polymer fluid to develop high viscosity and inhibitive properties against clay swelling. The advertised advantages of this product include the following: it (*a*) disperses easily with minimal shear, (*b*) is an efficient viscosifier and shale/clay stabiliser, (*c*) does not require a solids control unit to clean the fluid, (*d*) results in maximum skin friction and ultimate end-bearing capacity, (*e*) is non-fermenting, (*f*) breaks down chemically with bleach (sodium hypochlorite) and (*g*) is non-toxic. Most of these advantages are, in fact, common to almost all polymers. The claim regarding maximum ultimate end-bearing capacity does, of course, require that good construction practice is followed, and the importance of this will be elaborated in Chapter 5. The manufacturer also states that POLY-BORE is not designed to be used in conjunction with bentonite fluids, and the authors agree that the addition of bentonite to products that are designed to be used as pure polymer systems is to be avoided unless carefully proven by prior testing.

In terms of fluid preparation, the manufacturer proposes that the mix water should meet the following requirements: (*a*) a total chloride content of <1500 ppm (mg/l), (*b*) a total chlorine content of <50 ppm, (*c*) a total hardness of <150 ppm as calcium and (*d*) a pH of between 8.5 and 9.5. Typically, the last two requirements may be satisfied by adding soda ash at a recommended dose of 0.6–1.2 kg per cubic metre of make-up water. In terms of dosage, the recommended concentration of POLY-BORE ranges from 0.4 to 1.2 kg/m^3 (3 to 10 lb per 1000 US gallons) for construction applications and from 0.6 to 1.2 kg/m^3 for reverse circulation drilling. However, neither the expected nor the required range of Marsh funnel times of the resulting polymer fluid is given. In terms of the use of additives, it is interesting to note that this manufacturer does not promote the use of any additives to augment the properties of the polymer fluid, although a wide range of additives are available from this manufacturer for clay-based drilling fluids. Lastly, it should be mentioned that the manufacturer also sells a liquid emulsion PHPA known as EZ-MUD PLUS for foundation drilling. This polymer has a molecular weight of 15–17 million g/mol and an anionic charge (degree of hydrolysis) of about 30%. The active concentration in the emulsion is 30%. There are several other PHPA products provided by this manufacturer with similar names, such as EZ MUD (4–6 million molecular weight, 30% anionicity, emulsion), EZ-MUD GOLD (1–2 million molecular weight, 30% anionicity, dry granules) and EZ MUD DP (15–17 million molecular weight, 30% anionicity, dry granules); these products are not listed for use in foundation construction.

4.3.6 SlurryPro CDP

The SlurryPro CDP system, marketed by KB International (formerly KB Technologies), is one of the most commonly used and long-established polymer systems for foundation works; some of the case studies will be discussed in Chapters 7 and 8. It is called a polymer system because it comprises a base polymer coupled with a range of additives. When describing the creation of the SlurryPro system, the manufacturer mentions that one base product could not meet the requirements of all environments, and therefore extra products were developed for addition to the base system to create specific performance enhancements as required. This polymer system is also marketed as a vinyl polymer, and is referred to as such in some literature (e.g. Holmes, 1993; Goodhue and Holmes 1995; Lennon *et al.*, 2006); the meaning of the term 'vinyl' was explained in Section 3.3.1. According to the manufacturer, this polymer system possesses both positive and negative charges and has associative characteristics, thus making it easier for the polymer molecules to bind to soil particles to form a 'membrane' on exposed soil surfaces on both the walls of an excavation and the cut soil.

The base polymer, known as CDP, is a dry granular polyacrylamide with a molecular weight of approximately 20 million g/mol and an anionic charge density of nominally 40% (Goodhue and Holmes, 1995). KB International recommends that the CDP polymer should be mixed with water pretreated with a pH control agent known as ProTek 100, which is a concentrated potassium hydroxide solution. Figure 4.9 shows a drum of ProTek 100 and a bag of CDP. ProTek 100, when used, must be handled with great care as it has a very high pH and is thus very corrosive. The manufacturer has now developed a system that does not require added alkali (see Section 4.5). The purpose of ProTek 100 is to create an alkaline environment and to reduce water hardness so as to maximise the development of polymer properties. The general guideline for using ProTek 100 is to add the product slowly to potable water or recycled polymer fluid until

Figure 4.9 The pH controlling agent ProTek 100 and the base polymer CDP

the pH reaches a level of 9.5 to 11.5. The base polymer and other additives can then be added. The recommended dosages of the base polymer and various additives are summarised in Table 4.2. The typical Marsh funnel times of the fluid at various polymer concentrations are also given. It can be seen that the recommended average dosage of the CDP polymer ranges from 0.4 kg/m^3 for clay and shale to over 2.0 kg/m^3 for gravel and cobbles. In coarse permeable soils, a higher polymer dosage will result in a higher fluid viscosity and thus less fluid loss to the ground.

LA-1-D is a polymeric additive supplied in powder form. The manufacturer states that it is designed to work in conjunction with the CDP polymer to create a polymeric film on the exposed surfaces of an excavation. It is intended to be introduced slowly at the

Table 4.2 Recommended average dosages for the SlurryPro CDP polymer system[a]

Formation type	Dosage: kg/m^3				Marsh funnel time: s
	SlurryPro CDP	LA-1-D	Impac	KobbleBlok	
Clay and shale	0.4–1.0	0.02–0.08	0.02–0.05	0.01–0.075[b]	45–75
Silt and fine to medium sand	0.5–1.5	0.03–0.25	0.03–0.08	0.02–0.1	55–120
Coarse sand to pea gravel	0.7–1.8	0.04–0.50	0.04–0.10	0.03–0.2	65–150
Gravel to cobbles	0.9–>2.0	0.05–>1.0	0.05–>0.15	0.03–>0.3	75–>150

KB International (2013).
[a] Additives whose dosages are not dependent on the ground conditions are not listed here.
[b] Clay and shale are fractured.

excavation or into the mix tank. Rapid uncontrolled addition of this additive to the base CDP fluid will result in agglomeration of the powder and so material wastage. In terms of functionality, it is known to be an associative thickener that contributes to the associative characteristics of the polymer fluid. According to the manufacturer, since the polymer molecules contain hydrophobic pendant groups, in solution the pendant groups are associated with each other to form a three-dimensional network that increases the overall fluid viscosity and so reduces the potential for fluid loss to the ground. It may be noted that associative thickeners are widely used in the formulation of paints to improve their rheological properties, to reduce splattering and brush drags (e.g. Mussard 2003; Shay 2012).

MPA is a reactive additive for the CDP system. It is supplied in liquid form, and is stated to act as a catalyst for the CDP polymer, producing strands of a soft gel-like material that have a strong affinity with soil particles. MPA can therefore be used to remove dispersed soil from used support fluid. The soil particles stick to the strands, which, under the weight of these particles, settle through the fluid, collecting more soil as they sink, thereby greatly accelerating the soil removal process. The gel mass can also physically plug voids or coarse soils, so reducing fluid loss to the ground. Figure 4.10 shows soil particles held by some of the elastic strands produced by this additive.

Figure 4.10 Soil particles held together by the elastic strands formed by adding the polymeric additive MPA to the CDP polymer fluid. (Courtesy of KB International)

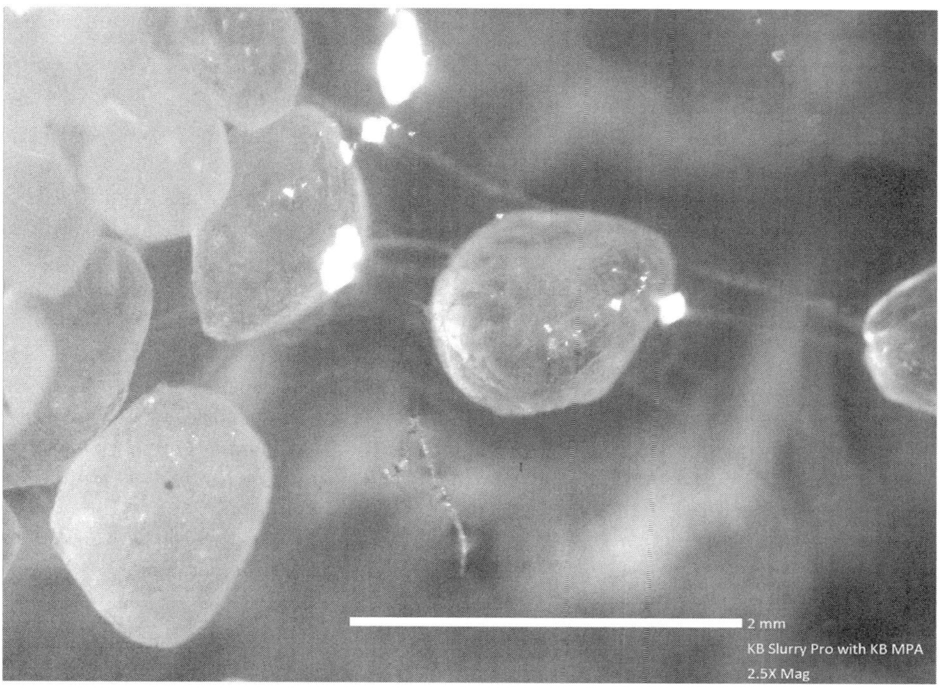

Figure 4.11 Adding dilute MPA solution in a weighted plastic bag to the polymer support fluid in a pile bore

MPA is supplied as a concentrated solution, and should be diluted to about 1–1.5% for use. During excavation, the dilute MPA solution may be added slowly to a stream of fluid entering the excavation, to clean the fluid and also to reduce fluid loss. After completion of the excavation of a hole and prior to concreting, it is often necessary to reduce the concentration of suspended solids in the fluid to prevent the formation of a 'soft toe'. To achieve this, dilute MPA solution may be introduced by pouring 4–10 litres into a weighted plastic bag and then dropping the bag into the excavation. The manufacturer colloquially refers to this procedure as a dropping an 'MPA bomb'. Figure 4.11 shows a photograph of this operation on a bored piling job site. The contents of the bag are released by lowering an excavation tool to the bottom of the excavation. The tool can then be slowly rotated and raised up and down a few times to disperse the MPA solution through the fluid column. A subsequent waiting period of at least 30 min is required to allow the soil particles to settle to the base of the excavation. Once the sand content of the fluid has reached below the specified limit, the settled solids collected at the bottom of the excavation are removed with a cleaning bucket (or grab) or by airlifting. MPA is a vigorous additive, and care should be taken not to overdose the system.

SandSeal is an inexpensive modified mineral supplied in dry powder form. The safety data sheet of the product identifies it as a natural kaolinite/silica mineral containing crystalline silica with a grain specific gravity of 2.56. As the name suggests, it is a fluid-loss

Figure 4.12 Addition of the mineral-based additive SandSeal to a polymer fluid entering a pile bore

control additive to help block and seal coarse formations. The manufacturer's guideline for SandSeal addition is to pour the product slowly, in 10–20 kg increments, into fresh or recycled fluid in a mixing tank or directly into the support fluid during transport to an excavation. Figure 4.12 shows a bag of SandSeal being added to a stream of CDP polymer fluid entering an excavation. Additional materials may be added until the desired effect is achieved. However, as the solid particles will increase the fluid density, the manufacturer recommends that the SandSeal dosage should not exceed 3 kg/m^3, or the fluid density may exceed the prescribed limit, although the actual effect on the fluid density at this dose rate will be very minor. As for the other additives, SandSeal should be introduced slowly into the fluid in motion, as rapid addition or a lack of fluid agitation can lead to the formation of unwetted clumps, and so wastage. The manufacturer reports that SandSeal, at the recommended dosage, does not significantly alter the fluid viscosity, as do some reactive colloids such as bentonite.

Impac is a polymeric additive supplied in dry granular form. It is listed as a polyacrylamide on the material safety data sheet. This product also acts as a catalyst for CDP. It is thought that Impac is a cationic polyacrylamide and so can interact with CDP – an anionic polymer. The manufacturer notes that Impac is a milder additive than MPA, and that in many formations it can be used in its place. However, in more challenging situations (e.g. very coarse grounds) dilute MPA solution should still be incorporated into the fluid system. The recommended dosage of Impac is given in Table 4.2. As for SandSeal and the other additives, Impac may be added to the fluid in a mixing tank, or directly at the hole – but this must be done slowly. It may be added as a small stream of dry powder or as a dilute solution in water.

KobbleBlok is a polymeric additive supplied in dry granular form. It is designed to clog coarse formations and to reduce bulk fluid loss to the ground by physically plugging coarse voids. The manufacturer recommends that it is used when other additives cannot provide sufficient control of bulk fluid loss. For use, the required amount (see Table 4.2) should first be mixed into a bucket of fresh water to allow the granules to hydrate for about 10–15 min. Then, when the swollen particles are introduced into the excavation, they will follow the fluid loss paths to clog the soil pores. Further, since this additive does not react with the base polymer fluid, it can be added until the desired control of fluid loss is reached. The preparation method and function of KobbleBlok are very similar to those of Aquasorb, an additive from PDS for the Super Mud fluid system. It is therefore possible that KobbleBlok is also a cross-linked polyacrylamide; that is, a common superabsorbent polymer. Figure 4.13 shows this additive before and after hydration.

InstaFreeze is a synthetic grouting additive supplied as a heavy, clear liquid. It is used to create a soft polymeric grout within the side walls of an excavation, thereby improving the stability of the excavation. The grouting action can be accelerated by the immediate follow-up addition of InstaFreeze C2, which acts as a setting agent. The general guideline for adding InstaFreeze is to add the product slowly in 50–190 litre increments directly to the fluid in the excavation until the desired effects are achieved. For spot addition, such as to plug a small broken utility pipe, InstaFreeze should be applied directly to the excavation, with the digging tool placed in the excavation at the desired point of application. During the addition, the tool should be slowly rotated to assist in delivering the additive to the desired treatment point. Since InstaFreeze is a heavy liquid, it will slightly increase the overall density of the fluid. As InstaFreeze does not contain any mineral colloids that can plaster the walls of an excavation, the shaft resistance of the completed foundation elements should, in theory, not be affected, although there is currently no laboratory or field evidence to support this.

WeightIt is a heavy, clear liquid. It may be used in situations where it is necessary to increase the density of the support fluid, such as excavation through aquifers under artesian pressure or through material containing flammable gas (e.g. Hagy, 2013). In such circumstances, it is necessary to increase the hydrostatic pressure of the support fluid in order to counteract the underground water or gas pressure, and increasing the fluid density could be an alternative to increasing the support fluid level by use of a long surface casing – a case study of the latter approach is mentioned by Brown *et al.* (2010). It is worth noting that the use of solid weighing agents such as barite (specific gravity 4.5 g/cm^3) is a common and well-established approach to densify drilling fluids in oil well drilling (Darley and Gray 1988). Since WeightIt does not add any mineral colloids as do traditional weighting agents such as barite and haematite, it will not form a filter cake on the excavation side wall, and therefore will not reduce the shaft resistance of the

Figure 4.13 A superabsorbent polymer for fluid-loss control in coarse grounds: (a) dry granules; (b) hydrated form

(a)

(b)

completed foundation elements. According to the manufacturer, WeightIt is applied in a similar manner to InstaFreeze; that is, the product is added slowly, in 50–190 litre increments, to the fluid in a mix tank or the excavation until the desired density is achieved.

4.3.7 Enhanced SlurryPro CDP

The Enhanced SlurryPro CDP system, as the name suggests, is an improved version of the original system. The major difference is that the number of additives has been reduced significantly from the original eight to just two: EnhancIt 100 and EnhancIt 200. Both products are supplied in dry granular form. Unlike the original system, no pH adjustment with liquid potassium hydroxide or soda ash (sodium carbonate) is required prior to mixing, except when extreme water hardness or salt water has to be used. If pH adjustment is required, the original pH-increasing agent Protek (potassium hydroxide) can be used. According to the manufacturer, EnhancIt 100 is designed to work in conjunction with the CDP base polymer to create a fluid with the capacity to control fluid loss in porous soils through the creation of a polymeric membrane or film on and within the soils of the excavation side-wall. In this respect, the EnhancIt 100 additive works in a similar way to the LA-1-D additive of the original system. In difficult grounds, EnhancIt 200 can be used to increase the strength of the polymeric gel membrane and increase the chemical grouting effect. Figure 4.14 shows a polymeric membrane formed on a pack of sand and then pulled by hand. The sand particles can be seen to adhere to a polymeric membrane that acts as an elastic web. In this respect, the EnhancIt 200 additive behaves

Figure 4.14 A polymeric membrane formed on the surface of a pack of sand. (Courtesy of KB International)

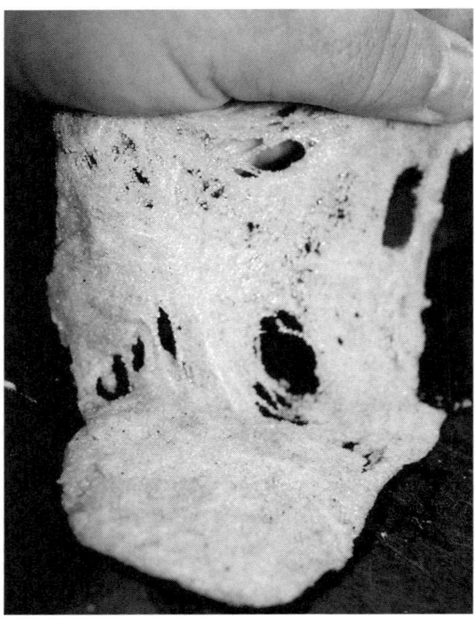

in a similar way to the original MPA additive, which also produces elastic strands when added to the base polymer fluid.

As both EnhancIt 100 and 200 are proprietary products, their specific chemical formulation is not disclosed. However, on the material safety data sheet, EnhancIt 100 is stated to be a mixture of sodium polyacrylamide, modified cellulosic ether and carboxylated vinyl polymer, and EnhancIt 200 is described as a mixture of polyacrylamide and organic cationic polyelectrolyte. Both products are therefore mixtures of chemicals, and this is why only two additives are needed in the new system. The manufacturer recommends that EnhancIt 100 is used at a dosage of about 0.2–0.5 kg/m^3 and that it is added to fresh fluid in a mixing tank using an in-line mixer or a splash plate. Used fluid recovered from an excavation can be supplemented with fresh EnhancIt 100 as required to achieve the required level of fluid loss control. Typical supplemental dosages range from 0.01 to >0.5 kg/m^3. For EnhancIt 200, the recommended dosage depends more highly on the ground conditions. In general, the greater the water sensitivity of the formation or the more fines in the formation, the higher is the dose, which typically ranges from 0.005 to 0.05 kg/m^3. EnhancIt 200 may be added to the fluid in a mixing tank or directly to a stream of support fluid entering an excavation. For the latter option, the product may be added as a dilute solution at a concentration of 1% or less. This method of addition may be used when the suspended fines are building up in the fluid so that the density is high or if sloughing of the side wall occurs, as the elastic strands formed in the fluid trap solids and so help to mitigate both situations.

4.3.8 Shore Pac

CETCO supplies a polymer fluid system comprising a base polymer known as Shore Pac and a wide range of additives for performance enhancement. In many ways, the CETCO system is similar to KB International's SlurryPro CDP system although there are some differences in terms of the number of additives available and their mode of application. The manufacturer suggests that, before mixing the Shore Pac fluid, the make-up water should be treated with soda ash to raise the pH to between 8 and 10. The manufacturer also notes that this typically requires 6 lb (3.6 kg) of soda ash per 1000 US gallons (3.8 m^3) of water; that is, a concentration of 0.95 kg/m^3. A granular dechlorination additive known as De-Chlor is also available for treating mix water containing residual chlorine. About 0.5 lb (0.23 kg) of De-Chlor can be used to treat 1000 US gallons (3.8 m^3) of mix water (about 0.06 kg/m^3). De-Chlor can be added at the same time as soda ash.

The base polymer Shore Pac is listed as a copolymer of sodium acrylate and acrylamide (i.e. a PHPA). The recommended dosage ranges from 0.4 to 0.5 kg/m^3 for clays and shales to 1.0–1.3 kg/m^3 for gravels and cobbles. At these concentrations, the Marsh funnel time ranges from 35 to 125 s. Table 4.3 lists the recommended polymer concentrations for a range of ground conditions and the expected ranges of Marsh funnel time. The manufacturer notes that the concentration of Shore Pac should be increased when brackish water, salt water or seawater may contaminate the prepared fluid or are used as make-up water. The manufacturer advises that Shore Pac can be mixed by sifting it carefully into a stream of water directed into an excavation or an agitator mix tank. The

Table 4.3 Recommended dosages for Shore Pac

Formation type	Suggested dosage: kg/m^3	Marsh funnel time: s
Clay and shale	0.4–0.5	35–50
Silt and fine to medium sand	0.5–0.6	45–55
Coarse sand to pea gravel	0.7–0.9	60–75
Gravel to cobbles	1.0–1.3	80–125

CETCO (2012).

authors have reservations about the first method, as dry polymer granules typically require at least 30–40 min to fully dissolve into the mix water to produce a viscous fluid. Therefore, if the polymer is mixed in situ in an excavation, the contractor may be digging with a low-viscosity fluid during the initial period after polymer addition. A liquid emulsion PHPA known as ACCU-VIS (previously called Shore Pac XX-Poly) is also available. This is better suited to mixing directly into an excavation, as in an emulsion the polymer molecules can uncoil much more quickly, and the manufacturer notes that it can also be used to instantly boost viscosity. The recommended dosage is 1 US gallon of the emulsion polymer per 1000 US gallons of premixed fluid (0.1%), or 1–2 US gallons if added to fresh water. This product works similarly to the Super Mud polymer described earlier, although the suggested mix concentration is slightly different.

Several additives are marketed by the manufacturer for use with the Shore Pac polymer to control fluid loss to the ground. These include Macro-Fill (previously known as Sure Seal), Sand Sealant, Multi-Seal and Stone Stop. Macro-Fill is a granular superabsorbent polymer described as a lightly cross-linked sodium polyacrylate that can rapidly absorb and retain up to 300 times its own weight of water. These characteristics enable the product to seal voids and fractures in highly permeable formations. The product is similar to Aquasorb from PDS and KobbleBlok from KB International, which are also superabsorbent polymers. As for other water-absorbent polymers, Macro-Fill should be allowed to hydrate in water before being added to the excavation. A typical addition increment might be about 15 lb (7 kg) added directly to an excavation after hydration.

The second fluid-loss control additive, Sand Sealant, is a dry powdered mineral described as a natural mineral containing magnesium aluminium silicate. It is similar to KB International's SandSeal, which is also a silicate-based mineral. Originally marketed as an additive to be used on its own, the manufacturer now recommends Sand Sealant to be used in combination with Multi-Seal, which is another additive for control of fluid loss. Multi-Seal is a dry reddish coloured blend of four types of material: a flake material, a granular material and fine and coarse fibres. The material data sheet describes it as a blend of wood fibres. The manufacturer suggests the use of this product to improve excavation stability in highly permeable formations and for those containing natural fractures or voids. For use, about 10 kg of Sand Sealant and 10 kg of Multi-Seal are mixed with 30 US gallons (110 litres) of water or polymer fluid to form a thick fluid of oatmeal consistency. The entire 30 US gallon mix is then added to the excavation when the drilling tool is out of the hole.

The last fluid-loss control additive is called Stone Stop. It is described as a granular sealant composed of dried minerals in various mesh sizes. On the material safety data sheet it is stated to be bentonite clay. This product is coarser in size than Sand Sealant, and is intended for use in more extreme conditions. When used, it can be poured directly into the polymer fluid in an excavation. For sand and gravel, 2–3 pails (50 lb or 22.7 kg each) can be added as a start. For formations containing large fractures and cobbles, 4–6 pails are normally required. This product is similar to Granular Seal supplied by PDS, as both are granular bentonite pellets. However, the application methods are slightly different: Granular Seal should be allowed to first hydrate in water before being introduced to an excavation, whereas there is no such requirement for Stone Stop. The first application method is probably preferable, as polymer molecules tend to absorb on the surface of bentonite and so retard its swelling, in turn compromising its efficiency in reducing fluid loss, although it may be argued that there is usually very little time to act when excessive fluid loss occurs and hence the need for immediate addition of additives.

Insta-Clear Dry is a dry granular product designed to react with the base polymer fluid to accelerate the settling of suspended solids (Figure 4.15). It is described as a blend of inorganic salts and an organic polymer on the material safety data sheet. Although detailed chemical information is not available, it is possible that this product may contain non-ionic polyacrylamide, which is a common flocculating agent and is also an ingredient of PDS's Quick Floc. Insta-Clear Dry is believed to include a component with an acid reaction, as it can lower the pH of the treated fluid and so reduce its viscosity. Probably for this reason, the manufacturer advises that addition of this product to excavations in unstable sands and gravels should be done cautiously. Handling of the product with bare, unprotected hands results in a burning sensation due to the acidic component. The authors have conducted a small-scale bucket trial, and found that the pH and the viscosity of the fluid are significantly reduced at high dosages. This additive also can be used to treat used fluid for disposal. For use, the manufacturer recommends adding 40 lb (18 kg) of the product to 3000 US gallons (11 m^3) of used fluid, and then circulating the contents of the storage tank for 10 min.

Slurry Buster and Slurry Buster Dry are two additives from CETCO for the treatment of used fluid for disposal. Slurry Buster is a sodium hypochlorite solution, which is the chemical commonly used in a bleach. Five US gallons (19 litres) of the product can treat about 6000 US gallons (23 m^3) of polymer fluid (0.8% v/v) to break the polymer viscosity by oxidation. The product is no longer promoted by the manufacturer as it has been replaced by the granular equivalent Slurry Buster Dry, which is listed as calcium hypochlorite on the material safety data sheet. Fifteen pounds (6.8 kg) of this product can treat about 10 000 US gallons (38 m^3) of used polymer fluid. The treated fluid must then be circulated by a pump to ensure oxidation full breaking of the polymer viscosity.

4.3.9 Neat Vis NV-220

Neat Vis NV-220 is a granular polyacrylamide-based polymer marketed by Tondalee Trading in Taiwan for civil engineering works. The material data sheet states that this

Figure 4.15 The solids-settling additive Insta-Clear Dry

product has a strong anionic charge so it can be deduced that this product is a PHPA (Tondalee Trading, 2013). The molecular weight of Neat Vis NV-220 is between 18 million and 25 million g/mol. The manufacturer recommends treatment of the make-up water with soda ash. The suggested polymer dosage is 0.2–0.3 kg/m^3, and the expected Marsh funnel time is 32–50 s – these values are rather low for PHPAs although they might be established based on local experience. This product is known to have been used in Taiwan, for example, for the excavation of some diaphragm wall panels at the Cailiao Station of the Taipei Metro.

4.3.10 G3

Recently, a polymer system known as G3 was introduced to the market by Ground Engineering Operations. The main constituent of this fluid system is a high-molecular-weight granular polymer known as PolyMud. Two liquid additives, known as AlfaBond and MicroBond, are used for viscosity boosting and fluid cleaning. The manufacturer claims that there are both positive and negative charges on the polymer backbones, and that the polymer fluid can form a 'membrane' on exposed soil surfaces. These are the same features that are advertised for the SlurryPro CDP system as discussed previously. Trindade (2014) described the use of this polymer for the construction of the bored piles and diaphragm walls for the Delhi Metro railway project. According to the manufacturer, this polymer system was also used for the construction of bored piles and diaphragm walls in Portugal, Spain and Brazil.

4.4. Blended polymer systems

4.4.1 SlurryShield and SlurrySMART

A blended polymer system known as SlurryShield was introduced by KB International in the mid-2000s. The major difference between this and its SlurryPro CDP system is that the additives and the viscosifying base polymer are now incorporated into a single product. This has considerably simplified mixing on site, as operatives no longer have to deal with many separate additives and remember their 'recipes' for each site situation. The liquid caustic additive has also been removed and replaced by a granular product in the dry blend, and so separate treatment of the make-up water is no longer required in most situations. However, the manufacturer notes that in about 15–25% of situations the more challenging ground conditions may still require extra additives. Additives such as SandSeal, MPA, InstaFreeze and WeightIt therefore have been retained for these situations. In salt water environments, the use of ProTek 100 (concentrated potassium hydroxide solution) may also be necessary.

In the early 2000s, SlurryShield was further developed, and is now called SlurrySMART. Figure 4.16 shows a scanning electron micrograph of the product as supplied. It can be seen that there are many different shapes and sizes of particles in the blend, indicating the use of many components. Although detailed information is not available, the material safety data sheet mentions sodium polyacrylamide, hydrophobically modified cellulose ether, CMC and several other proprietary products. According to the manufacturer, SlurrySMART has highly associative and dual charge chemical characteristics when dissolved in water. These allow it to form polymeric membranes on soil surfaces and at locations of high fluid loss. Figure 4.17 shows such a membrane formed by simply stretching the fluid between the fingers. A simple anionic PHPA fluid cannot achieve this kind of behaviour. In terms of application, the manufacturer recommends preparing fresh polymer fluid at a higher concentration and then using a reduced dosage for the refreshment of recycled fluid. Table 4.4 shows the recommended dosages for different soil types.

4.4.2 HydroCUT

Although SlurrySMART can be used to replace the SlurryPro CDP system for bored piling and diaphragm walling with grabs, it is not suitable for diaphragm walling using a hydromill (hydrofraise). This is because when a hydromill is used, the soil and rock are cut into small pieces by the teeth on the cutting wheels, and this significantly increases the area of exposed soil surfaces and hence the chemical demand on the fluid. A higher solids-carrying capacity and greater resistance to shear degradation are also required of the fluid, as the cut soils are removed by pumping the solids–fluid mixture with a powerful pump mounted on the hydromill. The soil-laden fluid must also be able to pass through the screens of the separation plant without sheeting over them. To tackle these issues, a polymer product known as HydroCUT was developed and marketed by KB International. Similar to SlurrySMART, HydroCUT is an all-in-one blended polymer

Figure 4.16 A scanning electron micrograph of the SlurrySMART polymer

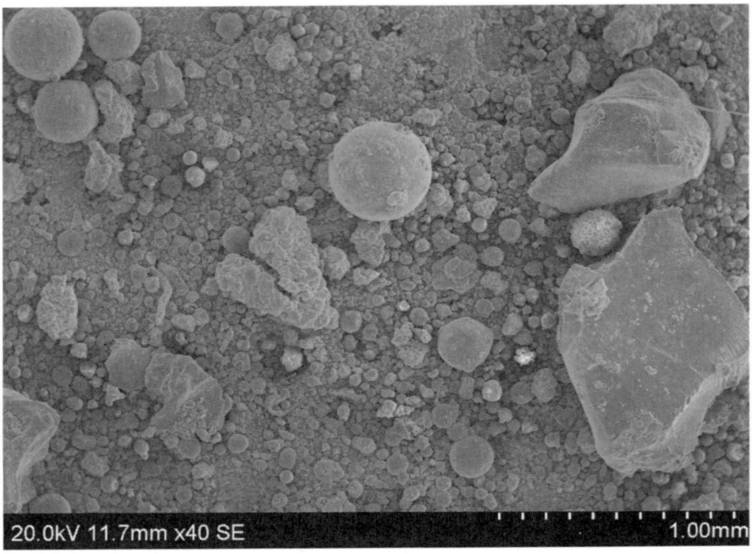

Figure 4.17 A polymeric liquid membrane formed by SlurrySMART. (Courtesy of KB International)

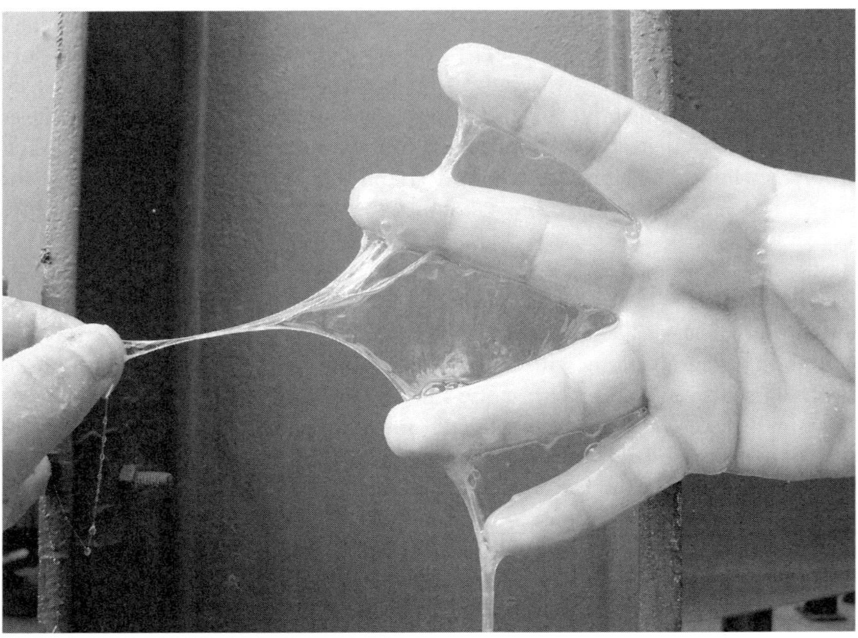

Table 4.4 Recommended initial and refreshment dosages for SlurrySMART

Formation type	Suggested initial dosage: kg/m^3	Suggested refreshment dosage: kg/m^3	Typical Marsh funnel time: s
Clay and shale	0.70–1.00	0.45–0.70	60–75
Silt and fine to medium sand	0.80–1.00	0.50–0.80	65–100
Coarse sand to pea gravel	0.90–1.15	0.60–1.00	75–100
Gravel to cobbles	1.00–1.40	0.75–1.25	85–>100

KB International (2013).

system consisting of a viscosifying base polymer coupled with a range of additives for enhanced fluid performance. The charge density of the HydroCUT polymer molecules is known to be higher than for SlurryPro CDP in order to satisfy the greater chemical demand from the soils. The membrane-forming property of the fluid enables it to coat the surfaces of the cuttings, so that when drilling in water-sensitive clays and shales the cuttings are inhibited from dispersing into the fluid as they come off the cutting face. This can prevent the excavated materials from gumming up or encasing the cutting teeth

Figure 4.18 A coal core in Mozambique, Africa, drilled using the CoreSMART polymer fluid. (Courtesy of KB International)

of the cutter wheels. Furthermore, as the cuttings are less likely to disintegrate into fines, the fluid can be kept at a relatively low solids content, and the demand on the fluid treatment plant thus also kept low.

4.4.3 CoreSMART

CoreSMART is a dry granular product designed and marketed by KB International for coring geological materials, especially for the minerals exploration market. The design of this product follows the same principles as for SlurrySMART and HydroCUT, combining a base viscosifying polymer with additives in a single package. The manufacturer states that the CoreSMART fluid can form a protective coating on the core surface as it is being cut so as to ensure a high core recovery. The polymer also acts as a lubricant to reduce torque and as a coolant to cool the drill bit. CoreSMART also has high-temperature stability for deep coring. In terms of usage, the manufacturer recommends a dosage ranging from 0.25 kg/m^3 for standard conditions to 1.0 kg/m^3 for drilling through highly fractured or permeable zones. The manufacturer states that in 90% of cases no additional additives are needed to achieve exceptional preservation of even very weak or highly water-sensitive materials, thus promoting good core recovery. Figure 4.18 shows a coal core with full recovery in Mozambique, Africa.

REFERENCES

Ashland (2013) Trade literature. http://www.ashland.com (accessed 10/04/2017).

Baroid Industrial Drilling Products (2013) Trade literature. http://www.baroididp.com (accessed 10/04/2017).

Beckwith RC, Rankin WJ, Blight IC and Harrison I (1995) Overview of initial ground investigations for the Channel Tunnel high speed rail link. *Advances in Site Investigation Practice: Proceedings of the International Conference Held in London on 30–31 March 1995*. Thomas Telford, London, UK, pp. 119–144.

Beresford JJ, Cashman PM and Hollamby RG (1989) The merits of polymeric fluids as support slurries. *Proceedings of the 3rd International Conference on Deep Foundations, London* (Burland J and Mitchell J (eds)). Balkema, Rotterdam, the Netherlands, vol. 1, pp. 3–10.

Brown DA, Turner JP and Castelli RJ (2010) *Drilled Shafts Construction Procedures and LRFD Design Methods*. US Federal Highway Administration, Department of Transportation, Washington DC, USA, Report FHWA NHI-10-016.

Bustamante M, Gianeselli L, Boato R and Conedera A (1998) Performance of polymer slurries in large diameter bored pile. *Proceedings of the 3rd International Seminar on Deep Foundations on Bored and Auger Piles (BAP III), Ghent*. Balkema, Rotterdam, the Netherlands, pp. 119–127.

Campanella RG, Stewart WP and Jackson RS (1990) Development of the UBC self-boring pressuremeter. *Proceedings of the 3rd International Symposium on Pressuremeters, Oxford*. Thomas Telford, London, UK, pp. 65–72.

CETCO (2012) Trade literature. http://www.cetco.com (accessed 10/04/2017).

Clayton CRI, Matthews MC and Simons NE (1995) *Site Investigation*, 2nd edn. Blackwell Science, Oxford, UK.

Darley HCH and Gray GR (1988) *Composition and Properties of Drilling and Completion Fluids*, 5th edn. Gulf Publishing, Houston, TX, USA.

Goodhue Jr KG and Holmes M (1995) *Earth Support Fluid Composition and Method for its Use*. US Patent 5 407 909.
Hagy M (2013) Polymer slurry used in methane gas suppression. *Foundation Drilling* **34(1)**: 45, 48.
Hardy B (2009) ExCel-lent adventure. *European Foundations* **43(summer)**: 26.
Holmes MM (1993) Drilled shafts and new slurry technology used instead of driven piles: sets record on the Ohio River. *Foundation Drilling* **32(5)**: 13–15.
Hosoi T and Matsushita S (2016) Recent diaphragm wall technologies and future challenges. *Geotechnical Engineering Journal of the SEAGS and AGSSEA* **47(2)**: 112–125.
Howsam P, Misstear B and Jones C (1995) *Monitoring, Maintenance and Rehabilitation of Water Supply Boreholes*. Construction Industry Research and Information Association, London, UK, CIRIA Report 137.
Ikuta Y (1974) Diaphragm walling in Japan. *Ground Engineering* **7(5)**: 39–44.
Inyang HI, Bae S, Mbamalu G and Park SW (2007) Aqueous polymer effects on volumetric swelling of Na-montmorillonite. *Journal of Materials in Civil Engineering* **19(1)**: 84–90.
Jannaire MM (1981) Récupération intégrale de carottes intactes en terrains meubles. *Revue de la Societé de l'Industrie Minérale* **63(1)**: 29–33. (In French.)
Johnson JGA, Newman RL, Paul TS and Pennington DS (1995) The measurement of strength, stiffness and in situ stress in the Thanet Beds using advanced techniques. *Advances in Site Investigation Practice: Proceedings of the International Conference held in London on 30–31 March 1995*. Thomas Telford, London, UK, pp. 887–899.
Kawachi T (1987) 滞水性地盤の止水法に関する土質化学的研究. PhD Thesis, Kyoto University, Kyoto, Japan. (In Japanese.)
KB International (2013) Trade literature. http://www.kbtech.com (accessed 10/04/2017).
Kheng H, Bloomquist D, Townsend FC, Thompson PY and Dobson E (1991) *Slurry Performance in Drilled Shaft Construction*. Department of Civil Engineering, University of Florida, Gainesville, FL, USA, Final Report. (Available from Association of Drilled Shaft Contractors, Irving, TX, USA.)
Kita D and Kawachi T (1979) Application of clay–polymer complex for slurry excavation of diaphragm wall. *Journal of the Clay Science Society of Japan* **19(1)**: 10–21. (In Japanese.)
Lam C and Jefferis SA (2015) Physical properties of polymer support fluids in use and their measurement techniques. *Geotechnical Testing Journal* **38(4)**: 490–500.
Lennon DJ, Ritchie D, Parry GO and Suckling TP (2006) Piling projects constructed with vinyl polymer in Glasgow, Scotland. *Proceedings of the 10th International Conference on Piling and Deep Foundations, Amsterdam*. Deep Foundations Institute, Hawthorne, NJ, USA, pp. 499–506.
Lesemann H and Vogt N (2012) Untersuchungen zur hydraulischen Stützung mit Polymerlösungen. *Geotechnik* **35(1)**: 11–21. (In German.)
Marsh HN (1931) Properties and treatment of rotary mud. *Transactions of the AIME* **92(1)**: 234–251.
Mussard I (2003) Practical aspects of formulating with associative thickeners. In *Additives in Water-borne Coatings* (Davison G and Lane B (eds)). Royal Society of Chemistry, Cambridge, UK, pp. 46–52.
Oliver B (2015) Solid answers to questions about drilling fluids disposal. *National Driller* **36(2)**: 28.

PDS (2013) Trade literature. http://www.pdscoinc.com (accessed 10/04/2017).

Petrind (2011) Trade literature. http://www.petrind.com (accessed 10/04/2017).

Pickles A and Everton S (1995) Aspects of the site investigation for the Waterloo International Terminal. *Advances in Site Investigation Practice: Proceedings of the International Conference held in London on 30–31 March 1995*. Thomas Telford, London, UK, pp. 1–12.

Richard MR (1979a) The organic drilling fluid controversy: Part I. *Water Well Journal*, Apr., pp. 66–74.

Richard MR (1979b) The organic drilling fluid controversy: Part II. *Water Well Journal*, May, pp. 50–58.

Scarrow JA and Gosling RC (1986) An example of rotary core drilling in soils. In *Site Investigation Practice: Assessing BS 5930, Engineering Geology Publication No. 2*. Geological Society, London, UK, pp. 357–363.

Shay GD (2012) Thickeners and rheology modifiers. In *Paint and Coating Testing Manual: 15th Edition of the Gardner-Sward Handbook* (Koleske JV (ed)). ASTM International, West Conshohocken, PA, USA, pp. 341–372.

Smith RA, Sowerby C, Knapman D, Myall D, May J, Lewis R, Bamfield B and Fox-Davies T (2005) *Feasibility of China Clay Secondary Aggregate Use. Final Report by TRL Ltd to Imerys Minerals Ltd.* Imerys Minerals, Par, UK, Report UPR ISS/031/05. (Unpublished.)

Süd-Chemie (2013) Trade literature. http://www.clariant.com (accessed 10/04/2017).

Tondalee Trading (2013) Trade literature (in Chinese). http://www.tondalee.com.tw

Trindade R (2014) Third-generation polymers. *GeoDrilling International*, May, pp. 17–19.

White Jefferis and Associates (1989) *An Investigation of the Effects of Viskeau Biopolymer and Bentonite on the Heat of Hydration of Ordinary Portland Cement and the Development of Strength in Concrete*. White Jefferis and Associates, Hertfordshire, UK, Report 179/236/F. (Unpublished.)

Xanthakos P (1979) *Slurry Walls*. McGraw-Hill, New York.

Polymer Support Fluids in Civil Engineering
ISBN 978-0-7277-5786-9

http://dx.doi.org/10.1680/psfce.57869.089

Chapter 5
Site equipment and procedures

5.1. Introduction

As polymer fluids are very different from their bentonite counterparts both physically and chemically, the site equipment necessary for their effective use is also very different. Over the past two decades, a much better understanding has been gained of the appropriate site equipment and procedures for polymer fluids. The purpose of this chapter is to discuss the choice of site equipment and procedures so that polymer fluids can be used properly on site. Photographs will be used extensively to illustrate the features of the equipment.

5.2. Polymer fluid mixing

The preparation of bentonite slurry requires high-energy mixing so that the clay platelets can be dispersed as finely as possible to maximise the development of slurry properties (see Section 2.3). However, most polymer fluids are sensitive to high shear stresses, and so require different mixing equipment and procedures. Furthermore, partially wetted polymer granules have a strong tendency to stick to one another to form lumps that do not dissolve, or do so very slowly. The lumps, which are transparent soft solids, are colloquially known as 'fish eyes'. The addition of dry polymer granules is therefore best done using either a splash plate or addition pan (Figure 5.1) or a Venturi eductor (Figure 5.2). A splash plate should provide a sufficiently large surface area onto which the polymer granules can be sprinkled to avoid their lumping. The splash plate requires manual powder addition. For more automated addition, a Venturi eductor can be used. This exploits the pressure reduction (vacuum created) at a constriction in pipe flow to draw polymer granules from a hopper evenly into the flow. Venturi eductors are simple and used in many industries to wet out powders. As Venturi eductors have no moving parts, they are largely maintenance free, requiring only a fluid flow. In contrast, rotary feeders used in some industries are less convenient for site use as they require mechanical drives.

After powder addition, to promote dissolution of the polymer granules, the mix water should be kept agitated for at least 30–40 min. For small projects requiring only a modest amount of fluid such as for drilling site investigation boreholes, the water can be agitated with a small motorised stirrer (paddle mixer), such as that shown in Figure 5.3. The stirrer speed should be kept low to limit shear damage to the polymer molecules (e.g. see, Nagashiro and Tsunoda, 1977; Nakano and Minoura, 1978). However, the stirrer also should not be run at too low a speed, otherwise polymer granules may build up on to the base and walls of the mix tank. A rotational speed of around 200–400 rev/min may

Figure 5.1 A splash plate for mixing polymer fluids. (Courtesy of KB International)

Figure 5.2 A Venturi eductor designed to suck polymer granules from containers via a hose. (Courtesy of KB International)

Figure 5.3 A three-blade motorised stirrer for preparing polymer fluid in a small vessel. (Courtesy of KB International)

be suitable, but this will depend on the dimensions of the stirrer and the vessel as well as on the volume of fluid being mixed. Some experimentation may be necessary to identify the optimum speed for a particular set-up. The stirrer should be of the axial-flow type designed to impart fluid motion without high shear. Stirrers with sharp edges or blades with acute angles should be avoided.

For projects requiring greater amounts of fluid (>20 m^3), it is advisable to agitate the water in a large open-top tank by aeration (Figure 5.4). A well-designed mixing tank has several lines of perforated metal pipes fixed at the bottom of the tank that are supplied with compressed air. Figure 5.5 shows an example. In situations where a purpose-built mixing tank is not available, perforated plastic hoses firmly held down by sand or gravel bags can be used. Figure 5.6 shows such a set-up on a site with mixing in progress. It can be seen that the fluid in the makeshift mixing tank is much less evenly agitated than that in a properly designed tank such as that shown in Figure 5.4. Increasing the number of perforations and the number of lines (hoses) would have produced a better result. The selection of a suitable compressed air pressure is also important, and this can be done on a trial basis. The overriding objective should be to agitate the full area of the tank.

If the required fluid volume is small and no mixing facilities are available, then as a last resort the fluid may be circulated through a pump. This method is difficult to control because too low a pumping rate will not produce the desired mixing effect, but if the pumping rate is too high the polymer molecules will be damaged and the fluid viscosity

Figure 5.4 A tank of polymer fluid well agitated by aeration

reduced (Lam *et al.*, 2010). Furthermore, if a larger volume is to be mixed, the flow is unlikely to be sufficient to fully agitate the fluid unless several pumps are used. In these circumstances, a better solution is to use liquid emulsion polymers, which can be mixed more easily although they are less economical (active concentration typically less than 40%) than their dry counterparts. Figure 5.7 shows a pail of liquid emulsion polymer being added to a stream of support fluid entering an excavation. In this instance, the polymer addition was a viscosity booster to compensate for the reduction in viscosity caused by the use of a centrifugal pump. After adding an emulsion polymer directly to the hole, the digging tool should be worked up and down the fluid column a few times to disperse the polymer; the tool should be moved slowly to avoid the generation of excess pore pressures or suctions in the surrounding soil.

5.3. Pumping

Centrifugal pumps, such as the one shown in Figure 5.8, are commonly used for transferring water and bentonite slurries, but can cause serious damage to polymer fluids due to scission of the polymer molecules. Their use is therefore discouraged but may be essential when high flow rates are required, for example when used fluid from an excavation is returned to storage tanks during concreting. Centrifugal pumps should not be used to supply polyacrylamide-based fluids to an excavation, as viscosity loss may occur even at low pump power settings. Circulating polymer fluids in a closed-loop system with a centrifugal pump, although regularly and beneficially used for bentonite slurries, can reduce the viscosity of polymer fluids to near that of water (Lam *et al.*, 2010). If a centrifugal pump has to be used, the viscosity of the fluid should be regularly

Figure 5.5 A purpose-built polymer-mixing tank with perforated air pipes. (Courtesy of KB International)

Figure 5.6 (a) A makeshift mixing tank. (b) Fluid mixing in process. (Courtesy of KB International)

(a)

(b)

Figure 5.7 Adding liquid emulsion polymer to a pile bore

Figure 5.8 A common centrifugal pump

checked and, if necessary, boosted by direct addition of emulsion polymer to the excavation.

Positive-displacement (PD) pumps, such as screw pumps and diaphragm pumps, should be used for pumping polymer fluid between tanks and to excavations. PD pumps work by trapping a fixed amount of fluid with each revolution (rotary type) or stroke (reciprocating type) of the pumping element and then moving that volume forward to the discharge. PD pumps are commonly used in place of centrifugal pumps to transfer shear-sensitive fluids, such as blood and emulsions. The peristaltic pumps commonly used to transfer blood in hospitals and donation centres are a good example of this type of pump. Since basic centrifugal pumps will not generate useful suction at the inlet unless primed, PD pumps can be used when suction at the inlet is required, such as when the support fluid is stored in a pond and thus a suction capacity is required to initiate the fluid flow. Finally, PD pumps are often the pump of choice on economic grounds for transferring fluids of variable or high viscosity, because, unlike centrifugal pumps, their flow rate is not affected by the viscosity of the fluid. Figure 5.9 shows a screw pump that was used to transfer polymer fluid from the storage tanks to the pile bores for a project in Hong Kong (Lam and Jefferis, 2014). As can be seen, a centrifugal pump was also used on the same site to transfer fluid back to the tanks because of the requirement for a higher flow capacity during tremie concreting. Typically, a 3 in (75 mm) (inlet/outlet diameter) centrifugal pump will have a flow rate of greater than 300 m^3/h, and over 1000 m^3/h can be achieved with a 4 in (100 mm) pump.

Figure 5.9 A screw pump (*right*) for pumping fluids into excavations and a centrifugal pump (*left*) for pumping fluids back to the storage tanks during concreting

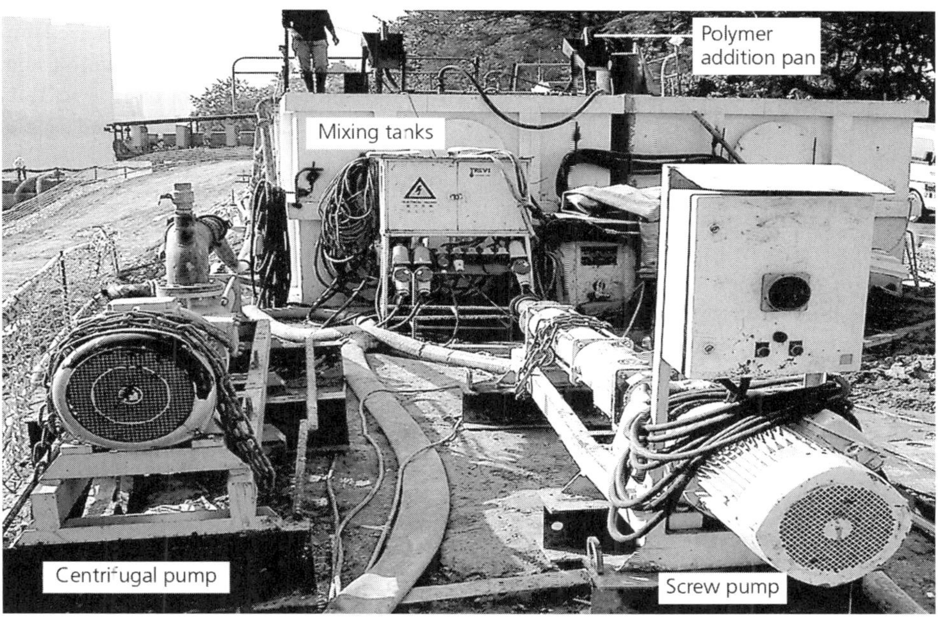

Air-operated double-diaphragm pumps are another type of PD pump that can be used for transferring polymer fluid around construction sites. These pumps have two liquid chambers and two plastic diaphragms that are driven by compressed air. Since the two diaphragms are linked to a single shaft, they move in unison to fill and empty the two liquid chambers repeatedly. The fluid is drawn into the chambers from a common inlet and discharged through a common outlet. The fluid flow is thus not continuous but pulsating, although this should not be an issue for most situations. The major advantage of these pumps is that if the outlet valve is closed the pump will stop automatically when the discharge pressure equalises with the inlet air pressure. This is an attractive feature because (*a*) the site workers can simply turn a ball valve next to the excavation on and off, depending on the fluid requirements, while keeping the pump in readiness at all times and (*b*) as the pump and the storage tank may be at some distance from the excavation, it eliminates the need for constant recirculation of the support fluid to avoid overheating, as would occur with a centrifugal pump. However, to take advantage of the self-stopping feature of diaphragm pumps, the operator must ensure that the rated pressure of the fluid lines is greater than the compressed air pressure. Typically, diaphragm pumps require no more than 9 bar to operate, and therefore as a minimum requirement the fluid hoses should have a rated pressure of 10 bar. However, flow capacities are more limited than for centrifugal pumps; a 4 in double-diaphragm pump may have a maximum flow rate of 100 m^3/h, and thus several would be needed to provide the flow of a typical 3 in centrifugal pump. Figure 5.10 shows three diaphragm pumps working in parallel on a construction site.

So-called trash pumps, which get their name from their ability to pass small debris, can be of the centrifugal or the single-diaphragm type. The latter should be requested whenever possible for the reasons described above. Trash pumps usually run on diesel for portability, but may also be powered electrically. Diaphragm-type trash pumps are commonly used to pump high-viscosity sludges and slurries. As the flow rate of diaphragm trash pumps is usually quite small, they are only suitable for smaller jobs, and more than one may be required. The maximum flow rate of a 3 in mechanically operated single-diaphragm trash pump is typically about 20 m^3/h.

5.4. Storage and self-cleaning

Although it may seem to be a trivial issue to some, the success of a polymer project often hinges on the design of the fluid-handling plant. Unlike their bentonite counterparts, polymer fluids are commonly self-cleaned by sedimentation, and so major plant items such as vibrating screens, hydrocyclones and decanting centrifuges are not required. This places a substantial demand on the fluid-handling plant, especially in terms of its ability to rejuvenate recycled fluids, which may be loaded with some solids. Although every project and its requirements are different, a properly designed plant should be able to perform two major functions: to provide sufficient storage and to allow self-cleaning. These twin demands are discussed below.

Firstly, the plant should have sufficient capacity so that the volume of fluid required for the excavation and return flows can be effectively managed. The required capacity will depend on many site-specific factors and constraints, including the space available, the

Figure 5.10 Air-operated double-diaphragm pumps used in parallel on a construction site. (Courtesy of KB International)

number of rigs operating at any one time, the volume of each excavation, the excavation speed per rig and the type of construction tool used (e.g. grab or hydromill for diaphragm walling). As a general rule of thumb, the plant should have a capacity of at least one and a half times the volume of fluid required for the maximum volume of excavation open at any one time. This requirement is especially important when drilling in coarse or fissured soils, where there is a potential for sudden large fluid losses, or for situations where exchange of the fluid in an excavation is necessary prior to concreting, for example as a result of fluid contamination. Figure 5.11 shows a row of open-top tanks used for a bored piling project in the UK using polymer fluid. Five tanks are shown in the photograph, but seven were actually used, giving a total capacity of almost 300 m^3. Details about this project are discussed in Section 8.3.

Secondly, the plant should be designed so that the recycled fluid can self-clean by sedimentation. This can be achieved in hole using treatment chemicals, if required (see Section 5.7 and Jefferis *et al.*, 2017). However, if the polymer system employed cannot achieve reliable in-hole cleaning then this can be achieved by having one or more designated sedimentation tanks. For smaller projects, it is recommended that there should be at least three tanks, one for each of the following: mixing, storage and sedimentation. Fluid returned from the excavations should not be allowed to mix with fresh fluid without first going through a sedimentation stage. For larger projects, at least two sedimentation tanks

Figure 5.11 Mixing and sedimentation tanks used for a bored piling project

should be available. The first is used as a primary sedimentation tank to receive fluid returned from the excavations. The role of this tank is to settle coarse sand and gravel-sized particles from suspension, a process that can take several hours depending on the viscosity of the recycled fluid. After this, the fluid should be transferred to the secondary sedimentation tank(s) to allow the fine to medium sand and coarse silt-sized particles to settle, a much slower process. If after a day the density of the recycled fluid is still high (>1.02 g/cm^3), the fluid probably contains a significant amount of yet finer material. A high fluid density can be a warning sign that the chemical activity of the fluid has been exhausted by sorption of the polymer onto fine soil particles. To put this into context, a density of 1.02 g/cm^3 represents 32 kg of suspended soil particles per cubic metre of solid-laden fluid. Recalling that the typical concentration of dry granular polymer is only around 1 kg/m^3, this would represent a solid-to-polymer mass ratio of about 32. If sedimentation under gravity alone is not sufficient for reducing the density of the used fluid, cleaning additives will be necessary to remove the suspended soil particles from the fluid before it is rejuvenated.

Lastly, it can be useful if the height of the sedimentation tanks is limited so that settled solids can be removed with a backhoe excavator. Sedimentation tanks that are too high to be cleaned by a backhoe may have to be tipped over to remove the sediment. Alternatively, on-site wastewater treatment systems may be used. Figure 5.12 shows a mobile unit known as AquaSed in use in Hong Kong. This system comprises a chemical dosing tank, a reaction tank and a large sedimentation tank, all incorporated into a single unit. When required, the sludge can be discharged from the bottom of the sedimentation tank

Figure 5.12 A compact mobile wastewater treatment system. (Courtesy of Environmental Pioneers and Solutions)

via an automatic valve, which eliminates the need for manual cleaning. It also facilitates the dosing of cleaning chemicals. This type of system has been used successfully for treating wastewater on construction sites in Hong Kong and similar systems are also available in other countries. However, the authors are not aware of the use of such plant on projects using polymer fluid.

The above review of plant was for excavation by grab, bucket or auger. For piling using reverse-circulation drilling or for diaphragm walling using a hydromill, the excavated material is removed from the cutting face by air lift or a suction pump. This means that there must be constant cycling of fluid and a flow of recycled fluid containing significant amounts of soil that must be passed through a desander. Typically this will consist of a set of vibrating screens to remove the coarser soil particles, which, if not removed, will build up very rapidly in the primary sedimentation tank. This type of desander effectively replaces a primary sedimentation tank. However, some polymer fluids, including those based on partially hydrolysed polyacrylamides, are known to sheet over the screens of the shakers and reduce their efficiency (e.g. Tallard, 1992). Therefore, if screens are to be used, the fluid should be checked for compatibility.

Even when desanding screens are used, if flow rates are low enough to permit, the treated fluid should still be allowed to further self-clean in a sedimentation tank without any external influences. Fluid movement in or out of a sedimentation tank should be minimised as it will disturb the particle settling process. This can be achieved by using more than one sedimentation tank, to allow one to be filled up while leaving the other undisturbed. Because of the constant circulation of the support fluid in this mode of excavation, a continuous supply of rejuvenated fluid will be required to maintain the stabilising pressure in the excavation. These extra requirements place a greater responsibility on the person in charge of running the fluid system, as they have to closely monitor the quantity and conditions of the fresh, recovered (prior to concreting) and rejuvenated fluids.

5.5. Managing excavations in soils with high groundwater pressures

Support fluids work by exerting a stabilising pressure on the base and side walls of an excavation (see Section 1.4). When groundwater is present, the net positive pressure will be the hydrostatic pressure of the fluid column less the free-field groundwater pressure. Polymer manufacturers commonly recommend a minimum fluid head difference of 1.5–2.0 m, which provides a net stabilising pressure of about 15–20 kPa, as the density of clean polymer fluids is essentially that of water. However, at sites where the groundwater table is near the surface, the stabilising pressure can be effectively reduced to zero or be negative. This is because, without taking special measures, the highest fluid level that can be consistently maintained in an excavation is about 0.5–1.0 m below the top of the steel casing for piles or the concrete guide walls for diaphragm walls. This reduced head is due to the volume of the fluid displaced by the digging tool as it travels in and out of the excavation. There have been cases where the reduced level has caused instability of the excavation. For example, Figure 5.13 shows a site in Dubai where a pile bore was excavated in cohesionless desert sand near a watercourse. The pile bore collapsed during excavation, and this caused substantial surface settlement around the bore. To manage the problem, the bore was backfilled, and a temporary reservoir (overflow basin) was built around the bore using site-won materials (Figure 5.14). The purpose of the reservoir was to collect the fluid displaced by the digging bucket during insertion with the fluid flowing back to the hole during extraction. The slinging holes near the top of the temporary steel casing allowed the fluid in the reservoir to flow back into the bore until equilibrium was reached – an extra notch could have been cut, but it was found to be unnecessary in this case. The use of the overflow basin allowed the support fluid level to be maintained at the ground level, and the pile was completed successfully.

An overflow basin can also be created using two concentric steel casings as shown in Figure 5.15. The outer casing should be capable of holding all the fluid displaced by the Kelly bar and the excavation tool without overflowing. A notch should be cut on the inner casing to facilitate fluid flow, especially during retrieval of the excavation tool when the stabilising pressure is at its lowest level. The relative height between the two casings is not important as long as the fluid can flow easily from one to the other to maintain the head in the bore. The outer casing does not need to penetrate deep into the ground provided that a seal can be formed at the base.

When drilling through an artesian aquifer, an extended casing above the ground level could be used to increase the support fluid level. In theory, there is no limit to the height

Figure 5.13 Collapse of a pile bore at a site in Dubai next to a watercourse. (Courtesy of KB International)

Figure 5.14 A fluid reservoir (overflow basin) built using site-won materials around a pile bore. (Courtesy of KB International)

Figure 5.15 An overflow basin created using two concentric steel casings. (Courtesy of KB International)

of the casing, but in practice it will be restricted by the height of the drilling rig and also by the stability of the casing itself. Brown *et al.* (2012) reported a case where a tall surface casing was required to overcome an artesian head of approximately 14 ft (4.3 m) above the ground level. Problems have been reported with drilling through artesian aquifers in South Georgia in the USA; using an extended casing or an overflow basin is typically recommended to mitigate the problem (Litke, 2013).

If it is found that the mitigating measures already discussed are insufficient to counterbalance the artesian groundwater pressure, the use of weighting agents should be considered. Only those designed for use with polymer fluids should be used, although the traditional mineral-based weighting agents for bentonite slurries (e.g. barite and haematite) might also be suitable, but this will require confirmation. It is worth adding that the use of weighting agents to densify drilling fluids is common practice in oil well drilling operations, although much less so in foundation drilling.

5.6. Excavation tools

The previous discussion on dealing with high groundwater pressures assumed that the full hydrostatic pressure of the support fluid exists within the fluid column. However, the actual stabilising pressure acting on the soil will be affected by the dynamic pressures induced by the insertion and retrieval of the excavation tools and the speed at which this is done. There have been many cases where the stability of the excavations in soft or loose soils was improved simply by digging at a lower speed. During tool insertion there

Figure 5.16 An internally vented digging bucket

may be a positive pressure pulse ahead of the tool, and during extraction there may be suction (negative pressure) behind it. The latter effect is analogous to the suction effect caused by pulling the piston of a syringe when the tip is closed. Both effects can cause soil disturbance and excess pore water pressure in the soil, which could result in a more unstable excavation and ultimately a lower-capacity foundation element. To mitigate these effects, specially designed digging tools such as vented buckets or twin-flight augers can be used. Figure 5.16 shows the inside of a digging bucket that has two internal vents on the side walls. The purpose of the vents is to allow passage of the support fluid through the bucket during insertion and retrieval of the tool, so that the hydrostatic pressure in the fluid column is maintained as fully as possible and pressure surges and suctions minimised. Figure 5.17 shows a flat-sided cleaning bucket that allows the passage of fluid outside the bucket.

Figure 5.18 shows a twin-flight auger loaded with silty fine sand. It can be seen that only one of the two flights is loaded with soil. This is as designed, the purpose of the second flight is to allow fluid passage. It is recognised that the use of this type of tool will inevitably reduce the volume of soil excavated per tool insertion, but the second flight allows faster tool insertion and, especially, tool removal, so that the benefits will outweigh any potential reduction in the soil excavation rate. In fact, considerable savings in time may be achieved with this tool when digging in unstable ground, as the risk of soil collapse is reduced due to relief of the suction effect. However, the use of this tool demands good skills and experience from the rig operator, as the second flight may become loaded with

Figure 5.17 A flat-sided cleaning bucket

Figure 5.18 A twin-flight auger

soil if too much downward thrust is applied to the auger. Inexperienced rig operators who do not understand the importance of relieving the suction effect may deliberately aim to load both flights with as much soil as possible.

5.7. Cleaning in excavations

Bentonite slurries, when left undisturbed, will develop a reversible gel due to the reorientation and linking of the clay platelets. This gel can be broken by shear but can redevelop with time – a material behaviour known as thixotropy. The physical strength of the gel gives the slurries the ability to suspend fine soil particles; the stronger the gel, the coarser the particles that can be suspended. Polymer fluids, on the other hand, do not form a gel, so they can only temporarily suspend soil particles, which will settle at a rate dependent on their particle size and the fluid viscosity. To prevent sediment build-up at the base of the excavation during the insertion of the reinforcing bar cage and tremie pipes, polymer fluid at the end of an excavation either has to be cleaned in situ or exchanged for fresh or rejuvenated fluid from the storage tanks. If neither of these processes is carried out, the sediment at the base may become what is known as a 'soft toe' for the completed foundation element. The soft toe may not cover the full base area of the foundation element, as some of the sediment may be flushed aside by the first pour of the concrete to enable some limited base contact. Yet again the sediment may mix with the wet concrete to form a weak material. During the concreting process, the settling soil particles may also contaminate the fresh concrete at the interface between the rising concrete and the support fluid. For example, Brown (2004) noted that on a project near the Atlantic coast in the USA, the contractor had to overpour the concrete by more than 2 m to remove concrete contaminated with silt particles. This should not have occurred if the support fluid had been properly cleaned or exchanged prior to concreting.

As the cleaning of polymer fluid prior to concreting is very important, many polymer manufacturers provide additives to accelerate the particle settlement process so that the sediment can be collected by cleaning bucket or by air lifting. Since most of these additives work by reacting with the base polymer fluid, the manufacturer's recommendation on the dosage should be strictly followed, or else the fluid may be damaged while still in the excavation. If an additive is used for the first time on a project, it is always advisable to test its effect on the polymer fluid in a bucket before applying it to the excavation. Figure 5.19 shows a bucket of used polymer fluid after treatment with a cleaning additive at a dosage much higher than the manufacturer's recommended value. Although the additive was effective in settling the soil particles from the fluid, as shown in the figure, the viscosity of fluid was actually reduced to that of water due to the overdose. This type of simple, yet instructive, small-scale trial should be carried out on all additives for every project so as to minimise the risk of an unintended overdose or unexpected consequences from the treatment. It is now clear that some of the less successful polymer projects in the past were compromised by improper use of cleaning additives; adding too much or too little can both cause unsatisfactory results. After introduction to an excavation, a cleaning additive needs to be dispersed in the fluid column, to promote contact and chemical interaction. This can be achieved by simply traversing the excavation tool slowly up and down the fluid column a few times. Following this, a waiting period must be allowed, for the flocculated soil particles to settle to the base of the excavation. Some polymer

Figure 5.19 Soil particles collected from a bucket of used polymer fluids treated with a chemical additive at a very high dosage

manufacturers recommend a waiting period of 30 min, but the actual time will depend on the depth of the excavation, the solids content of the used fluid, and the type and dosage of the additive used, which will influence the floc size and hence the settlement rate. Note that actually treatment agents may cause soil in the fluid to lump into coagulated masses as shown in Figure 5.19, or form strings; for convenience these processes are referred to as flocculation but the mechanics are more complex. It must be emphasised that the waiting period must not be skipped even when time is constrained, otherwise the results will be worse than if nothing at all had been done. After the waiting period, the soft sediment at the base can be removed by a cleaning bucket or air lift. The density and sand content of the fluid should also be checked at the specified depths to ensure compliance with the specified limits. The whole cleaning process should be repeated if the results do not meet the project requirements.

At sites where the soils are dominated by fine to medium sands and silts, polymer fluids may become more difficult to clean, and the situation may become more problematic. This is because while particles coarser than medium sand usually settle out of suspension relatively quickly, those finer than coarse silt can stay in suspension for much longer. Therefore, fine to medium sand and silt particles are the most likely be settling for many hours after completion of an excavation. To tackle this tricky problem, a possible strategy is to use treatment chemicals and to prolong the waiting period, to allow the settlement of the finer soil particles. During this period, the approximate thickness of the base sediment can be checked with a weighted tape measure to monitor the progress of sedimentation. In grounds with coarse voids, during the waiting period the support fluid level may need to be topped up to maintain the necessary stabilising pressure. A successful bored piling case study using this simple yet effective technique will be discussed in Chapter 8. Tallard (1992) has also described the successful application of this approach to clean the base of diaphragm wall panels supported by polymer fluid. Tallard's recommendation was to allow the suspended solids in the fluid to settle for a full day after excavation, and prior to concreting the sediment collected at the base can be removed by air lifting or pumping. If the project specification does not permit an extended waiting period, the used fluid, at the end of excavation, will need to be exchanged for fresh or rejuvenated fluid to minimise the risk of a 'soft toe' and concrete contamination.

Finally, it should be mentioned that under no circumstances should strong oxidising agents such as bleach be used to clean polymer fluid in an excavation. These will destroy the polymer molecules, thus reducing the support fluid to no more than muddy water. Bleach should be permitted only for the destruction of unwanted polymer fluid, for example prior to the disposal of used fluid, as will be discussed later.

5.8. Recycling and rejuvenation for reuse

At the end of an excavation cycle, polymer fluid should be rejuvenated for reuse. Care should be taken not to pump any heavily contaminated fluid back to the storage tanks. With some fluids, the last 1.0 to 1.5 m of fluid above the concrete column may be heavily loaded with soil particles or contaminated by the calcium ions from the concrete, or both. The heavily contaminated fluid should therefore be diverted to a waste tank for disposal or, if permitted by the local environmental regulations, allowed to spill onto the ground and then immediately mixed with dry soil. However caution is necessary, as polymer fluids are very slippery, so if spilled onto a hard surface they should be immediately mixed with dry soil for safe disposal.

Before a polymer fluid can be rejuvenated for reuse, it must first be cleaned to remove the suspended soil particles, which, if left in the fluid, will sap the chemical activity of any freshly added polymer materials. As discussed previously, since polymer fluids can be self-cleaned by sedimentation, shakers and screens are seldom required except for excavations carried out by the reverse-circulation method. Fluid recycled from an excavation if not cleaned in hole, therefore has to go through one or two sedimentation stages. If after a long period of sedimentation (e.g. over 24 h) the density of the fluid is still above, say, 1.02 g/cm^3, then it probably contains a significant amount of fines. Polymer fluids that look like a clay slurry are likely to be completely exhausted and should not be rejuvenated straight away, if at all. This is because adding new polymer materials to a fluid that is heavily loaded with soil will only serve to satisfy the chemical demand of the soil particles; the properties of the polymer fluid will not develop until this demand is fully satisfied. In this situation, cleaning additives will need to be used to help flocculate/coagulate the fine particles so that the heavier flocs settle more rapidly from the fluid. As for cleaning the fluid in an excavation, overdosing with additives must be avoided, as otherwise any residual additives will react with the newly added polymer materials during rejuvenation. In theory, decanting centrifuges, which use a high rotational speed

Figure 5.20 A decanting centrifuge

to separate components of different densities, can be used to clean used polymer fluids, but the authors are not aware of any case studies as high usage may be necessary to recover the hire costs. This equipment might therefore be more suitable for excavation projects using the reverse-circulation method where there is a high volume of returned fluid but only if the dispersion of fine soil has not been effectively prevented by good polymer fluid management. Figure 5.20 shows a decanting centrifuge.

Since the pH and the viscosity of polymer fluids are both reduced by chemical interactions with soil and by dilution with soil pore water during excavation, after the sedimentation process the pH of the recycled fluid first needs to be checked and, if necessary, raised to the desired level, typically around 10. Following this, new polymer materials can be added to the fluid to raise the viscosity back to the desired level. Additional procedures and mixing methods should be the same as those described for preparing fresh fluids (see Section 5.2). As for fresh fluids, rejuvenated fluids can be used immediately after complete dissolution of the polymer materials. If either the demand for support fluid has increased or the total volume has decreased due to fluid loss to the ground, one may choose to prepare fresh polymer fluid at a higher dosage and then blend it with recycled fluid after sedimentation and pH adjustment. Additional polymer materials can be added to the blended fluid if it still does not have the viscosity required. Due to the increased fluid volume, the density of a blend of fresh and used fluids will be lower than that of the used fluid, and this can help mitigate the effect of solids build-up in the fluid due to repeated reuse. Which rejuvenating method to use at a particular time will depend on the expected fluid demand and the available storage capacity. For most projects, a combination of these two methods is likely to be appropriate to effectively control the quality and total volume of the fluid.

5.9. Disposal

Polymer fluids, unlike their bentonite counterparts, can be treated on site and discharged into a local sewer if the consent of the local water undertaker (sewer authority in some jurisdictions) can be obtained. To obtain the necessary permit, the contractor needs to apply to the water undertaker before or near the start of the project. A generic discharge permit covering the entire local area may also be applied for and obtained in advance if there is provision for this service. For the permit application, laboratory test results confirming the quality of the treated fluid may be required, to demonstrate compliance with the undertaker's discharge requirements. During the construction period, inspectors from the water undertaker may visit the site to obtain fluid samples for further laboratory analyses. It should be noted that under no circumstances should polymer fluids (treated or untreated) be discharged into a sewer or a nearby watercourse without prior treatment to the required standards and the consent of the authority.

Although the water undertaker may specify limits for a wide range of parameters, the critical ones typically are the pH, the solids content and either the biological oxygen demand (BOD) or the chemical oxygen demand (COD). Laboratory test results reported by Hossain *et al.* (2011) showed that the BOD and the COD for two polyacrylamide-based polymers were so low that they were both below the detection limit of the method, confirming that the material was not easily biodegradable. However, some chemical oxygen demand is to be expected and it would seem that the oxidising agent used did not oxidise the polymer. They concluded that these polymers are thus unlikely to pose a threat to groundwater quality. However, if natural biodegradable polymers are used, then the BOD and COD will become the key parameters, as they measure the amount of oxygen that is needed to break down the organic material in the fluid. Zhong *et al.* (2013) found that the COD of a xanthan fluid prepared at 1.5 kg/m^3 was over 3000 mg/l when fresh, but that it may reduce considerably after a prolonged period of contact with soil.

In practice, to satisfy the discharge requirements of the water undertaker, the contractor may need to reduce the pH to about neutral (pH 7) and break the viscosity back to that of water. Since used fluids may contain a large amount of suspended soil particles, breaking the fluids to reduce viscosity will promote solids settlement, so that the requirement on the solids contents may be automatically satisfied. To reduce the viscosity of the fluid, calcium or sodium hypochlorite solution (bleach) is commonly used to break down the long-chain polymer molecules in order to reduce their viscosifying effect. Many manufacturers' own-brand cleaning additives for disposal contain hypochlorites. Another type of chemical that may be used to reduce the viscosity of polymer fluids is a salt such as sodium or calcium chloride. When added to anionic polymer fluids such as partially hydrolysed polyacrylamides and carboxymethyl celluloses, the cations (Na^{+}- and Ca^{2+}) released from these salts will compress the ionic double layer around the charged polymer molecules, so reducing their hydrodynamic volumes and viscosifying effects (Jefferis and Lam, 2013). Obviously, salts will only work with charged polymer systems, and will have no effect on non-ionic polymers such as xanthan gum, guar gum and hydroxyethyl cellulose (see Chapter 3). When salts such as sodium or calcium chloride are used, the chloride content in the treated fluid will increase, so the contractor will need to make sure this does not exceed any limits specified by the water undertaker. Finally, if a centrifugal pump is available on site, it may also be used to circulate the fluid to mechanically degrade the polymer molecules, but the obvious drawback of this method is that it is very energy inefficient. It is therefore advisable to combine the use of chemicals and the centrifugal pump to maximise the effect, as the circulation can also help to promote contact between the chemicals and the polymer. After reducing the viscosity, the contractor will also need to adjust the pH back to neutral typically by adding acid to the fluid. However, it is always advisable to check the consent limits before proceeding to pH treatment, as some water authorities may allow fluids with a relatively wide range of pH to be discharged into their sewers. If the pH of the recycled fluid already falls within the specified limits, no pH adjustment will be required.

REFERENCES

Brown DA (2004) Zen and the art of drilled shaft construction: the pursuit of quality. *Proceedings of GeoSupport 2004 Conference, Orlando*. American Society of Civil Engineers, Reston, VA, USA, pp. 19–33.

Brown DA, Turner JP and Castelli RJ (2010) *Drilled Shafts Construction Procedures and LRFD Design Methods*. US Federal Highway Administration, Department of Transportation, Washington, DC, USA, Report FHWA NHI-10-016.

Hossain A, Cofer W and Yonge D (2011) *Impact of Drilled Shaft Synthetic Slurries on Groundwater Quality. Phase I: Literature Survey. Final Report to Washington State Department of Transportation*. Washington State University, Pullman, WA, USA, Report WA-RD 773.1.

Jefferis SA and Lam C (2013) Polymer support fluids: use and misuse of innovative fluids in geotechnical works. *Proceedings of the 18th International Conference on Soil Mechanics and Geotechnical Engineering, Paris*. Presses des Ponts, Paris, France, pp. 3219–3222.

Jefferis SA, Ouyang Y, Wiltcher P, Suckling T and Lam C (2017) The on-site management of polymer support fluids for the construction of drilled shafts and diaphragm walls. *Proceedings of Grouting 2017, the 5th International Conference on Grouting, Honolulu, Hawaii*. American Society of Civil Engineers, Reston, VA, USA, pp. 523–532.

Lam C and Jefferis SA (2014) The use of polymer solutions for deep excavations: lessons from Far Eastern experience. *HKIE Transactions* **21(4)**: 262–271.

Lam C, Jefferis SA and Goodhue Jr KG (2010) Observations on viscosity reduction of PHPA polymer support fluids. *Proceedings of Sessions of the GeoShanghai 2010 International Conference, Shanghai*. American Society of Civil Engineers, Reston, VA, USA, pp. 184–191.

Litke SS (2013) ADSC/FHWA virtual seminar part II, slurries in drilled shaft construction. *Foundation Drilling* **34(2)**: 27–28, 31–34.

Nagashiro W and Tsunoda T (1977) Degradation of polyacrylamide molecules in aqueous solutions by high-speed stirring. *Journal of Applied Polymer Science* **21(4)**: 1149–1153.

Nakano A and Minoura Y (1978) Degradation of aqueous poly(acrylic acid) and its sodium salt solutions by high-speed stirring. *Journal of Applied Polymer Science* **22(8)**: 2207–2215.

Tallard GR (1992) New trenching method using synthetic bio-polymers. In *Slurry Walls: Design, Construction, and Quality Control* (Paul DB, Davidson RR and Cavalli NJ (eds)). American Society for Testing and Materials, Philadelphia, PA, USA, pp. 86–102.

Zhong L, Oostrom M, Truex MJ, Vermeul VR and Szecsody JE (2013) Rheological behavior of xanthan gum solution related to shear thinning fluid delivery for subsurface remediation. *Journal of Hazardous Materials* **244–245**: 160–170.

Polymer Support Fluids in Civil Engineering
ISBN 978-0-7277-5786-9

http://dx.doi.org/10.1680/psfce.57869.113

Chapter 6
Fluid properties and field testing

6.1. Introduction

The physicochemical properties of support fluids are of paramount importance to the effectiveness of the excavation process and the quality of the completed foundation elements, regardless of the fluid type used. For example, fluids of insufficient viscosity are less likely to be able to stabilise an excavation in coarse grounds, and fluids containing excessive solids will be more prone to the creation of 'soft toes'. The required functions and properties of an ideal polymer support fluid have been summarised in Table 1.1. As can be seen, the relevant fluid properties are viscosity (rheological properties), density, sand content and chemical reactivity. The techniques that can be used to measure these properties are discussed in this chapter. Both conventional and new techniques developed specifically for polymers are discussed. Since polymer fluids commonly do not form a layer of filter cake on the exposed soil surface, the American Petroleum Institute (API) filter loss test is not applicable and has not been included in this discussion.

As polymer fluids are active chemical systems consisting of charged polymer molecules, the quality of the mix water is very important, and it can have a significant impact on the development of the solution properties. Key mix water parameters are pH, water hardness, and total dissolved solids. The last of these can be assessed via electrical conductivity. These parameters and associated measurement methods are discussed below.

6.2. Dissolved solids and electrical conductivity

6.2.1 Basics

The water available for polymer solution preparation is likely to contain some dissolved salts – even potable water can contain significant amounts. For example, calcium is commonly found in water due to the dissolution of calcium-containing minerals. Although some polymer manufacturers claim that brackish water and even seawater may be used for mixing polymer fluids, the dissolved salts in these waters may significantly reduce the performance of the fluids by interacting with the polymer molecules. This statement is certainly true for anionic polymers such as partially hydrolysed polyacrylamides (PHPAs) and carboxymethyl celluloses (CMCs). Non-ionic polymers such as hydroxyethyl celluloses (HECs), xanthan and guar gums are not affected by the dissolved ions, so they may be considered if brackish water or seawater has to be used. However, there is considerably less industry experience with these latter types of polymer, and their effects on excavation stability and foundation performance are little known. Whenever possible, fresh water should be used to prepare polymer fluids, as brackish water and seawater will have significant salt concentrations compared with the polymer concentration. For example, typical seawater will have a salt concentration of about 35 000 ppm (35 g/l) whereas a typical synthetic polymer may be used at a concentration of 1 g/l or less. Water from ponds or private wells and so on should be checked prior to use.

To assess the quality of water, one can conduct a comprehensive chemical analysis, but for polymer fluid work a much simpler, and usually sufficient, approach is to measure the electrical conductivity (EC). Pure water has a very low EC, which is significantly increased by any ions in solution. This indirect approach is helped by the fact that, for many salts at modest concentrations, there is a near-linear relationship between the EC and the ion concentration, so the former can be seen as a direct indication of the latter. Conductivity is commonly expressed in microsiemens per centimetre (μS/cm). The EC of pure water is about 0.05 μS/cm. For high-conductivity solutions such as seawater, the results may be presented in millisiemens per centimetre (mS/cm). Seawaters typically have an EC of about 50 mS/cm. The unit 'siemens' is named after the German industrialist Ernst Werner Siemens, who founded the electrical engineering company Siemens. It may be noted that 'siemens' can also be referred as 'mho'; that is, the inverse of the word 'ohm', to indicate the inverse relationship between conductivity and resistivity (i.e. 1 S is equal to 1 mho, and the reciprocal of 1 Ω).

6.2.2 Measurement techniques

Figure 6.1 shows a combined EC and pH meter that can be used on site. Stand-alone EC meters are also available at a lower cost. Before use, the EC meter has to be calibrated using standard conductivity solutions, which can be obtained from the instrument manufacturer or a scientific equipment supplier. Typically, standard solutions are provided with conductance values of 1413 and 12 880 μS/cm at 25°C. For mix waters used for typical polymer fluid preparation, the lower EC standard is likely to be the more suitable, as the EC of potable water is typically between 50 and 500 μS/cm. It must be emphasised that the EC of a solution is strongly temperature dependent, and an adjustment must be made if calibration is performed at a temperature other than that specified for the standard solution – although conductivity data over a wide range of temperatures are provided with the solution bottle for many standard solutions. To measure the EC of a fluid, simply immerse the EC probe to the prescribed mark on the electrode casing and allow a few seconds for the reading to stabilise. As mentioned before, the EC of a fluid is a direct indication of the dissolved ion content of the fluid, the latter commonly termed the total dissolved solids of the solution. The unit is typically parts per million (ppm), which for dilute solutions, is numerically equal to milligrams per litre (mg/l). The conversion factor between the EC and the total dissolved solids depends on the particular ions in solution and their concentration, but typically an EC of 1 μS/cm corresponds to about 0.5–0.7 mg/l of dissolved ions.

6.3. pH

6.3.1 Basics

pH denotes the 'power' of the hydrogen ion, and is commonly used as a measure of the acidic or alkaline nature of a solution. The standard pH scale is from 0 (most acidic) to 14 (most alkaline). Values outside this range are possible but are not relevant to civil engineering practice. Pure water has a neutral pH (7). Due to the presence of the dissolved minerals, the pH of tap water may be slightly above or below 7. The relationship between pH and the hydrogen ion concentration is

$$pH = -\log_{10}[H^+] \tag{6.1}$$

where $[H^+]$ is the concentration (strictly the activity) of hydrogen ions in the solution in moles per litre (mol/l); a concentration of 1 mol/l is equivalent to 1 molar (1 M). For example, a 0.1 M hydrochloric acid (HCl) solution will fully ionise in water to give $[H^+] = 0.1$ M. Substituting this into Equation 6.1 will give a pH of 1.

Water dissociates to produce hydrogen and hydroxyl ions, and the relationship between their concentrations is

$$[H^+][OH^-] = 10^{-14} \tag{6.2}$$

and thus, for a neutral solution,

$$[H^+] = [OH^-] = 10^{-7} \tag{6.3}$$

or

$$pH = pOH = 7 \tag{6.4}$$

where $[OH^-]$ is in moles per litre and pOH defined as for pH. It follows from the above equations that pH + pOH = 14, so that the pH of an alkaline solution can be obtained if one knows the value of $[OH^-]$. For example, sodium carbonate (soda ash) when dissolved in water dissociates to release sodium and carbonate ions as follows:

$$Na_2CO_3 \leftrightharpoons 2Na^+ + CO_3^{2-} \tag{6.5}$$

In the absence of other cations in the system, the carbonate ion then undergoes hydrolysis to accept a proton from the water molecule, releasing bicarbonate and hydroxyl ions as shown below:

$$CO_3^{2-} + H_2O \leftrightharpoons HCO_3^{2-} + OH^- \tag{6.6}$$

Equation 6.6 explains why soda ash has an alkaline reaction. The equilibrium constant is the ratio of the concentration of products to the concentration of reactants for a reversible reaction, and for the hydrolysis of carbonate ions it is

$$K_b = \frac{[HCO_3^-][OH^-]}{[CO_3^{2-}]} \tag{6.7}$$

Since the hydrolysis releases bicarbonate and hydroxide ions in equimolar amounts, Equation 6.7 can be reduced to

$$K_b = \frac{[OH^-]^2}{[CO_3^{2-}]} \tag{6.8}$$

K_b has been determined to be about 2×10^{-4}, and so the value of $[OH^-]$ can be obtained if $[CO_3^{2-}]$ is known. For example, in a 0.001 M sodium carbonate (soda ash, Na_2CO_3) solution, if fully dissociated, $[CO_3^{2-}] = 0.001$ M, and $[OH^-]$ can therefore be shown to be 4.5×10^{-4} M. Taking the negative logarithm of $[OH^-]$ gives a pOH of 3.35, and hence a pH of 10.65. This analysis assumes that the water does not contain any reactive species or other salts at concentrations sufficient to affect the dissociation of the sodium carbonate. However, drinking water may contain calcium ions (the cause of calcium hardness), which can react with carbonate ions to precipitate calcium carbonate, with the release of sodium ions into solution. This explains why manufacturers typically recommend using soda ash at a high concentration ($\approx$1 kg/m^3) to achieve a mix water pH of around 10 (see Section 4.3). For sodium carbonate, which has a molecular weight of 106 g/mol, a concentration of 1 kg/m^3 is approximately equal to 0.01 M, which is higher than the theoretical concentration required to increase the pH of pure water to 10.65 as shown above. The extra sodium carbonate is to allow for the presence of other reactive species (e.g. calcium ions) in the mix water. The use of sodium carbonate to remove water hardness will be discussed in more detail in Section 6.4.

6.3.2 pH measurement techniques

Most polymer manufacturers currently recommend indicator strips such as those shown in Figure 6.2 for pH monitoring. These indicator strips are cheap to buy, but the result is usually only accurate to the nearest integer. The result is also operator-dependent, as different people see colours slightly differently. It is important to recognise that indicator strips must be kept dry in a sealed container: they can be rendered useless if they are exposed to damp conditions.

A much more accurate approach is to use a properly calibrated electronic pH meter, such as that shown in Figure 6.1. A stand-alone pen-type pH meter is shown in Figure 6.3. A pH meter consists of a glass sensing bulb at the base of an electrode. The glass in the bulb separates the external test fluid from the internal reference solution. The glass electrodes used for pH meters are designed to be sensitive only to hydrogen ions, so that the electrochemical potential that develops across the glass membrane is proportional to the pH of the test fluid. The glass bulb needs to be kept moist at all times, and should not be left to dry out in air. If the electrode has dried out, follow the manufacturer's recommendation for reconditioning the electrode. As a minimum, soak the electrode in pH 7 standard solution or tap water for at least an hour before use and then recalibrate the electrode using two pH buffer solutions spanning the range of interest. For optimal performance, the glass electrode should be kept in a specially formulated storage solution when not in use (see Figure 6.1), which can be obtained from most scientific equipment suppliers. Never store an electrode in distilled or deionised water, as this will shorten the life of the electrode. If electrode storage solutions are not available, use standard calibration solutions or, less preferably, tap water.

Before a pH meter is used, it must be calibrated using standard solutions to establish the relationship between the electrical potential and the pH – this relationship can be held in the electronics of the pH meter. Reference (buffer) pH solutions can be obtained from instrument suppliers and most scientific equipment suppliers. The majority of pH meters

Figure 6.1 A portable combined electrical conductivity and pH meter. From left to right: digital readout, conductivity electrode and pH electrode in a storage bottle

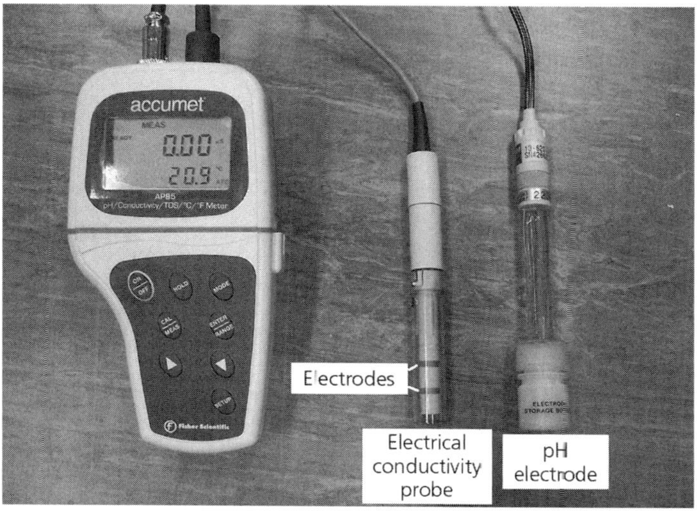

require a two-point calibration. Indeed, it is always advisable to calibrate at two pH points, as electrodes tend to lose sensitivity as they age, and a single-point calibration will not identify a loss response. Standard buffer solutions are commonly available for pH 4, 7 and 10; note that the pH of buffer solutions is somewhat temperature sensitive (temperature data are normally supplied with the solutions). Since polymer fluids are

Figure 6.2 pH indicator strips

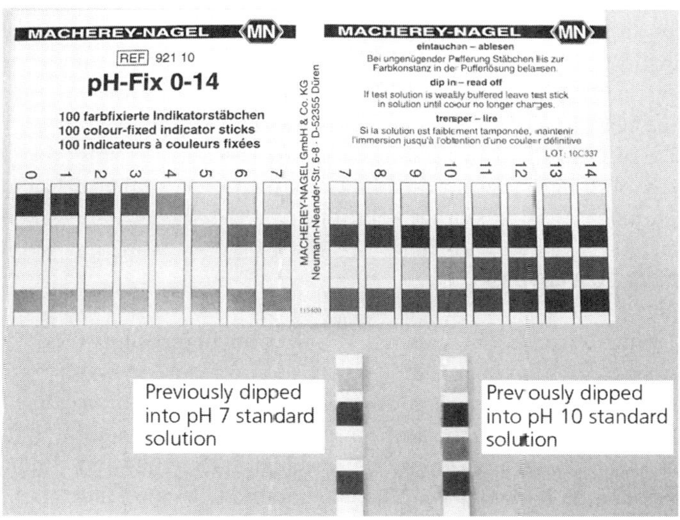

Figure 6.3 A pen-type pH meter

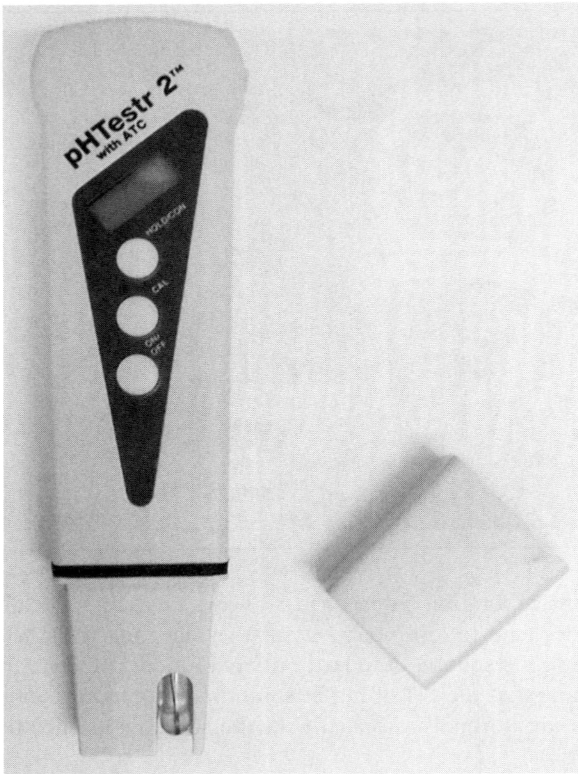

always used in an alkaline environment, use pH 7 and 10 solutions for the calibration. The frequency of calibration will depend on the usage and the accuracy required. Typically, for support fluid testing, weekly calibration may be sufficient for the required accuracy of the results. To make a pH measurement, simply dip the electrode into the test fluid and wait until the reading stabilises. After use, always rinse the electrode with water or buffer solution to prevent the build-up of material on the surface of the sensing bulb, and then place it in the storage solution.

6.4. Water hardness

6.4.1 Basics

Hardness in water is caused by calcium (Ca^{2+}) and magnesium (Mg^{2+}) ions. Other multivalent cations such as iron (Fe^{3+}) can also contribute to water hardness, but in most natural waters they are present at much lower concentrations than calcium and magnesium. Ions such as calcium and magnesium will enter natural waters from the dissolution of minerals such as limestone (calcium carbonate) and dolomite (calcium magnesium carbonate). Large parts of the UK and the USA have hard municipal waters. The following chemical equations show how calcium and bicarbonate ions are released from limestone or chalk by the action of dissolved carbon dioxide. The dissolution of carbon dioxide in water produces carbonic acid, which can then dissolve calcium carbonate:

$$CO_2 + H_2O \leftrightharpoons H_2CO_3 + OH^- \quad (6.9)$$

$$CaCO_3 + H_2CO_3 \leftrightharpoons Ca^{2+} + 2HCO_3^{2-} \quad (6.10)$$

These chemical reactions can be reversed by heating (boiling) the water, and therefore this type of carbonate hardness is referred to as temporary hardness. In hard water areas, calcium carbonate limescale will deposit on the heating elements of kettles, boilers and so on. Non-carbonate or permanent hardness, on the other hand, cannot be removed by boiling, as it is caused by the presence of dissolved magnesium salts and calcium salts such as calcium sulphate and chloride. These salts are unaffected by heating the water. Both temporary and permanent calcium hardness can be removed by adding soluble sodium carbonate, which precipitates the calcium ions as calcium carbonate and is effectively insoluble in alkaline solutions:

$$Ca^{2+} + Na_2CO_3 \rightarrow 2Na^+ + CaCO_3 \quad (6.11)$$

As can be seen, the calcium ions in solution are replaced by twice the number of sodium ions. The replacement of calcium ions by sodium ions is recommended for the preparation of anionic water-soluble polymer solutions. This is because monovalent ions have much less impact on polymer performance than polyvalent ions, and the exchange reduces the ionic strength of the solution. Note that the ionic strength of a solution is the sum of the concentrations of each ion multiplied by the square of its charge. Some polymer manufacturers recommend adding soda ash (sodium carbonate) to the mix water until pH 10 is achieved. This is good practice, as it serves the dual purposes of removing water hardness and creating an alkaline environment for the polymers. Water hardness can also be removed by passing the water through an ion exchange column containing resin beads with active sodium charge sites. Calcium ions in the water are absorbed by the beads, and twice the number of sodium ions are released – the same result as the precipitation process. However, the process is substantially more expensive than simple sodium carbonate dosing.

For situations where there is high magnesium hardness in the water, hydrated lime (calcium hydroxide) can be used to precipitate the magnesium, but the lime treatment must be followed by sodium carbonate treatment to remove the resulting dissolved calcium from the solution. This process is known as lime soda softening in the treatment of water for potable supply.

Water hardness is typically reported in terms of the equivalent concentration of calcium carbonate. Numerically, it is calculated as the calcium ion concentration multiplied by 100 and divided by 40, plus the magnesium ion concentration multiplied by 100 and divided by 24, where 100 is the molecular weight of calcium carbonate and 40 and 24 are the atomic weights of calcium and magnesium, respectively. Different countries have different classification system for water hardness, but, as a general guide, water that has less than 100 g/ml of calcium carbonate can be considered soft to moderately soft, between 100 and 200 mg/l as slightly hard to moderately hard and over 200 mg/l as hard to very hard. A large part of the UK has very hard water. The tap water in the south-east of England, for example, typically has a total hardness of over 250 mg/l of calcium carbonate.

6.4.2 Measurement techniques

Despite its potentially significant effect on polymer fluids, water hardness is not usually measured on construction sites. This is because local contractors usually have a good knowledge of the level of water treatment required, and water hardness information may also be available from the water supplier. If soda ash is used to raise the pH of the mix water, this will automatically ensure that calcium hardness is removed, so the risk is already greatly reduced. If it is necessary to confirm the effectiveness of sodium carbonate treatment or the remaining (magnesium) hardness, there are several methods that can be employed, but the simplest is the use of water hardness indicator strips. Similar to pH indicator strips, these indicator strips are cheap and easy to use. They usually have a working range of up to 400 or 500 mg/l (or ppm) with a resolution of between 50 and 100 mg/l – the lower the value the better the resolution. Since accurate measurement of water hardness is usually not required, the indicator strips are sufficient for most situations. Combined pH and water hardness indicator strips are also available.

For better measurement accuracy, a water hardness testing kit can be used. Typically, these are based on colorimetric titrations. The specific procedure will depend on the system, but it usually involves adding a reagent dropwise to a fixed volume of water until a colour change occurs. The number of drops up to the colour change is multiplied by a specified factor, to provide the water hardness in mg/l (or ppm). For example, if ten drops of reagent are used and the multiplication factor is 20, then the water hardness is 200 mg/l. Before the titration, the more sophisticated test kits may require the use of buffer solutions to maintain the pH of the test fluid at 10. This ensures the reactivity of the reagent chemical, which is typically ethylenediaminetetraacetic acid (EDTA). Detailed information about the chemical reactions between EDTA and Ca^{2+} and Mg^{2+} ions can be found in many analytical chemistry textbooks.

6.5. Viscosity

6.5.1 Basics

The viscosity of a fluid is defined as the ratio of the shear stress to shear strain rate; that is,

$$\eta = \frac{\tau}{\dot{\gamma}} \tag{6.12}$$

where η is the viscosity, τ is the shear stress and $\dot{\gamma}$ is the shear (strain) rate. Note that in some publications μ is used instead of η to denote viscosity, but the latter is adopted in this book. Newtonian fluids are those that have a constant viscosity regardless of the shear rate. In other words, there is a linear (proportional) relationship between the shear stress and the shear rate at all shear rates, as indicated in the Equation 6.12. Water is a Newtonian fluid and has a dynamic viscosity of 1 millipascal-second (mPa s) at 20°C: 1 mPa s is equivalent to 1 centipoise (cP). Fluids that do not display Newtonian behaviour are collectively known as non-Newtonian fluids. Both bentonite slurries and polymer solutions are non-Newtonian fluids, and an understanding of their flow behaviour is key to their successful use. In the civil engineering industry, the flow properties of these fluids have traditionally been assessed with the Marsh funnel, but instruments such as the Fann and Brookfield viscometers are used for the determination of shear stress versus the shear rate behaviour. In the following sections, the operating principles and test procedures for these three instruments are discussed.

6.5.2 Marsh funnel

The Marsh funnel was invented by the US petroleum engineer Hallan N. Marsh in the early 1930s for the field control of the viscosity of clay-based drilling fluids in the oil industry (Marsh, 1931). When the bentonite slurry technique was 'borrowed' from the oil industry for civil engineering works in the 1940s and 1950s, this tool was also adopted for viscosity control, and has been used ever since. More recently, it has been accepted for the management of polymer support fluids. Figure 6.4 shows a Marsh funnel mounted on a workshop-made stand and the accompanying graduated measuring cup.

Figure 6.4 A Marsh funnel and cup

The funnel has an internal diameter of 6 in (152 mm) at the top, reducing to 3/16 in (4.75 mm) at the bottom over a length of 12 in (305 mm). An orifice that has a 2 in (50.8 mm) long tube with a diameter of 3/16 in is attached to the bottom of the funnel. Near the top of the funnel there is a mesh screen with openings of approximately 1/8 in (3.2 mm) covering half of the funnel. The screen allows the operator to exclude any coarse particles that could block the discharge orifice during a test. The Marsh funnel is still used in the oil industry, and it can be obtained from equipment suppliers servicing that sector.

To run a test, cover the orifice with a finger and fill the funnel with the test fluid to the level of the metal screen (a volume of 1.5 litres), then with the graduated cup placed below the funnel, remove the finger and simultaneously start a stop watch. Record the time required for a discharge of 1 US quart (946 ml) from the funnel. The 1 US quart level is clearly marked on the graduated cup. The recorded efflux time in seconds is the Marsh funnel time (MFT). It is suggested that the test is repeated at least twice. The results should agree to within ± 0.5 s, and testing should be repeated if they do not, although a greater difference may be acceptable for fluids with long flow times. For gelling mineral slurries, the MFT will be sensitive to the standing time of the slurry between sampling and testing. It is recommended that the slurry samples are vigorously stirred prior to testing.

The funnel should be checked with water prior to use. The MFT for water should be 26 ± 0.5 s at 21°C. If the measured MFT for water is outside this range, the funnel should be checked for cleanliness and damage, especially around the spigot. After use, always thoroughly clean the funnel and the graduated cup with a water jet and/or a long wire brush, because soil particles may build up around the orifice tube during a test.

Although this test is deceptively simple to run, there have been instances where operators failed to follow the test procedure. Examples of wrong procedures include filling the funnel to the top rather than to the marked line at the screen, and recording the time for the funnel to drain completely instead of 1 US quart (946 ml). It may be noted that although it is now customary to conduct a Marsh funnel test with an initial volume of 1.5 litres and measuring the efflux time for 946 ml, the original procedure proposed by Marsh (1931) was to use 500 ml of test fluid and to record the efflux time for the full 500 ml, for which the time should be about 18.5 s for water. The 500 ml in and 500 ml out procedure was followed for the measurement of the Telmarch fluid viscosities shown in Figure 4.2. A further complication is that some operators, especially in the UK and other European countries, record the efflux time for 1 litre rather than 946 ml, although for the same initial volume of 1.5 litres. The efflux time for water should 28 ± 0.5 s for this procedure. The results previously shown in Figure 4.6 were based on a 1-litre discharge. Industry standards and specifications setting out the dimensions and current test procedure for the Marsh funnel include those by the British Standards Institution (BSI, 2001), the API (2003), the Federation of Piling Specialists (FPS, 2006) and the American Society for Testing and Materials (ASTM, 2009). If in doubt, always refer to these documents for authoritative advice.

Note that the terms 'Marsh cone' and 'Marsh funnel' seem to be used interchangeably in the literature. Some texts, including the early FPS bentonite fluid specifications (FPS, 1975, 1977, 1985) and the *ICE Specification for Piling and Embedded Retaining Walls* (ICE, 2017), uses 'cone'. Flow funnels (cones) are used for testing fluids other than excavation fluids, including structural and geotechnical grouts. Sometimes these funnels are erroneously referred to as Marsh funnels. For example, although the US standard ASTM C939 (ASTM, 2010) specifies the dimensions of a flow cone and correctly does not describe it as a Marsh cone, some equipment suppliers and researchers (e.g. Roussel and Le Roy, 2005) still refer to it as a Marsh cone.

Lastly, it is worth noting that the result of a Marsh funnel test is often referred to as Marsh funnel viscosity (e.g. FPS, 2006). However, this term is discouraged by the authors because a Marsh funnel test is only a measurement of the efflux time of the test fluid.

6.5.3 Direct-indicating viscometer

A direct-indicating viscometer designed specifically for testing oil well drilling fluids has been adopted by the civil engineering industry and used for the testing of bentonite support fluids for many decades (e.g. FPS, 1977, 1985; Hodgson, 1977; ICE, 2017). The instrument was designed by Savins and Roper (1954), and the first commercial model was produced by the Fann Instrument Company, and as a result the instrument is commonly known as a Fann viscometer. The instrument is referred to as 'direct indicating', as it is designed so that the viscosity of a Newtonian fluid (in mPa s) can be read directly from the dial at a certain speed of rotation (300 rev/min). Although in regular use for bentonite slurries, this instrument is less used for polymer fluids. This is not due to technical unsuitability but rather because there is a lack of industry guidance on procedures and specifications for the test results. As will be shown in this section, the direct-indicating viscometer is a much more powerful tool than the Marsh funnel for the characterisation of non-Newtonian fluids such as polymer fluids and bentonite slurries.

The direct-indicating viscometer is essentially a co-axial cylinder viscometer with an outer rotating sleeve and an inner bob on which the torque is measured. Figure 6.5 shows a photograph and a schematic diagram of the instrument. The test fluid is contained in the narrow gap between the two cylinders. To conduct a test, the operator fills the sample cup with sufficient test fluid such that the co-axial cylinders can be immersed in the fluid to the scribed line on the rotor sleeve. There are several bleed holes on the sleeve to allow air to escape as the fluid level rises in the narrow gap between the bob and the sleeve. On some instruments, fine adjustment of the level of immersion in the fluid can be made with an adjustment screw at the back of the instrument; others have an adjustable table on which the sample cup is mounted. After the initial set-up, the required test speed is selected from the control knob at the top. When the outer rotor sleeve is rotated, a torque is exerted on the bob, which is counter-balanced by and measured against a torsion spring mounted on the bob shaft. The dial reading, which represents the shear stress of the fluid, is then read through a magnifying lens on the top of the instrument. The reading may fluctuate slightly at high speeds, and in those cases simply take the middle value. Most direct-indicating viscometers are hand-cranked two-speed (300 and

Figure 6.5 The direct-indicating viscometer: (a) variable speed model by the Fann Instrument Company; (b) schematic diagram

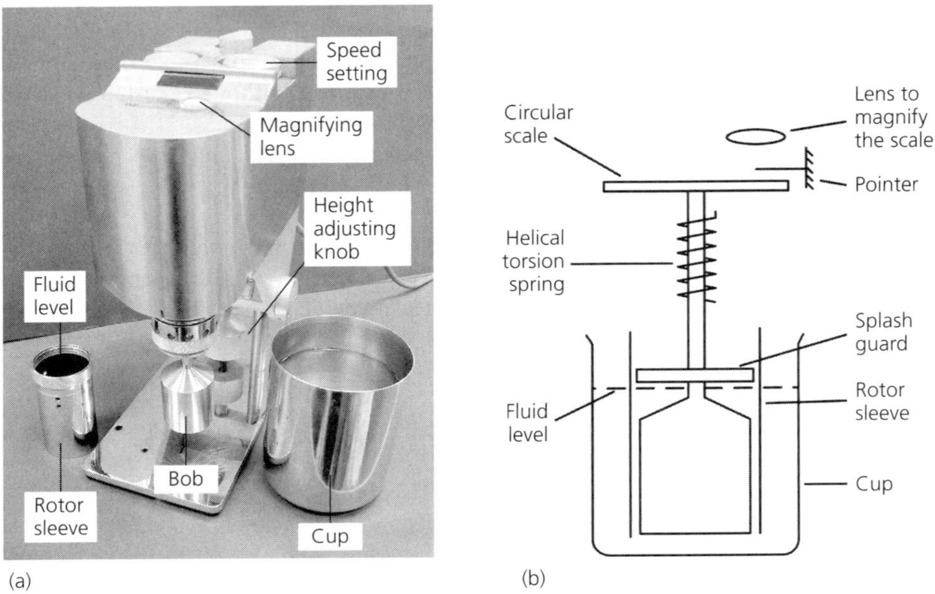

600 rev/min) or electrically driven six-speed (3, 6, 100, 200, 300 and 600 rev/min) models, but variable-speed models are also available, such as the one shown in Figure 6.5. The speed setting for 3 rev/min is labelled as ‘gel’ on many models, since it can be used for the measurement of the gel strength of bentonite slurry. As the fluid temperature is not controlled in this test and viscosity is sensitive to temperature, it is recommended that the temperature of the test sample is measured immediately before or after a test.

After a test, the dial readings can be converted to pounds per 100 sqaure feet (lb/100 ft^2) by multiplying by 1.065 (Darley and Gray, 1988), although this conversion factor is often ignored and the dial reading is assumed to give lb/100 ft^2. Since 1 lb/ft^2 equals 0.48 Pa, the shear stresses in pascals (N/m^2) can be obtained by multiplying the dial readings by 0.51 (=0.48 × 1.065). For the standard bob and rotor geometry (R1-B1), the shear rate at the bob in reciprocal seconds (s^{-1}) can be obtained by multiplying the rotor speed (in rev/min) by 1.703. The common speed settings of 3, 6, 100, 200, 300 and 600 rev/min therefore correspond to 5.1, 10.2, 170.3, 340.6, 510.9 and 1021.8 s^{-1}. Note that the above calculation procedure is only relevant to the standard version of the instrument equipped with rotor R1, bob B1 and torsion spring F1. There are also other sizes of rotor, bob and torsion spring available (e.g. Fann Instrument Company, 2013).

By using the shear stress and shear rate conversion factors given above, the apparent viscosity of the test fluid, at any test shear rate, can be obtained by dividing the shear stress (in Pa) by the shear rate (in s^{-1}). A multiplying factor of 1000 will be required

Table 6.1 Viscosity test results for a PHPA fluid

Raw data		Interpretation		
Rotor speed, ω: rev/min *A*	Dial reading, θ *B*	Shear rate, $\dot{\gamma}$: s^{-1} $C = A \times 1.703$	Shear stress, τ: Pa $D = B \times 0.51$	Viscosity, μ: mPa s $E = D/C \times 1000$
3	1.5	5.11	0.77	150.0
6	2	10.22	1.02	100.0
100	7	170.30	3.58	21.0
200	11	340.60	5.62	16.5
300	14	510.90	7.15	14.0
600	21	1021.80	10.74	10.5

if the results are to be presented in millipascal-seconds (mPa s, numerically equal to centipoise (cP)). To illustrate the calculation procedure, Table 6.1 shows a set of viscometer results obtained from a PHPA polymer fluid and the associated calculation. The apparent viscosity values are calculated from the dial readings and rotational speeds. To illustrate these results graphically, Figure 6.6 shows plots of the shear stress and the apparent viscosity against the shear rate.

It is worth noting that, in the oil industry, viscometer results are still commonly presented in the instrument units (i.e. shear stress in lb/100 ft^2 and rotor speeds in rev/min; Caenn *et al.*, 2011), although drilling fluid standards prefer SI units (API, 2003). In the construction industry, the situation is more confused, as currently there are no agreed and standard units for the presentation of viscometer results (e.g. Ata and O'Neill (1997) have reported shear stress results in lb/100 ft^2 and shear rates in s^{-1}, but also used Pa and s^{-1} (Ata and O'Neill, 1998)). Although the choice of units is not in itself a problem, it does make the use data from different sources cumbersome, especially for those who are less experienced, and invites risks of error. For these reasons, it is recommended that in site laboratories the instrument readings are recorded on the laboratory sheets for later conversion to SI units, as previously described.

For a generalised instrument, the apparent viscosity also can be determined from

$$\eta = kf\frac{\theta}{\omega} \tag{6.13}$$

where η is the viscosity (in mPa s), θ is the dial reading, ω is the rotor speed (in rev/min), k is the overall instrument constant and f is the torsion spring factor. The value of k is 300 for the standard rotor–bob combination R1-B1, and the value of f is 1 for the standard torsion spring F1 (the actual spring constant is 386 dyne-cm/(degree of deflection)), and so Equation 6.13 reduces to

$$\eta = \frac{300\theta}{\omega} \tag{6.14}$$

Figure 6.6 Test results for a polymer solution obtained from a direct-indicating viscometer and their interpretation

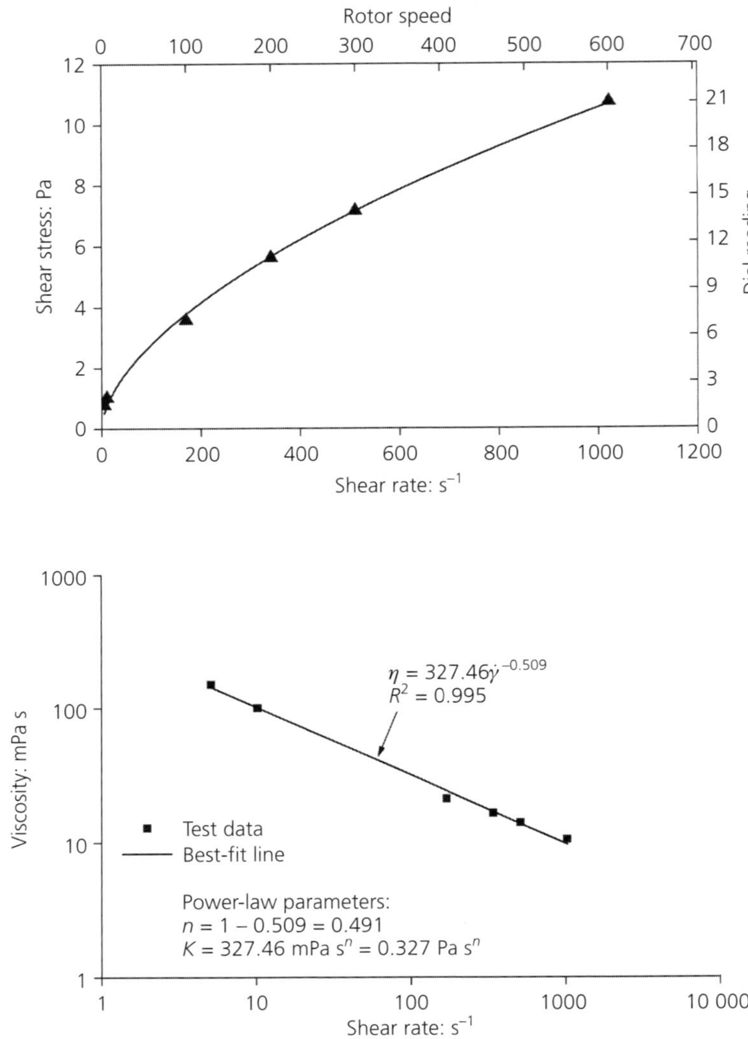

As can be seen, the apparent viscosity of the fluid (in mPa s) at a rotational speed of 300 rev/min is numerically equal to the dial reading; this relationship is demonstrated in Table 6.1. It is this property that gives the instrument its description of 'direct indicating'. It is important to note that for non-Newtonian fluids the viscosity will be shear rate dependent, and the viscosity determined from the dial reading at any rotational speed (shear rate) is the apparent viscosity at that shear rate. For results from tests with the direct-indicating viscometer, the term 'apparent viscosity' (AV) is reserved for the viscosity determined at a rotor speed of 600 rev/min unless stated otherwise. Substituting $\omega = 600$ into Equation 6.14 gives

$$\mathrm{AV} = \frac{300\theta_{600}}{600} = \frac{\theta_{600}}{2} \tag{6.15}$$

Equation 6.15 shows that AV (in mPa s) can be obtained simply by dividing the dial reading at 600 rev/min by 2. The dial reading and the viscosity for 600 rev/min given in Table 6.1 confirm this numerical relationship.

If it is assumed that the fluid flow can be modelled with a particular rheological model, for example the Bingham plastic model or the power law model then the direct-indicating viscometer can be used to determine the model parameters. The Bingham plastic model (discussed in Chapter 9) requires the determination of two parameters, namely the plastic viscosity (PV) and the yield point (YP). The standard practice adopted from the oil industry is to derive these parameters from test results at 300 and 600 rev/min. When these two rotational speeds are used, the equations for the PV and YP are

$$\mathrm{PV} = \theta_{600} - \theta_{300} \tag{6.16}$$

$$\mathrm{YP} = \theta_{300} - \mathrm{PV} = 2\theta_{300} - \theta_{600} \tag{6.17}$$

where θ is the dial reading and the numerical subscripts indicate the rotor speed (in rev/min). Note that the PV is in mPa s and the YP is in pounds per lb/100 ft^2. Since no graphical plots and curve fitting are required, Equations 6.16 and 6.17 greatly simplify the interpretation of viscometer results for fluids that follow, or are assumed to follow, the Bingham plastic model for a shear rate range between 510.9 and 1021.8 s^{-1} (achievable at 300 and 600 rev/min).

However, as will be shown later in Chapter 9, polymer fluids do not have a yield stress, and hence the power law model is more suitable. There are two parameters in the power law model, namely the flow behaviour index n and the consistency index K. If these parameters are to be determined from the 300 and 600 rev/min readings, then it can be shown that (Darley and Gray, 1988)

$$n = \frac{\log \theta_{600} - \log \theta_{300}}{\log \omega_{600} - \log \omega_{300}} = 3.32 \log_{10} \frac{\theta_{600}}{\theta_{300}} \tag{6.18}$$

and

$$K = \frac{\theta_{600}}{1022^n} \tag{6.19}$$

where θ is the dial reading (in lb/100 ft^2), n is dimensionless and K is in lb/100 ft^2 s^n. The value of K can be converted into Pa s^n by multiplying 0.51, as discussed above. To illustrate how these equations can be used to interpret test results of polymer fluids, the data set previously given in Table 6.1 has been interpreted using these equations and by the

Table 6.2 Power law parameters for a PHPA fluid

Calculation method	Flow behaviour index, n	Consistency index, K: Pa s^n
Directly from dial readings	0.585	0.187
Curve fitting	0.491	0.327

curve-fitting method; the latter better reflects the full data set, whereas the direct method uses only the 300 and 600 rev/min data. The results of interpretations by both methods are given in Table 6.2, and the best-fit curve is shown in Figure 6.6. As can be seen, although there is reasonable agreement between the n values determined by the two methods, the difference between the K values is much greater. The use of just the 300 and 600 rev/min results, although useful for rapid field interpretations, is therefore not recommended when a fuller picture is required.

6.5.4 Brookfield viscometer

Although the direct-indicating viscometer has a long history of use for testing drilling and support fluids, it is arguably not the best tool to use if the in-hole behaviour of support fluids is to be predicted. This is because the shear rates achievable on this instrument (3–1022 s^{-1}) are more representative of fluid conditions during high-shear operations such as pumping, fluid circulation and mechanised cleaning. In excavations and in settling tanks, support fluids principally work in a low shear environment, and therefore the relevant shear rates are typically much lower than those of the direct-indicating viscometer. In the authors' opinion, the Brookfield viscometer, which is named after its designer Don Brookfield, can be a more appropriate tool for testing support fluids because it can measure fluid viscosity over a wider range of shear rates, which can be less than 0.1 s^{-1}. Although the Brookfield viscometer is portable and requires only electricity for operation, at the time of writing it has yet to be applied to support fluids on construction sites, possibly because of lack of relevant research, guidance and, most especially, awareness. However, it has been used for support fluids in laboratories (Abichou *et al.*, 2004; Lam *et al.*, 2014) and for testing chemical grouts (ASTM, 2014). In contrast, in other industries, the Brookfield viscometer is regularly used as a 'shop floor' instrument for the quality control of fluid products such as lubricants, resins, paints and adhesives.

Figure 6.7 shows a digital Brookfield viscometer. As for the direct-indicating viscometer, the torque is measured on a central bob. However, in contrast, the central bob rather than outer sleeve is rotated. This allows a number of test geometries to be used, including the 'infinite sea' or wide-gap geometry shown in Figure 6.7. This permits a fluid to be tested without prior screening – as required for the direct-indicating instrument. Analysis of data with a wide gap typically requires the assumption of an infinite cup boundary (Chhabra and Richardson, 1999), a condition that can be satisfied if a narrow spindle is used with a 600 ml low-form beaker. The operating principle of the instrument is straightforward: the spindle is driven by a motor through a torsion spring, and the viscous drag exerted by the fluid is determined from the deflection of the spring. Rotational speeds can be up to 100 rev/min, and a range of interchangeable spindles

Figure 6.7 A Brookfield viscometer

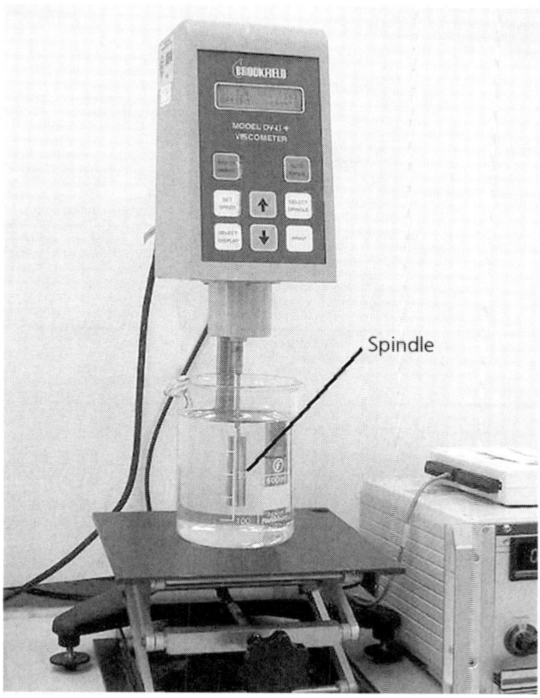

is available, as is a range of models with different spring constants. For example, the one shown in Figure 6.7 is a low-viscosity (LV) model equipped with a soft spring; the spring constant is 0.0673 mN m (millinewton-metre). For testing polymer fluids, the LV model has been found to be the most suitable. The medium- and high-viscosity models are more appropriate for measuring other industrial chemicals such as creams, paints and gels.

To set up a test with the Brookfield viscometer, the operator first needs to attach a spindle to the instrument head and then fill the beaker with the test fluid up to the scribed mark on the spindle stem. The test speed can be selected via a control knob on a basic electrical model or via the keypad on digital models. The shear stress is read either from the dial or on a digital display. On the digital models, the viscosity of the fluid can be shown directly on the display if the appropriate spindle code is entered on the keypad before the test. This simplifies interpretation of the results and minimises the potential for human error.

Unlike the direct-indicating viscometer, the Brookfield viscometer is, by design, a relatively low-shear instrument. The rotational speed typically ranges from 0.3 to 60 rev/min, and can be up to 100 rev/min on some models; however, the shear rate range depends on the test geometry employed, but is usually less than 100 s^{-1}. Some spindles, such as those with a disc-shaped geometry, cannot provide a defined shear rate, and, although suitable

Table 6.3 Constants for Brookfield low-viscosity spindles

Spindle designation	Geometry	Shear rate constant (SRC)	Spindle multiplier constant (SMC)	Maximum viscosity[a]: mPa s
LV-1	Cylindrical	0.220	6.4	20×10^3
LV-2C	Cylindrical	0.212	32	100×10^3
LV-3C	Cylindrical	0.210	128	400×10^3
LV4	Cylindrical	0.209	640	2×10^6
LV5	Cylindrical	0.209	1280	4×10^6
LV-2	Disc	–	32	100×10^3
LV-3	Disc	–	128	400×10^3

Brookfield Engineering Laboratories (2014a, 2014b).
[a] Calculated based on a rotational speed of 0.3 rev/min. Torque constant taken as 0.09373 for the LV model.

for routine quality assurance, are not satisfactory for situations where an actual viscosity value is required. Cylindrical spindles, on the other hand, can give a defined shear rate for a given rotational speed; that is,

$$\dot{\gamma} = \text{SRC} \times \theta \tag{6.20}$$

where $\dot{\gamma}$ is the shear rate (in s^{-1}), θ is the rotation speed (in rev/min) and SRC is the shear rate constant, which is a function of the chosen test geometry. Table 6.3 summarises the values of the SRC for common low-viscosity spindles. For example, when equipped with the low-viscosity spindle LV1, the Brookfield viscometer can measure fluid viscosity over a shear rate range of between 0.066 and 13.2 s^{-1} for a rotational speed range of between 0.3 and 60 rev/min, provided that the limit of the torsion spring is not exceeded. For fluids with higher viscosity, smaller cylindrical spindles such as LV-2C, LV-3C and LV-4 may be used to keep the torque within the operating range. Brookfield uses the letter 'C' to denote 'cylindrical' spindles, as the original LV-2 and LV-3 spindles have a disc-shaped geometry.

For a given viscometer–spindle combination, the full-scale viscosity range can be calculated using the following equation:

$$\eta = \text{TK} \times \text{SMC} \times \frac{10\,000}{\theta} \tag{6.21}$$

where η is viscosity (in mPa s), TK is an viscometer torque constant (=0.09373 for the LV model), and SMC is the spindle multiplier constant. Table 6.3 lists the values of SMCs for common low-viscosity spindles and the corresponding full-scale viscosity values.

To illustrate how the Brookfield viscometer can be used for polymer fluids, Figure 6.8 shows a viscosity plot for the same polymer fluid as was used for Figure 6.6. The low-

Figure 6.8 Viscosity curve of a PHPA solution measured using a Brookfield viscometer

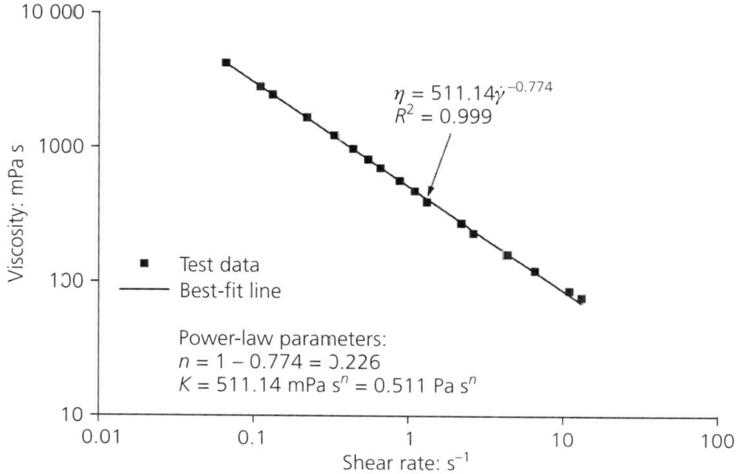

viscosity spindle LV1 shown in Figure 6.7 was used for the tests. It can be immediately seen that the shear rates are much lower than those of the direct-indicating viscometer. As discussed, these lower shear rates are more relevant to the typical in-use situation of support fluids. It can also be seen that the power law parameters ($n = 0.226$, $K = 0.511$ Pa s^n) given in Figure 6.8 are rather different from those previously obtained at the higher shear rates of the direct-indicating viscometer and shown in Figure 6.6 ($n = 0.491$, $K = 0.327$ Pa s^n). This should not come as a surprise, as the power law is merely a model, and different parameters may be needed to describe the flow behaviour of a polymer fluid over different shear rate ranges. Further discussion about the rheological properties of polymer fluids will be given in Chapter 9.

6.6. Density

6.6.1 Basics

Density is the mass per unit of volume of a material; that is, $\rho = m/V$, where ρ is the density, m is the mass and V is the volume. At 20°C, water has a density of 998 kg/m^3 or .998 g/cm^3; the latter unit is more commonly used for expressing the density of drilling and support fluids. The measurement of density is important, as it gives a direct indication of the suspended solids in a fluid (Jefferis and Lam, 2017), and such solids can affect the performance of the fluids in several ways (see Table 1.1). Since used polymer fluids may contain settling soil particles, instruments such as hydrometers are not suitable. The immersed-body method (BSI, 2011a) is also unsuitable. The presence of suspended solid particles in the fluids also renders the oscillation method (BSI, 2011b) unsuitable for fluids containing coarse particles. Figure 6.9 shows a portable density meter utilising the oscillating U-tube principle. Since the internal diameter of the glass tube is only about 2 mm, any coarse solids in the fluids may block or damage the tube. This type of density meter is therefore unsuitable for use on sites containing materials coarser than fine sand. Although electronic density meters based on other principles are

Figure 6.9 An electronic density meter

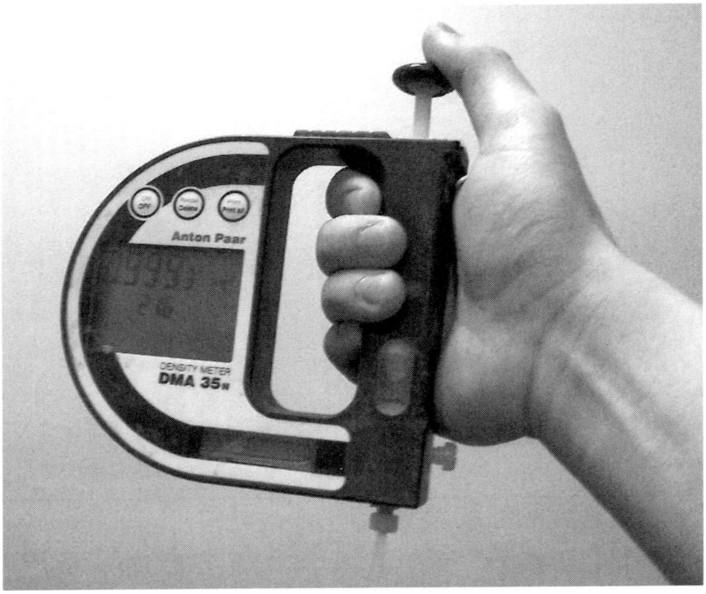

also available (e.g. the tuning fork or resonance principle), they are commonly affected by the high viscosity of the fluids and/or the particle-settling problem. This is probably the reason why the conventional mud balance remains as the industry standard equipment for measuring the density of drilling and support fluids. This instrument, together with the lesser known aerometer and density bottle methods, will be the subject of discussion in the following sections.

6.6.2 Mud balance

The mud balance is undoubtedly the most commonly used instrument for measuring the density of drilling and support fluids. This instrument is rather like a steelyard balance but with the scale pan replaced by a fixed-volume sample cup. The cup is rigidly fixed to a balance arm fitted with a sliding counterweight or a rider. There is a knife edge and a bubble level attached to the balance. When in use, the whole unit is mounted on a fulcrum, which may be set in a base stand or, for some designs, in the plastic carrying case for the instrument. Once mounted on the fulcrum, the rider is moved along the balance arm until the whole instrument is in balance, as shown in Figure 6.10. The density of the fluid then can be read from an engraved scale. Since this instrument was developed in the oil industry to measure the density of drilling fluids, it has a much wider measuring range (0.72–2.88 g/cm^3) than is necessary for civil engineering works, as noted by Jefferis (1977). This is especially true for polymer fluids, as their densities are usually no higher than 1.03 g/cm^3 when in use. Due to the wide measuring range, the mud balance has rather limited resolution. As shown in Figure 6.10, the smallest scale division is 0.01 g/cm^3, although with care the instrument may be read to 0.005 g/cm^3.

Figure 6.10 A conventional mud balance

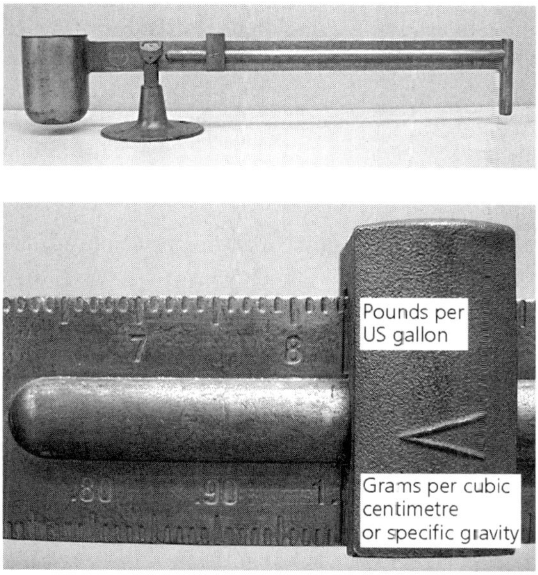

However, the repeatability between readings is seldom better than 0.01 g/cm^3, and this is clearly insufficient for polymer fluids. It should be noted that the instrument may have three scales in addition to the specific gravity scale. These are: (*a*) pounds per cubic foot (lb/ft^3), (*b*) pounds per US gallon (lb/gallon), and (*c*) pounds per square inch per 1000 foot depth (lb/in^2/1000 ft). None of these scales is required for civil engineering work, and only the specific gravity (numerically equivalent to g/cm^3) scale should be used.

There are several common sources of error in a mud balance test, with lack of proper calibration being the principal one. At the start of a new project, the instrument should be checked using fresh water, which should give a reading of 1.00 on the specific gravity scale at room temperature. If the balance does not give the correct reading, it should be calibrated using water or a liquid with a known density. The calibration can be done by adding lead shot to or removing it from the weight adjustment chamber, which is located at the remote end of the balance arm to the cup. If no spare lead shot is available, sand particles may be used instead – but never use a liquid as it will evaporate over time. Another common source of error is due to excess fluid clinging to the surface of the sample cup or the balance beam. Even a small amount of excess fluid can cause incorrect readings, so the external surfaces of the balance should be carefully dried before a measurement is taken. Less experienced operators are recommended to familiarise themselves with the instrument by doing several tests on a single batch of unused fluid; the results should agree to ± 0.01 g/cm^3. If agreement is not achieved, the source of the error must be identified and eliminated. With soil-laden fluids, it may be difficult to get repeatability to ± 0.01 g/m^3, as coarser soil particles can settle in the fluid sample; such fluids should be well stirred before a sample is taken for testing.

6.6.3 Aerometer

The aerometer is a lesser-known instrument than the mud balance, although it is also designed to measure the density of drilling fluids. In the literature, the word 'aerometer' is often used interchangeably with 'hydrometer', but here they are considered distinct instruments although with a common purpose. A hydrometer is an instrument consisting of a cylindrical glass stem at the top and a weighted bulb at the bottom. In use, it is allowed to float in the fluid to be measured. The density of the fluid can then be read off against a calibrated scale within the glass stem.

The aerometer, which is based on the same operating principle, is filled with the test fluid, and is then immersed in water. Figure 6.11 shows an aerometer floating in a water-filled plastic cylinder. As shown in the photograph, at the bottom of the device there is a small detachable compartment (cup), which is filled with the fluid to be measured. This design allows fluids containing settling particles to be measured, as the mass of the fluid inside the compartment remains the same even if particles settle to the base of the cup. After the aerometer has reached equilibrium in water, the density of the test fluid can be read from

Figure 6.11 (a) An aerometer and (b) a close-up of the graduation marks

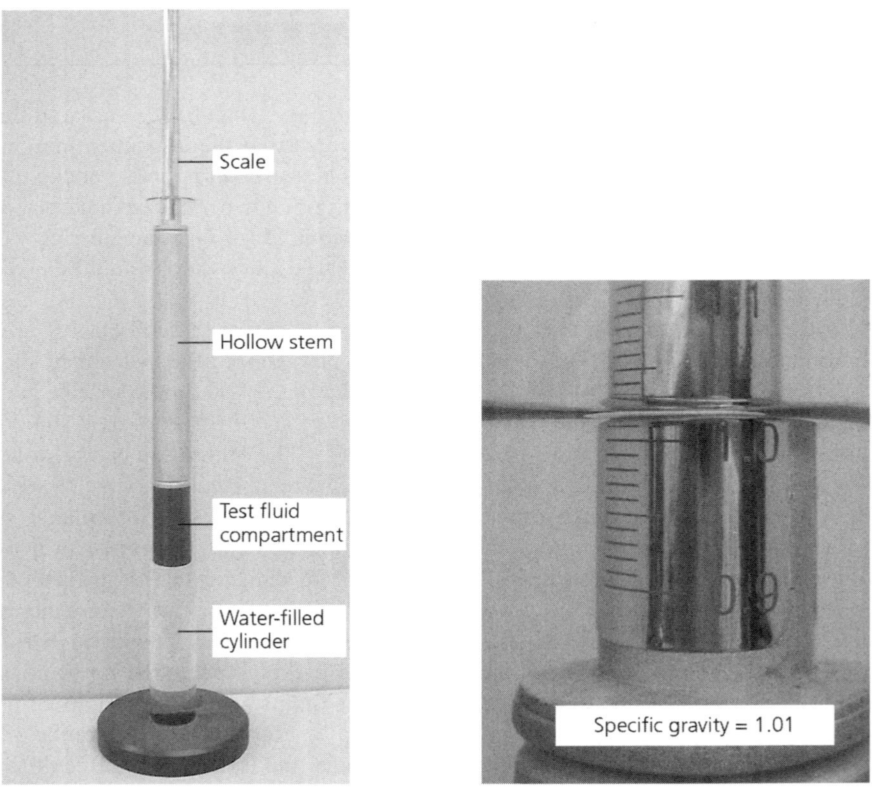

(a) (b)

a scale engraved on the barrel of the instrument, as shown in Figure 6.11. In this case, the specific gravity (density) of the fluid as indicated by the aerometer is 1.01 (g/cm^3). It is interesting to note that, similar to the mud balance, the aerometer also has a very wide density range of between 0.9 and 2.4 g/cm^3 (7.5 to 20 lbs/gal) and that its resolution is the same as that of the mud balance (0.01 g/cm^3). This instrument is therefore likely to provide similar results to the mud balance if used correctly.

Although the aerometer is a very simple instrument, it still needs to be checked regularly, using water to ensure the accuracy of the results. The most common source of error is probably due to the cup not being completely filled with the test fluid. To eliminate this potential source of error, always fill the cup to the top with the test fluid, and let some overflow when the cup is screwed to the main body of the instrument. Wash the surface of the instrument afterwards before insertion into the water-filled cylinder to avoid contamination of the water. If the correct test procedure has been followed but the instrument still gives an incorrect result during a check, then the weight of the instrument should be carefully adjusted by adding lead shot to or removing it from the hollow stem, which can be accessed by unscrewing the cap at the top. Lastly, it should be remembered that the detachable cup should not be changed for a different one, as the whole instrument is calibrated to the original one in the factory.

6.6.4 Density bottle

A density bottle or pyknometer is a simple device that can be used to measure the density of fluids or solids. As density is mass per unit volume, if the volume is fixed, then the density of a material will be directly proportional to its mass. The density bottle method is one of the standardised measurement methods in the petroleum and chemical industries for products such as paints, varnishes and bitumens (BSI, 2004, 2011c). This method is also used in civil engineering for the density measurement of soil grains (BSI, 1990) in the laboratory and of bentonite–cement slurry in the field (ICE, 1999). So far, this method has not been regularly used for the control of drilling and support fluids despite the notoriously low resolution of the mud balance (FPS, 2006; Jefferis, 1977). The procedure requires only a fixed-volume container and an electronic balance. As the latter can be battery powered, there is no restriction on the use of the procedure on site.

Although the concept is simple, the key is to find the right combination of a balance and a container. According to the Institution of Civil Engineers (ICE, 1999), the electronic balance should be accurate to ± 0.1 g and the vessel should have a volume of no less than 1.0 litres to achieve a measurement accuracy of ± 0.002 g/cm^3. Laboratory tests carried out by the authors under temperature controlled conditions support this claim, and further show that the measurement error can be reduced to less than 0.001 g/cm^3 if the vessel volume is increased to 2.0 litres (Lam, 2016). It was also found that a measurement accuracy of 0.005 g/cm^3 can be obtained by using a smaller vessel (e.g. 100 or 200 ml) and an electronic balance accurate to ± 0.01 g or better. Figure 6.12 shows a typical 100 ml narrow-necked glass bottle filled with a bentonite fluid. Although this type of bottle may be suitable for use in a laboratory, it is fragile and is difficult to clean and dry, so it may not be convenient for use in the field. The Hubbard-type pyknometer (BSI, 2011c) may be more appropriate, as it has a wide mouth and is available in metal. The downside is that

Figure 6.12 A laboratory glass bottle used for fluid density measurement

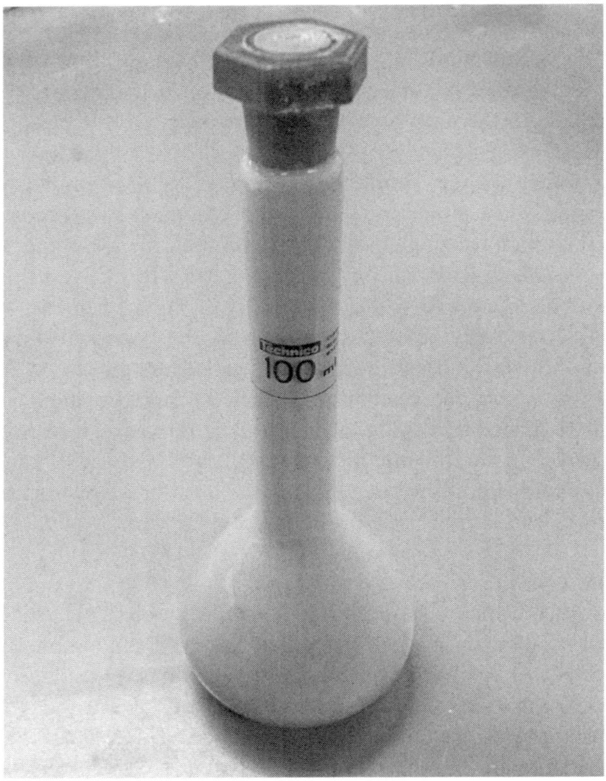

it only has a volume of up to 100 cm^3, so if it is used, a more accurate balance will be required. It is interesting to note that the Hubbard-type metal pyknometer is very similar to the mud balance cup, as both have a cylindrical metal body and a flat lid with a central bleed hole. Recent research by the authors has found that spring top bottles are convenient fixed volume containers for use on site (Jefferis and Lam, 2017).

To run a density bottle test, the following procedure can be followed:

(*a*) Carefully clean and dry a fixed-volume container and the top cap or lid. Weigh them on a suitable electronic balance and record the mass as M_1.
(*b*) Slowly fill the container with water to the top. Tap or vibrate the container if there are any bubbles (e.g. attached to the side walls). Put on the top cap or lid carefully so that there is no trapped air. Remove the excess water, and dry the outside surface with a cloth or paper towel. Weigh the whole assembly, and record the mass as M_2.
(*c*) Empty and dry the container and then fill it with the support fluid to be determined. Take the same precautions given in step (b) to ensure that there is no

trapped air inside the container and that the outside surface is dry. Weigh the whole assembly again, and record the mass as M_3.

(*d*) Calculate the specific gravity of the fluid using the equation $(M_3 - M_1)/(M_2 - M_1)$.

6.7. Sand content

6.7.1 Basics

The sand content test employs a sieving technique to determine the amount of sand particles in a drilling or support fluid. Since the test was originally developed in the oil industry in the USA, sand is defined as material coarser than 75 μm (No. 200 mesh US) but finer than 4.75 mm (No. 4 mesh US). However, particles coarser than the upper limit are not removed in this test, so in theory they are included in the result but in practice they are rarely found, as such large particles rapidly settle from the fluid column. The result of the test, the sand content, is the bulk volume of >75 μm material by volume of the test fluid.

When polymer fluids are to be tested, extra attention must be paid to the sampling procedure, to ensure the sample to be tested is representative of the fluid in the excavation. This is because polymer fluids have a much lower particle-holding capacity than bentonite, so it is necessary to ensure that all suspended soil particles, at the time of initial sampling, are counted. For example, if a sample of used fluid is allowed to stand even for, say, 1 min, then some of the soil particles will have already settled. These settled particles must be reintroduced into the fluid by agitation before a sample can be used for the testing of the sand content.

6.7.2 Measurement technique

This test is commonly carried out using a sand content kit, which can be purchased from most drilling fluid equipment suppliers. Figure 6.13 shows the key components: a glass measuring thimble, a plastic funnel, a No. 200 mesh screen and a stirring rod. A wash bottle, not shown in the photograph, is also commonly required. The measuring tube has a scale that is graduated from 0 to 20% (of the volume to the 'mud to here line' engraved on the thimble), as shown in Figure 6.13.

The procedure to run a test is:

(*a*) Vigorously stir the fluid to be tested in the sample container, to ensure that all soil particles are in suspension.
(*b*) Fill the glass measuring thimble with the fluid to the 'mud to here' line.
(*c*) Top up the thimble with household bleach (dilute sodium hypochlorite solution) to the 'water to here' line. The exact amount of bleach added is not important.
(*d*) Cover the mouth of the tube, and shake vigorously.
(*e*) Pour the mixture onto the mesh screen a little at a time. Wash the <75 μm fines through the screen using a wash bottle filled with bleach. Discard the liquid passing through the screen. If necessary, tap the side of the screen holder to accelerate the flow through the screen. Repeat until all the solids in the thimble have been washed on/through the screen.
(*f*) Turn the funnel upside down and fit it over the top of the sieve, invert the assembly and insert the tip of the funnel into the mouth of the measuring tube.

Figure 6.13 (a) A sand content kit and (b) a close-up of the graduated measuring tube

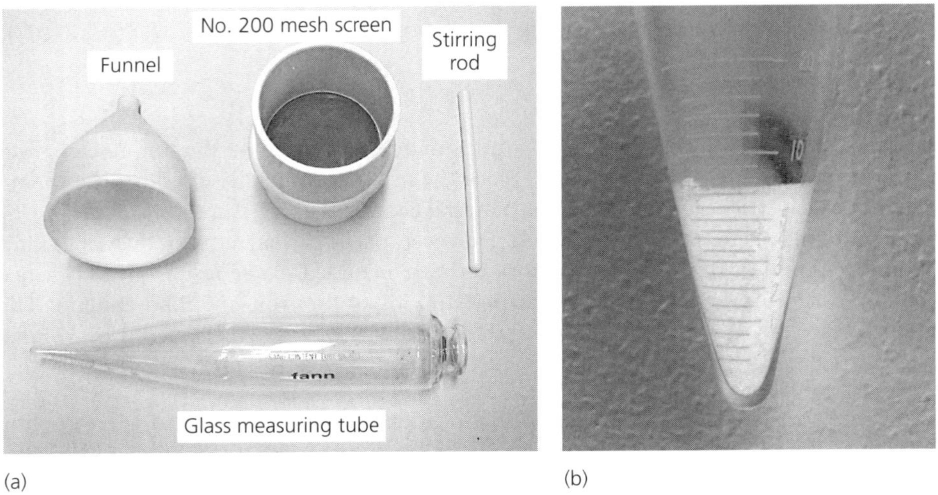

Wash the material retained on the sieve back into the tube with a fine spray of water. Take care not to overfill the tube.

(*g*) Allow time for the sand to settle, and read the volume of the sand against the graduations on the tube. Report the result as the sand content (percentage by volume).

This procedure is essentially the same as that commonly used for measuring the sand content of bentonite fluid, except that dilute bleach is used instead of water for the initial mixing and the subsequent wet-sieving process (steps (*c*) and (*e*)). This is because synthetic polymers such as PHPAs are strong flocculants when diluted – typical dosages for water treatment are 0.05–0.01% by mass according to the British Standards Institution (BSI, 2008). Therefore, if water is used the diluted polymer solutions will flocculate the fine particles to form agglomerates that will appear to be sand. The substitution of bleach for water breaks down the polymer molecules and thus prevents the flocculation from happening. Also, the viscosity of the polymer can make it very difficult to get through the screen; hence the use of bleach is now commonly recommended by polymer manufacturers. Typically, 5% hypochlorite household bleach is suitable, although some manufacturers recommend 10% bleach, which should achieve more rapid polymer breakdown. Thick bleach should be avoided because it has a high viscosity. Bleach can be harmful to the skin and eyes, and careful attention must be paid to health and safety: gloves and goggles should be worn, and clothing should be protected from accidental splashes and spills of bleach.

6.8. Residual polymer concentration

6.8.1 Basics

Although polymer fluids are very different from their bentonite counterparts in terms of their physicochemical properties, none of the tests that have been discussed so far is specifically designed for polymers. Test equipment such as the Marsh funnel, direct-indicating viscometer, mud balance and sand content kit were all developed in the oil industry for testing drilling fluids many decades ago. These tests were directly adopted by civil engineers for testing first bentonite and then polymer support fluids. Although these tests can provide some useful information about the physical conditions of the fluid, the chemical condition of the fluid remains unknown. Polymer fluids, being active chemical fluid systems, rely on their chemical activity for many of their principal functions, including viscosity development and soil encapsulation. However, fine soil particles can significantly reduce the amount of active polymers via a process known as sorption. Therefore, it is very important for all those involved in the management of polymer fluids to know the amount of residual polymer in a used fluid for both quality assurance and fluid management purposes.

Three possible conditions of a polymer fluid (i.e. totally clean (unused), under working conditions and completely exhausted) are briefly discussed here. When a polymer fluid is used in an excavation, the polymer molecules and the soil particles can form bonds with each other by sorption of part of a polymer molecule onto the soil. This allows the polymer fluid to encapsulate cuttings and to 'coat' the side-wall surfaces. When insufficient active polymer molecules remain in solution, the fluid becomes completely exhausted and ceases to function as a polymer fluid. In this text, the term 'flipped' is used to describe a polymer fluid that has changed from an active polymer solution to a clay water suspension with no active polymer – a state that is not identifiable by any of the conventional tests. If flipping is allowed to occur, soil cuttings can freely disintegrate, thus exacerbating the sorption situation, and promoting sloughing and swelling of the side-wall soils if the formation is water sensitive. To prevent flipping from occurring, a method is needed to determine the amount of active polymer in the system. A crude way to determine if there are any free polymer molecules in the system is to observe how the fluid drains from the digging tool upon its retrieval from an excavation. If the polymer fluid forms strings when draining off the tool, then it is an indication that there is still free PHPA polymer in the system.

6.8.2 Measurement techniques

The authors have developed a test that may be used to quantify the amount of free polymer in a used fluid (Lam *et al.*, 2014). This test is based on the assumption that if the suspended soil particles are removed from a used polymer fluid, then the properties of a solids-free fluid can be used as a measure of the residual active polymer. The removal of suspended soil particles is necessary because their presence can interfere with the measurement of the underlying polymer fluid properties In theory, solids removal can be done by either pressure filtration or centrifugation. However, if the API filter loss press is used, then the filtrate will contain entrained bubbles, which can make the filtrate unsuitable for immediate testing. A bench-top micro-centrifuge was found to effectively remove solid particles to produce solids-free liquid (supernate), and so it is recommended for the preparation of samples in the following test.

In terms of accuracy and repeatability, the viscosity method has been found to be the most satisfactory technique for active polymer determination. The major advantage of

Figure 6.14 The set-up for the measurement of the residual polymer concentration

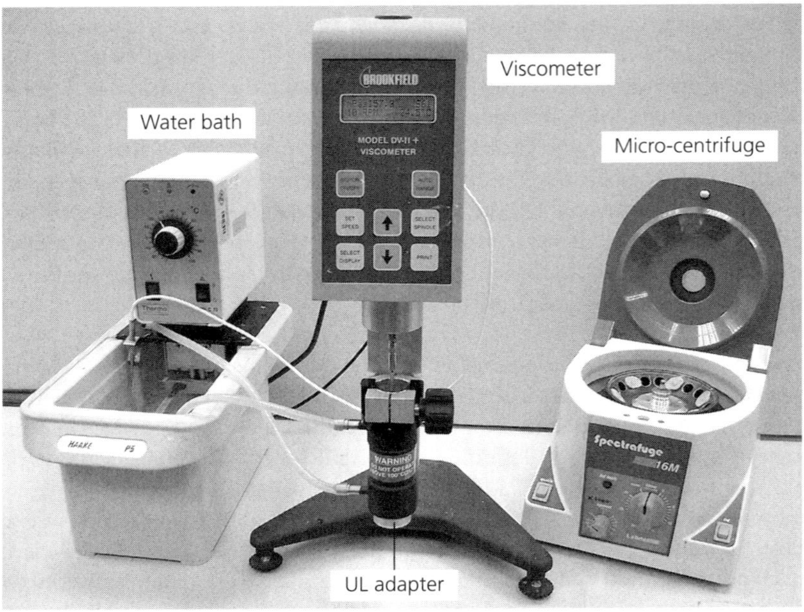

this method is that, because viscosity is a physical property, it will work for different types of polymer as long as they produce a significant increase in fluid viscosity over that of water. A Brookfield viscometer fitted with the UL (ultra-low) viscosity adapter has been found suitable for viscosities as low as 1 mPa s, which is the viscosity of water. For a used polymer fluid that has no active polymer, the supernate is expected to have the viscosity of water, so resolution to 1 mPa s is important. Once the viscosity of a supernate has been determined, the residual polymer concentration can be obtained from a viscosity–concentration calibration plot. Figure 6.14 shows the test set-up, including a micro-centrifuge, a viscometer and a water bath. The water bath is optional, but is recommended for more accurate work as the viscosity of the polymer solution is temperature sensitive.

A step-by-step procedure for the test is set out below.

(*a*) Prior to site testing, prepare a set of viscosity–concentration calibration curves for the polymer to be used. This can be done by testing the viscosity of polymer fluid prepared at a range of known concentrations. Viscosity is dependent on temperature, so a standard temperature of, say, 25°C may be used to prepare the calibration plots. In hotter climates, a slightly higher temperature may be appropriate. It is generally convenient to adopt a temperature slightly above that expected on site, so that only heating and not cooling is required. Note that as the viscosity of polymer solutions is shear rate dependent, it may be necessary to prepare calibration curves over a range of viscometer test speeds so that calibration data are available over the full range of the polymer concentrations expected.

(*b*) Sample the test fluid from the shaft, trench, settlement tank or other location as appropriate.

(*c*) If the fluid contains a significant amount of suspended solids, allow the sample to stand for 1–2 min to settle the coarsest particles.

(*d*) Centrifuge about 25 ml of the fluid using a micro-centrifuge to obtain at least 16 ml of supernate. This centrifugation process should take no more than 15 min to complete for a centrifuge operating at an acceleration of 15 000*g*.

(*e*) Transfer the supernate to the UL adapter of the Brookfield viscometer, and attach the adapter unit to the viscometer head.

(*f*) Prepare a water bath at the temperature corresponding to that of the calibration curves. If the water bath is equipped with a built-in circulatory pump, connect the pump to the UL adapter. Otherwise, simply immerse the adaptor in the water bath. In either case, allow several minutes for the temperature of the specimen to equilibrate. If a water bath is not available, use a wide-mouthed vessel with water temperature adjusted to that required for the test.

(*g*) Measure the viscosity at the rotational speed(s) to which the calibration curves(s) apply. For the best signal-to-noise ratio with low-viscosity specimens, make measurements at the highest speed for which calibration data are available. Similarly, make measurements at a low speed for high-viscosity specimens, to avoid the torque exceeding the upper limit of the instrument.

(*h*) Read off the polymer concentration corresponding to the measured viscosity from the calibration curve for the chosen viscometer speed.

(*i*) For improved confidence in the result, repeat the viscosity test at a different speed.

The whole test procedure, excluding steps (*a*) and (*b*), should take an experienced operator no longer than 30 min. As an example of this test, Figure 6.15 shows a set of calibration curves produced for a high-molecular-weight PHPA polymer fluid prepared using deionised water. Viscosity curves for different rotational speeds are given over a range of polymer concentrations. Any of these curves can be used to determine the residual polymer concentration for this polymer during a test. Figure 6.16 also shows how the active polymer concentration reduces when different amounts of clay are added to the fluid. London Clay was used in this example. It can be seen that the starting concentration of the fresh polymer was 0.4 kg/m^3; a low concentration such as this is recommended by some manufacturers for drilling in clay and shale. As can be seen, very little active polymer is left when the added clay concentration reaches 20 kg/m^3. At this clay concentration, the fluid density is only about 1.012 g/cm^3. At higher clay concentrations, the polymer concentration approaches a very low value, which for all practical purposes can be regarded as zero. For drilling, polymer fluids in this state are physically no different from a slurry of the in situ clay or muddy water.

Finally, it should be noted that this test is best suited to products containing a single anionic polymer such as the common PHPAs. Blended polymer systems may contain some non-ionic polymers that are not reactive to soil but may still contribute to viscosity.

Figure 6.15 Calibration curves for the measurement of the residual polymer concentration

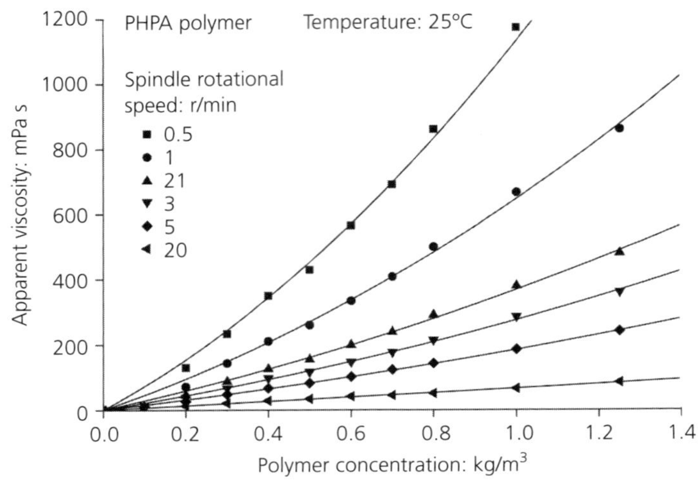

Figure 6.16 Changes in the active polymer concentration with clay concentration

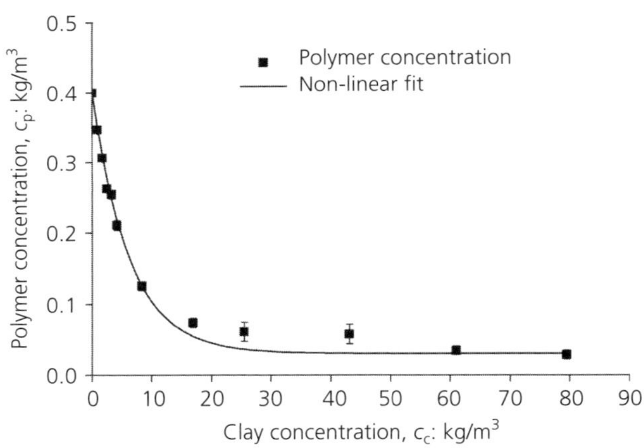

In use, the non-reactive polymers may accumulate in the system, leading to a false (high) residual active polymer indication, even though charged (cationic and anionic) species have been completely exhausted.

REFERENCES

Abichou T, Tawfiq K and Abdelrazig Y (2004) Using electrical conductivity to estimate properties of mineral slurries used in drilled shaft construction. *Geotechnical Testing Journal* **27(6)**: 532–539.

API (American Petroleum Institute) (2003) API RP 13B-1. Recommended Practice for field testing water-based drilling fluids, 3rd edn. API, Washington, DC, USA.

ASTM (American Society for Testing and Materials) (2009) D6910/D6910-09. Standard test method for Marsh funnel time of clay construction slurries ASTM International, West Conshohocken, PA, USA.

ASTM (2010) C939-10. Standard test method for flow of grout for preplaced-aggregate concrete (flow cone method). ASTM International, West Conshohocken, PA, USA.

ASTM (2014) D4016-14. Standard test method for viscosity and gel time of chemical grouts by rotational viscometer (laboratory method). ASTM International, West Conshohocken, PA, USA.

Ata A and O'Neill M (1997) *Characterization of the Effect of POLY-BORE Polymer on the Construction and Performance of Drilled Shaft Foundations. Phase I: Field Tests*. Department of Civil and Environmental Engineering, University of Houston, Huoston, TX, USA, Report to Baroid Drilling Fluids.

Ata AA and O'Neill MW (1998) Side-wall stability and side-shear resistance in bored piles constructed with high-molecular-weight polymer slurry. *Proceedings of the 3rd International Seminar on Deep Foundations on Bored and Auger Piles, Ghent*. Balkema, Rotterdam, the Netherlands, pp. 111–117.

Brookfield Engineering Laboratories (2014a) *Brookfield Digital Viscometer Model DV-II + Operating Instructions*. Brookfield Engineering Laboratories Middleboro, MA, USA.

Brookfield Engineering Laboratories (2014b) *More Solutions to Sticky Problems*. Brookfield Engineering Laboratories, Middleboro, MA, USA.

BSI (British Standards Institution) (1990) BS 1377-2:1990. Methods of test for soils for civil engineering purposes. Part 2: classification test. BSI, London, UK.

BSI (2001) BS ISO 10414–1:2001. Petroleum and natural gas industries. Field testing of drilling fluids. Part 1: Water-based fluids. BSI, London, UK.

BSI (2004) BS EN ISO 3838:2004. Crude petroleum and liquid or solid petroleum products. Determination of density or relative density. Capillary-stoppered pyknometer and graduated bicapillary pyknometer methods. BSI, London, UK.

BSI (2008) BS EN 1407:2008. Chemicals used for treatment of water intended for human consumption. Anionic and non-ionic polyacrylamides. BSI, London, UK.

BSI (2011a) BS ISO 2811-2:2011. Paints and varnishes. Determination of density. Part 2: Immersed body (plummet) method. BSI, London, UK.

BSI (2011b) BS ISO 2811-3:2011. Paints and varnishes. Determination of density. Part 3: Oscillation method. BSI, London, UK.

BSI (2011c) BS ISO 2811-1:2011. Paints and varnishes. Determination of density. Part 1: Pyknometer method. BSI, London, UK.

Caenn R, Darley HCH and Gray GR (2011) *Composition and Properties of Drilling and Completion Fluids*, 6th edn. Gulf Publishing, Oxford, UK.

Chhabra RP and Richardson JF (1999) *Non-Newtonian Flow in the Process Industries*. Butterworth-Heinemann, Oxford, UK.

Darley HCH and Gray GR (1988) *Composition and Properties of Drilling and Completion Fluids*, 5th edn. Gulf Publishing, Houston, TX, USA.

Fann Instrument Company (2013) *Model 35 Instruction Manual*. Fann Instrument Company, Houston, TX, USA.

FPS (Federation of Piling Specialists) (1975) Specification for cast in place piles formed under bentonite suspension. *Ground Engineering* **8(2)**: 50–52.

FPS (1977) Modifications to Specification for cast in place concrete diaphragm walling (published 1973) & Specification for cast in place piles formed under bentonite suspension (published 1975). *Ground Engineering* **10(8)**: 45.

FPS (1985) Specification for cast in place piles formed under bentonite suspension. *Ground Engineering* **18(4)**: 11–15.

FPS (2006) *Bentonite Support Fluids in Civil Engineering*. 2nd edn. FPS, Beckenham, UK.

Hodgson FT (1977) Design and control of bentonite/clay suspensions and concrete in diaphragm wall construction. *A Review of Diaphragm Wall: Seminar*. Institution of Civil Engineers, London, UK, pp. 79–92.

ICE (Institution of Civil Engineers) (1999) *Specification for the Construction of Slurry Trench Cut-off Walls*. Thomas Telford, London.

ICE (2017) *ICE Specification for Piling and Embedded Retaining Walls*, 3rd edn. ICE, London, UK.

Jefferis SA (1977) Contribution to the discussion of 'Design and control of bentonite/clay suspensions and concrete in diaphragm wall construction'. *A Review of Diaphragm Wall: Symposium*. Institution of Civil Engineers, London, UK, pp. 88–89.

Jefferis SA and Lam C (2017) Using density to determine the solids content of construction slurries. *Proceedings of the 19th International Conference on Soil Mechanics and Geotechnical Engineering, Seoul, South Korea*, pp. 1695–1698.

Lam C, Martin PJ, Jefferis SA and Goodhue, KG Jr (2014) Determination of residual concentration of active polymer in a polymeric support fluid. *Geotechnical Testing Journal* **37(1)**: 46–59.

Marsh HN (1931) Properties and treatment of rotary mud. *Transactions of the AIME* **92(1)**: 234–251.

Roussel N and Le Roy R (2005) The Marsh cone: a test or a rheological apparatus? *Cement and Concrete Research* **35(5)**: 823–830.

Savins JG and Roper WF (1954) A direct-indicating viscometer for drilling fluids. *API Drilling and Production Practice*. American Petroleum Institute, Washington DC, USA, pp. 7–22.

Polymer Support Fluids in Civil Engineering
ISBN 978-0-7277-5786-9

http://dx.doi.org/10.1680/psfce.57869.145

Chapter 7
Applications of polymer fluids to the construction of deep foundations around the world

7.1. Introduction

This chapter aims to give a summary of the use of polymer support fluids for deep foundation works around the world based on published case studies. Only field studies are considered, and the focus is on excavation stability and the performance of completed foundation elements. Because construction practice differs considerably between countries, the case studies are categorised according to their geographical location rather than the application, type of polymer used or geological conditions. By doing so, a historical account of the use of polymer fluids is developed.

7.2. Europe

7.2.1 UK

NORFOLK

Corbet *et al.* (1991) reported one of the first case studies of bored piles constructed using polymer fluids. The site was in Norfolk in the UK, and the ground consisted of 5 m of peat overlying 4 m of sand and gravel and then a very thick layer of weathered chalk. The groundwater table was located near the surface. The standard penetration test (SPT) blow count (N) ranged from less than 10 blows near the top to 20 blows at a depth of 45 m. The cone penetration test (CPT) cone resistance (q_c) ranged from 0.2 MPa at the top to 2 MPa at a depth of 32 m. To assist in the design and in the selection of a suitable construction technique, three 40 m-long bored piles were constructed using three types of support fluid: water, polymer and bentonite. Unfortunately, no information about the polymer is given in the article, and so its type and properties are not known. However, as the test piles were constructed in 1989 and at this time the polymer Viskeau was being promoted in the UK (see Section 4.2.3), it is possible that this polymer was used.

Figure 7.1 shows the load–settlement curves for the three piles, and Table 7.1 summarises the pile stiffness at different stages of loading during the tests. From the test results, Corbet *et al.* (1991) showed that the pile constructed using the polymer fluid significantly outperformed the other two in terms of load–settlement characteristics, even though this pile (P3) was installed with an artificial 'soft toe' made of polystyrene foam that was not included in the other two piles. The term 'soft toe' is commonly used to describe a layer of soft sediment at the base (toe) of a pile typically caused by insufficient base or fluid

Figure 7.1 Load–settlement curves for test bored piles in very weak chalk in Norfolk, UK. (Adapted from Corbet *et al.* 1991)

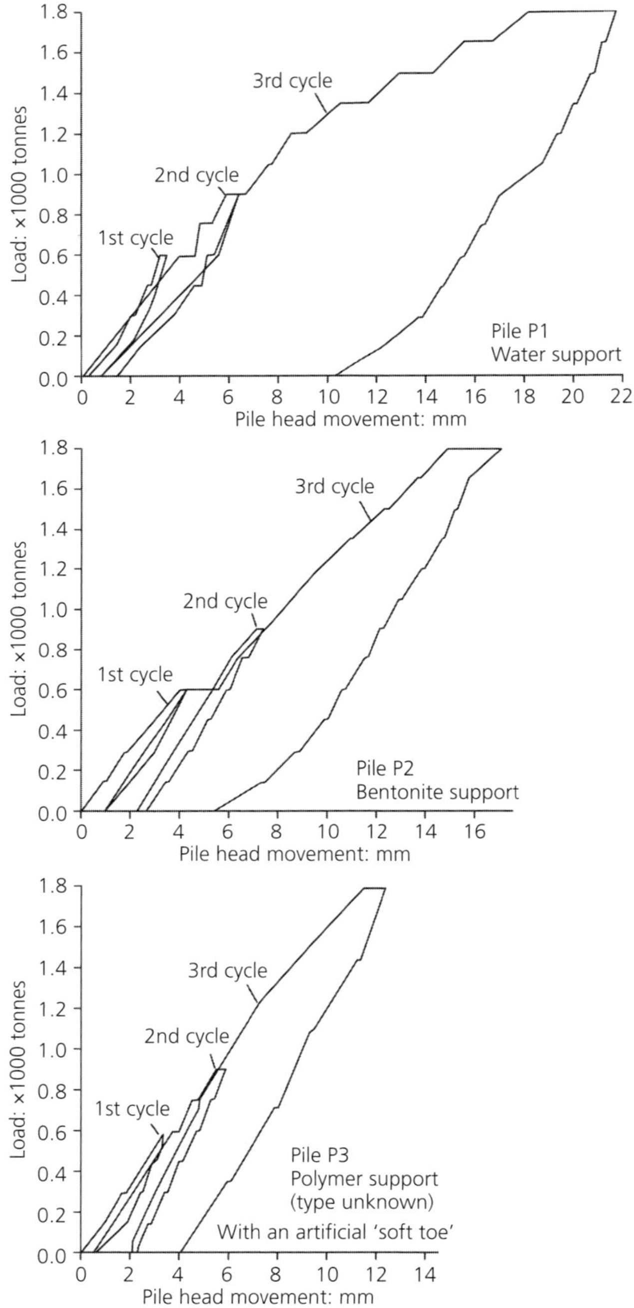

Table 7.1 Stiffness of test piles in very weak chalk in Norfolk, UK

Loading cycle: tonnes	Stiffness: tonne/mm		
	Pile P1 (water)	Pile P2 (bentonite)	Pile P3 (polymer)
0–600	177.0	141.2	169.5
0–900	148.7	138.7	167.6
0–1200	143.5	161.3	234.4
1200–1800	47.1	80.4	114.9[a]

Data after Corbet *et al.* (1991).
[a] Constant rate penetration test.

cleaning prior to concreting. A 'soft toe' is always undesirable, but it was employed for pile P3 to eliminate the effect of base resistance from the overall pile response. The superior performance of pile P3 indicates that the shaft resistance of this pile was very good, and this was probably due to successful stabilisation of the pile bore during excavation and better concrete-soil interface resistance.

There are two other findings from this study that are worth mentioning. First, although the water-supported pile bore had the most irregular side-wall profile and the largest actual diameter as measured by calipers, the completed pile showed the largest settlement under load. This was not expected by Corbet *et al.* (1991), as irregular side-wall profiles are commonly thought to lead to higher capacity due to better interlocking between the pile concrete and the adjacent soil. What probably happened in the water-supported bore was that, as water was unable to stabilise the pile bore, significant stress relief occurred in the adjacent soil, which led to softening of the side walls. Second, Corbet *et al.* also mentioned that a temporary casing of at least 15 m was necessary to ensure the stability of the very soft alluvium and weathered chalk near the surface (SPT $N < 10$ blows/0.3 m) at this site. This suggests that support fluids, regardless of their type, could not adequately support the very soft soil and that temporary casing may be necessary in very soft soils.

CANARY WHARF, EAST LONDON

Wheeler (2003) reported another site trial involving test piles constructed using different types of support fluid. Additional information has been obtained from the design consultant involved. The test site was located in Canary Wharf in East London, where the ground conditions consisted of interbedded clay and sand layers from three different formations, namely the Lambeth Group (stiff clay and sand), the Harwich Formation (alluvial sand and gravel) and Thanet Sand (very dense fine sand). The groundwater conditions at this site are believed to be similar to those described by Troughton (1992) for a nearby site. The aim of the trial was to see whether a pile design, based on the assumptions appropriate to bentonite fluid, could be adopted for piles formed under polymer fluid. To this end, three 0.75 m-diameter bored piles were constructed using polymer fluid (piles TP1, TP2 and TP2R) and one with bentonite (pile TP4). The polymer used was SlurryPro CDP provided by KB International (see Section 4.3.6). To assess the effect of construction time, pile TP1 was constructed over 37 h and the others in 12 h. All the piles were load tested to at least 220% of the design working load.

Figure 7.2 shows the load–movement curves for the piles. The results offer useful insights into the behaviour of piles formed under polymer fluids. First, the measured resistances of all the piles were found to be very good. Due to the high shaft resistances, there was little mobilisation of pile base resistances, and all the piles failed structurally near the top rather than geotechnically. This finding reassured the design consultant that the use of polymer fluids did not lead to a reduction in the pile shaft resistance, so that the existing design practice could be adopted for polymer fluids. Second, in terms of the possible effect of construction time, the pile constructed using polymer fluid in 37 h (TP1) was found to behave similarly to the other piles that were completed in 12 h (TP2, TP2R and TP4). This led to the conclusion that the bores of the working piles could be left open overnight to allow better use of site resources. This decision was reported to have led to an increase in productivity of an extra half a pile per day, since the contractor could start excavating a new bore in the afternoon and then concrete it the next day. This would not have been allowed if bentonite slurries were used, because they are known to reduce the pile shaft resistance with increasing construction time (Thasnanipan *et al.*, 1998). Due to the promising test results, a 4 m reduction in the pile length was allowed in the design of the working piles for this project. However, it was also noted that one of the polymer-supported pile bores collapsed during production due to the polymer behaving differently from the bentonite. The contractor had to change its production methods when using the polymer, although it was not stated what changes were made. Several other practical issues were noted by Wheeler (2003), including the possible adverse effects of some proprietary additives on the piles.

STRATFORD, EAST LONDON

Lam *et al.* (2010) reported another comparative study conducted in East London. The objectives of this study were threefold: (*a*) to compare polymer and bentonite fluids in terms of their effect on pile performance, (*b*) to study the effect of an extended pile bore open time and (*c*) to study the effect of polymer fluids on the quality of concrete. The trial involved testing three instrumented piles, two of which were constructed using a polymer fluid and one using bentonite slurry. The only difference between the two polymer-supported bores was the duration of the soil–fluid exposure time. The load test results showed that the piles constructed using polymer fluid significantly outperformed the one constructed under bentonite slurry, and that there was virtually no difference between the two piles constructed using polymer fluid, a finding that shows an apparently negligible effect of an extended pile bore open time. The polymer fluid when intermixed with the concrete was found to adversely affect the strength of the hardened concrete, but no more than bentonite did. This case study will be described in detail in Chapter 9.

ROYAL VICTORIA DOCK, EAST LONDON

Hardy (2009) reported one of the first uses of polymer fluids for the construction of marine piles in the UK. In addition to the article by Hardy, information has also been obtained from the contractor's website (Bachy Soletanche, 2010). The piles were for the

Figure 7.2 Load–movement curves for four test piles constructed using polymer (SlurryPro CDP) and bentonite fluids at Canary Wharf, London, UK. (From Lam and Jefferis, 2016)

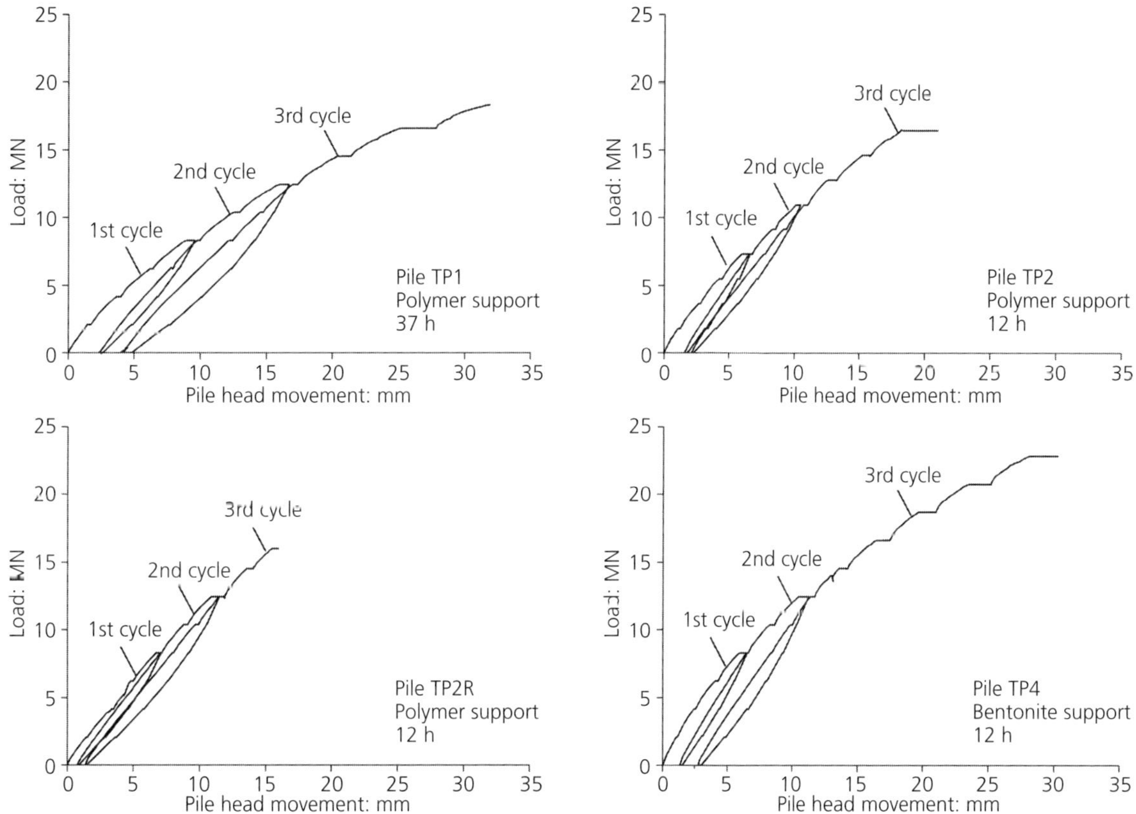

construction of a deck in the Royal Victoria Dock for the extension of the ExCeL conference centre in east London. The project required 148 marine piles, 85 of which were installed from the quay and 63 from two barges. The piles had diameters of between 0.9 and 1.05 m and lengths of between 34 and 44 m; the top 2–10 m was through water. Pile construction started by installing a 20 m-long permanent casing into the ground with approximately 2 m above the water level. The water inside the casing was then pumped out and replaced with Super Mud polymer fluid. Super Mud is a partially hydrolysed polyacrylamide (PHPA) supplied as a liquid emulsion (see Section 4.3.4) but photographs found on the manufacturer's website (PDS) suggest that Super Mud Dry was actually used. Super Mud Dry is also a PHPA, but it is supplied in dry powder form. No information about the polymer fluid was given in the article, and no problems (pile bore instability, etc.) were reported. One of the key messages of the article is that traditional large-diameter auger piling techniques can benefit a variety of projects both on land and in water.

GLASGOW, SCOTLAND

Lennon *et al.* (2006) reported some case studies of the successful use of polymer fluid for bored piling projects in central Glasgow, Scotland. The ground conditions in Glasgow typically consist of loose becoming very dense silty fine sand for the top 25 m, and then the bedrock. The groundwater table is associated with the River Clyde, and is typically at a depth of 4–5 m. It has been reported that continuous-flight auger piling is normally the preferred technique for urban sites in Glasgow, but flighting of the superficial (alluvial) deposits has on occasion resulted in building damage in the past. Polymer fluid was used to overcome this problem and to avoid using the bulky ancillary plant for bored piling with bentonite. The polymer used was the SlurryPro CDP system from KB International (see Section 4.3.6). To illustrate the effects of polymer, the load–settlement curves for test piles from two different sites in Glasgow (Jamaica Street and Pacific Quay) were presented by Lennon *et al.* The curves showed that both test piles had a near-linear load–settlement relationship up to the proof load, and it was concluded that the piles had very good contact with the soil, both along the shaft and at the base. A similar load–settlement response (unpublished) was obtained by the piling contractor at several other sites, such as the Anderston site in Glasgow (SlurryShield) and the University College Hospital site in central London (SlurryPro CDP).

Lam *et al.* (2015) reported another case study of polymer use in Glasgow. The ground conditions at the site led to the polymer fluid being heavily laden with suspended silt and fine sand particles at the end of digging each pile. Fluid cleaning using chemical additives alone was unable to reduce the sand content and density values to below the specified upper limits. The pile bores were therefore simply left open overnight, to allow the particles more time to settle out of suspension, and the bores cleaned in the morning. This strategy proved to be successful, as both the density and the sand content fell to within the specified limits. The overnight waiting period considerably reduced the risks of 'soft toe' and concrete contamination due to the settling soil particles – both are common problems with polymers. Structural integrity and dynamic loading testing showed that the completed piles were of good quality despite the additional waiting period. This case study will be discussed in more detail in Chapter 8.

7.2.2 Portugal

VASCO DA GAMA BRIDGE

Several published case studies of polymer use in other European countries have been found, although they are sometimes reported in the local language (e.g. French, German or Italian). For these publications the reviews are briefer. The first relates to the Vasco da Gama Bridge across the Tagus River in Lisbon, Portugal, where polymer fluid was used to construct some deep bored piles. Probably due to its high-profile status, several articles have been written about this project (Bustamante *et al.*, 1998; Guadagnini, 2001; Manuel Correia and Sêco e Pinto, 1999; Sêco e Pinto and Oliveira, 1998). The ground conditions below the river bed consisted of 35 m of very soft silty clay (layer 1), which was underlain by 9 m of medium dense to dense silty sands (layer 2), 6 m of hard clay (layer 3) and then at least 7 m of very dense sandy gravel with pebbles.

The polymer used in the project was a high-molecular-weight PHPA marketed as Geomud-15. A brief introduction to this polymer was given in Section 4.3.2. Some inconsistent information has been found between the published articles about the quality of the mix water: Bustamante *et al.* (1998) mentioned that the polymer was mixed with fresh water, while Sêco e Pinto and Oliveira (1998), Manuel Correia and Sêco e Pinto (1999) and Guadagnini (2001) stated that salty water from the Tagus River was used. According to Sêco e Pinto and Oliveira (1998) and Guadagnini (2001), the Geomud-15 polymer was used at a concentration of 2 kg per 1000 litres of water, and the Marsh funnel time was 40 s. This viscosity value is rather low for typical synthetic polymers, although it would be within the expected range if salty river water was indeed used for mixing the fluid. However, there are apparently conflicting data on the polymer, as Bustamante *et al.* (1998) stated that the polymer concentration was 1.2 kg per 1000 litres and the Marsh funnel time was 60 s.

As regards pile test results, Manuel Correia and Sêco e Pinto (1999) presented load–settlement curves for three 1.2 m-diameter instrumented piles (P8, P31 and P79) constructed at three different locations (the main bridge, central viaduct and south viaduct), and compared the results with predictions. The measured failures loads, defined as the applied load at a head settlement of 10% of the pile diameter, were found to be lower than predicted for piles P8 and P31. The difference was thought to be due to the lower pile shaft resistance. Bustamante *et al.* (1998) and Guadagnini (2001) presented the results of another test pile at pylon south (PS), which had a diameter of 1.2 m and a total length of 60.8 m, of which 52 m was embedded in the river bed. To evaluate the effect of toe (base) grouting, after the first load test to 17.5 MN, the base of the pile was grouted in three stages, and the pile was load tested again. Figure 7.3 shows the load–movement curve for the pile. It can be seen that base grouting resulted in a much stiffer response and also possibly higher ultimate capacity. From the load test results, it was concluded that polymer fluid did not lead to any reduction in the ultimate unit shaft resistance (skin friction), and polymer fluid was therefore allowed to be used for the construction of the working piles, which had a much larger diameter of between 2.0 and 2.2 m. A total of 278 working piles were constructed by the contractor Trevi using polymer fluid, and another 548 piles by a different contractor using bentonite. There were also 552 driven piles. Of the 278 piles constructed under polymer fluid, only three were found to have defects at the pile

Figure 7.3 Load–movement curve for a test pile constructed using polymer fluid (Geomud-15) at Vasco Da Gama Bridge, Portugal; intermediate unload–reload curves have been omitted for clarity. (Adapted from Guadagnini, 2001)

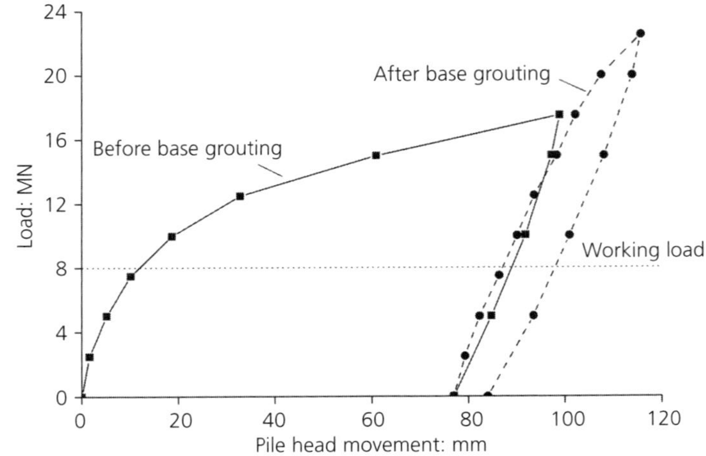

toes, which is equivalent to 1%. On the other hand, of the total 548 piles constructed using bentonite, five were found to have defects at the pile toes, and eight at the pile shafts, giving a total defective rate of 2.4%. These results appear to suggest that polymer fluid did not adversely affect the structural integrity of the completed piles.

Despite the satisfactory structural integrity of the piles constructed using polymer fluid, it was also reported that two of the working piles had to be redrilled using SlurryPro CDP fluid because the Geomud-15 fluid became contaminated by the soils. The manufacturer of the SlurryPro CDP polymer, in one of its advertising documents, explains that its fluid was contaminated by a large amount of calcium and magnesium ions from the ground due to the nearby salt flats. In the authors' opinion, the two collapses might have been prevented if the Geomud-15 polymer fluid had been mixed at a higher concentration, or potable water had been used rather than brackish water, to prevent the adverse effect of dissolved ions in the water. The second point also applies to bentonite slurry, because dissolved salts can inhibit the dispersion of bentonite clay particles. This case study highlights the importance of polymer selection and maintenance to ensure compatibility with the ground.

7.2.3 Italy

ITALIAN HIGH SPEED TRAIN PROJECT

In addition to the Vasco da Gama Bridge project, Bustamante *et al.* (1998) also reported the use of polymer fluid for the construction of the Naples–Rome section of the Italian High Speed Train (TAV) project. A total of 1732 bored piles were required to support the viaducts, although about half of these were constructed using bentonite. The polymer used was Geomud-15, and the support fluid was mixed at a concentration of 0.6 kg/m^3, which gave a Marsh funnel time of 48 s. Compared with the Vasco da Gama Bridge

Figure 7.4 Load–movement curve for a test pile constructed using polymer fluid (Geomud-15) at Peccia-1, TAV project, Naples–Rome section, Italy. (Adapted from Bustamante *et al.*, 1998)

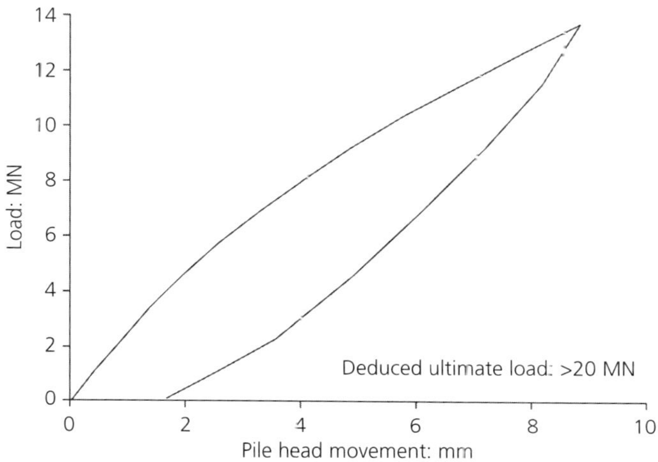

project, the dosage of polymer was lower but the viscosity was higher; this was probably due to the use of potable rather than brackish water for mixing. Two pile load tests were carried out at two different sites, namely Peccia-1 and Cassino South-2. The test pile at Peccia-1 had a diameter of 1.2 m and a total length of 33 m; Figure 7.4 shows the load–movement curve for this pile. Due to the limited reaction load, the pile was loaded to 13.8 MN, although the deduced ultimate capacity was believed to be over 20 MN. The derived unit shaft resistance (f_s) curves showed that, except near the pile top, the shaft resistance of this pile was not fully mobilised during the test. The second test pile at Cassino South-2 also did not reach its ultimate resistance during the load test. Nonetheless, by comparing the mobilised unit shaft resistance curves with those obtained for continuous-flight auger piles, it was concluded that polymers did not lead to any reduction in shaft resistance in the pyroclastic soil (pumice and lapilli).

Following the satisfactory performance of the test piles, the use of polymer fluid was extended to the working piles for eight different viaduct sites, all of which were on pyroclastic soils. Bentonite slurries were used for sites (e.g. Gari, Aquino and Rio Verne) underlain by alluvial deposits (silty clay, silty sand, clayey sand, sand and gravel). At the eight sites where polymer fluid was used, a total of 951 piles of a diameter of 1.2 m and up to 42 m deep were constructed, although it was reported that, at one site (Pisciarello), bentonite fluid had to be used instead due to the markedly cohesionless nature of the soil. In the authors' opinion, this was probably due to the relatively low Marsh funnel time of the polymer fluid used (48 s), as a funnel time of at least 60 s for PHPAs is commonly recommended for drilling in coarse soils (see Tables 4.1–4.3).

All the piles constructed under polymer fluid were tested for integrity, and about 10% were found to have defects at the pile tip, although none of the defects was such as to require any remedial work. Construction details, including the exposure time, total installation time and concrete overbreak, were recorded, but no direct correlation was found between these parameters and the occurrence of defects. In the authors' opinion, the defects may possibly have been due to sediment accumulating at the bases of the bores. Given the relatively low Marsh funnel time (48 s) of the polymer fluid used, it was possible for some sediment to build up at the bases over the period from the end of excavation to the first pour of concrete. The importance of base cleaning has already been discussed in Section 5.7.

7.2.4 Germany

MUNICH

Lesemann (2010) and Lesemann and Vogt (2012) reported the results of a field trial near Munich Airport in Germany. The ground conditions consisted of sand and gravel mixtures with an average permeability (k) of approximately 5×10^{-3} m/s. At the test site the groundwater table was about 2.0 m below ground level. A detailed description of the site, including borehole logs, in situ tests, photographs and particle-size distribution curves, has been provided by Lesemann (2010). The trial was carried out to assess the performance of polymer support fluids in highly permeable grounds. To this end, six 0.6 m-diameter bored piles (P1–P6) with a length of 10 m were constructed. Three types of polymer were tested: a polyacrylamide (PAA) marketed as SC MUD P, a carboxymethyl cellulose (CMC) marketed as SC VIS HV P and a xanthan gum marketed as SC XGUM. A bentonite with the trade name Tixoton was also used to construct the piles. The polymers used were provided by the chemical company Süd-Chemie, and their properties have been discussed in Sections 4.2.6 and 4.3.3. The support fluids were prepared at different dosages for the six piles: two piles with PAA at concentrations of 2 and 6 g/l, two with xanthan gum at 2 and 4 g/l, one with CMC at 4 g/l, and one with bentonite (Tixoton) at 50 g/l. Table 7.2 summarises the support fluid and test pile information. Note that the abbreviation PAA was used by the report authors to denote PHPA.

To assess the stability of the bores, profiling of the side walls was carried out using the ultrasonic technique after excavations. Negligible differences were found between the side-wall profiles of the piles. This finding confirms that polymer fluids are, in principle, capable of supporting coarse-grained highly permeable soils such as those encountered at the test site, and also disproves the common perception that only bentonite slurry can stabilise excavations in gravelly soil. However, it was noted that a problem (details are not given) occurred during the concreting of pile P1, which was constructed using PAA (PHPA) fluid prepared at the very high concentration of 6 g/l; the typical concentration range for a polyacrylamide-based polymer fluid is about 0.8–1.2 g/l. The problem was stated to have been solved by replacing the PAA fluid with water, and so the higher than necessary polymer concentration, and hence the fluid viscosity, may have been the reason for the problem.

After the construction of the piles, piles P3 (bentonite at 50 kg/m^3) and P4 (xanthan gum at 2 kg/m^3) were load tested in compression using the other four piles to provide the reaction (tension) forces. Figure 7.5 shows the load–movement curves for the six piles. It can be seen that pile P3 settled 10 mm less than pile P4 did under the maximum load of about 5 MN. This finding is contrary to those obtained from several other case studies

Table 7.2 Support fluid and test pile information for the test in Munich, Germany

Pile designation	Support fluid type[a]	Concentration: kg/m^3	Marsh funnel time s[b]	Loading direction	Maximum load: kN	Pile movement under maximum load: mm
P1	PAA	6	190–288 (225)	Tension	2,475	−19.8
P2	PAA	2	60–67 (63)	Tension	2,460	−11.3
P3	Bentonite	50	32–33 (32)	Compression	4,950	15.5
P4	XAN	2	37–41 (38)	Compression	4,930	23.3
P5	CMC	4	49–55 (52)	Tension	2,475	−12.4
P6	XAN	4	48–52 (50)	Tension	2,460	−9.5

Data after Lesemann (2010).
[a] PAA, anionic polyacrylamide (PHPA); CMC, carboxymethyl cellulose; XAN, xanthan gum.
[b] Numbers in parentheses are average values.

Figure 7.5 Load–movement curves for six bored piles constructed using polymer (Geomud-15) and bentonite support fluids at a test site in Munich, Germany; intermediate unload–reload curves have been omitted for clarity. (Adapted from Lesemann, 2010)

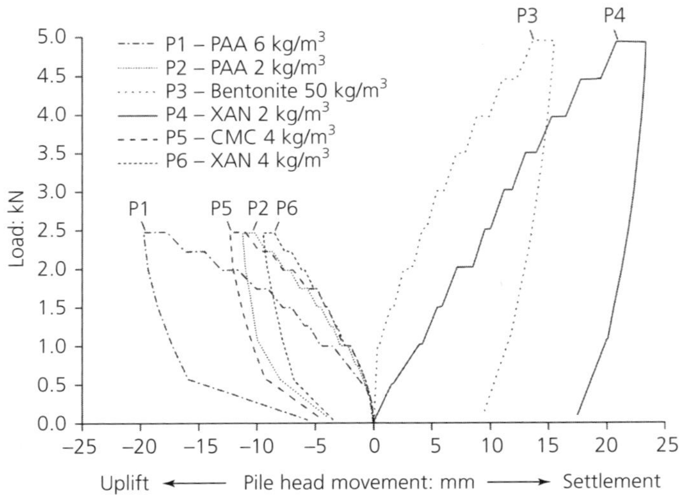

that generally show lower settlements with piles constructed using polymers (e.g. see Figures 7.1 and 7.2). The reason for the relatively poor performance of pile P4 at this site was possibly due to the low polymer concentration used. Typically, a concentration of 3 kg/m^3 would be used for a natural xanthan polymer for foundation drilling (Beresford *et al.*, 1989), whereas only 2 kg/m^3 was used for the production of pile P4. As a result, the average Marsh funnel time for pile P4, which was 38 s, was the lowest among all the piles constructed under polymer fluid (see Table 7.2).

The effect of polymer concentration can also be seen by comparing the settlement of piles P4 (tested in compression) and P6 (tested in tension), both of which were excavated under xanthan fluid but at different concentrations (2 and 4 kg/m^3, respectively). When the applied load is small, the effect of the pile base resistance is often small or negligible, so the response of these two piles can be compared. Figure 7.5 shows that the initial load–movement response of pile P6 was much stiffer than that of pile P4. This suggests better excavation support for the former, although the original researchers also cited a flaw in pile P4 as a reason for its inferior performance. Piles P1 and P2 are another pair that are worthy of discussion, as they were also excavated under the same type of polymer fluid (PAA) but at different concentrations (6 and 2 kg/m^3). As shown in Figure 7.5, pile P1 showed much larger movement than pile P2, although the former was excavated under higher-concentration (viscosity) fluid. The inferior performance of pile P1 can be explained by the fact that concreting of this pile was only successful on the third attempt, when the support fluid was replaced with water. During the first two failed attempts, the pile bore was supported by high-viscosity polymer fluid (Marsh funnel time 190–288 s), and the tremied concrete stiffened prematurely. The original researchers postulated several reasons for the failed attempts, including chemical interaction between the concrete and the high-viscosity fluid. In the light of this experience, it may be prudent not to use polymer fluids prepared at a very high concentration or viscosity. After the static load tests, an inspection of the top 2 m of the completed piles revealed good shaft profiles and also good bonding between the pile concrete and the soil. A defect was, however, found in pile P4 (xanthan gum at 2 g/l) near the pile cap, but the soil at this level would have been supported by a temporary casing, not polymer. Note that this case study was originally reported in German, so this summary might not exactly reflect the views of the original authors – readers should consult the original documents for fuller details.

7.3. Africa

At the time of writing, only two published case studies of polymer use in Africa have been found, although this apparent very low level of publication is by no means representative of the actual volume of polymer use in the region. The two case studies relate to the construction of a new bridge over Lake Nokoué (a lagoon) in Benin (Bustamante and Boato, 2005; Schwarz and Lange, 2004) and the João Landim Bridge in Mansôa, Guinea-Bissau (Bustamante and Boato, 2005). Note that these case studies were first reported in German and French, so the summaries below might not perfectly reflect the views of the original authors – readers should consult the original documents for specific details.

7.3.1 Benin

LAKE NOKOUÉ

For the bridge over Lake Nokoué, 72 piles of length 56–73 m and diameter 1.2 m were required to support the proposed structure. The working loads were between 2600 and 4300 kN. The ground conditions were loose to medium-dense sand and gravel (0–15 m, below ground level), soft becoming stiff, silty organic clay (15–55 m) and fine to medium sand (>55 m). A geological cross-section can be found in Schwarz and Lange (2004). The results of SPTs and CPTs and particle–distribution curves of the soils have been reported by Bustamante and Boato (2005). Two 1.2 m diameter trial piles (Nos. 1 and 2) were constructed using Geomud-15 polymer fluid, and two (Nos. 3 and 4) were constructed using Novagel polymers. The Novagel polymer is supplied in granular form, and is believed to be an anionic PHPA, according to the information given by the manufacturer. Collapses occurred during the excavation of all four piles when the bores reached between 48 and 66 m. Some information about the polymer fluids can be found in Table 7.3. As can be seen, the Marsh funnel time and pH of the fluids dropped considerably after use in each of the four polymer piles. Further laboratory investigations revealed that the soil pore water was very rich in sodium, potassium, chloride and sulfate. It was therefore believed that the collapses were caused by the contamination of the polymer fluids by salts from ground. It can also be seen in Table 7.3 that the viscosity of the bentonite fluid increased from the initial 37 to 60 s after use. This is a typical result of fluid contamination by salts that can flocculate bentonite clay platelets and increase the gel strength and viscosity of the fluid (Darley and Gray, 1988). Due to the poor performance of the polymer fluids, CMC-enhanced bentonite (Bentosund 120E) slurries were used to construct a further test pile (No. 5), and no stability problems were reported. A total of four completed piles were later load tested by the Osterberg cell (O-cell) method, and the results were considered satisfactory.

Table 7.3 Summary of support fluid and test pile information at Lake Nokoué, Benin

Pile No.	Support fluid type	Dosage: kg/m^3	Additive	Temporary casing length: m	Depth reached: m	Standing time after drilling: h	pH	Marsh funnel time: s	Before use: pH	After use: Marsh funnel time: s
1	Geomud-15	1.5–2.0	Soda ash	8	58	12	9.5	60	7.2	47
2	Geomud-15	2.0	—	8	66	2	7.5	75	7.5	59
3	Novagel	1.0	Soda ash	10	48	0	11.0	75	8.0	45
4	Novagel	2–2.5	Soda ash	22	54	0	11.0	95	7.5	50
5	Bentonite	45–50	Soda ash and CMC	24	68	36	8.7	37	7.0	60

Data after Schwarz and Lange (2004) and Bustamante and Boato (2005).

It is interesting to note that this case study is rather similar to the Vasco da Gama Bridge project in Portugal where Geomud-15 appeared to have been adversely affected by salts in the ground. These two case studies highlight the fundamental need to respect the specific ground conditions of a site when using a any support fluid system.

7.3.2 Guinea-Bissau

JOÃO LANDIM BRIDGE

In Guinea-Bissau, Geomud-15 fluid was used for the construction of 55–75 m-long, 1.8–2.0 m-diameter bored piles for the João Landim Bridge (Bustamante and Boato, 2005). The site was underlain by about 50–70 m of clay, followed by the limestone bedrock. The strength of the clay was described as soft at the top and gradually becoming hard near the bottom. The Geomud-15 fluid was prepared at a dosage of 2 kg/m^3, and the Marsh funnel time ranged between 55 and 60 s. A trial undertaken on the river bank showed that the polymer fluid was unable to support the pile bore at 64.5 m depth, and this was believed to be due to the presence of numerous coarse sand inclusions in the hard clay. Further attempts at boring with different dosages of Geomud-15 also failed. As a result, the contractor decided to support the pile bore with a full-length casing, and used the polymer fluid only as a lubricant to aid the installation of the casing through the clay.

7.4. Asia

7.4.1 Taiwan

TAIWAN HIGH SPEED RAIL PROJECT

In Taiwan, Bustamante and Boato (2005) and Corbet *et al.* (2005) reported the use of polymer support fluid for Contracts 260 and 270 of the Taiwan High Speed Rail project, which were undertaken between 2000 and 2004. Large-diameter bored piles were required to support the railway viaduct, which runs along much of the route of the two contracts. The ground conditions were alluvial sand, gravel and clay to a depth exceeding 100 m at some locations. SPT blow counts (N) ranged from between 6 and 20 for the first 20 m to refusal at a depth of 28 m, probably due to the presence of large boulders. SlurryPro CDP polymer was mixed at a concentration of between 0.4 and 0.6 kg/m^3, and the Marsh funnel time was between 52 and 65 s. Trials of different pile construction methods were undertaken at the beginning of the project. A test pile was first constructed using a grab for the excavation and water to counterbalance the groundwater which was at about 3 m below the ground level. The results of load tests using a single O-cell set at about 20 m below the pile head showed that the mobilised skin friction (unit shaft resistance) was lower than the values used in the tender design. Polymer fluid was therefore used for the second test pile, which was excavated using a rotary auger. However, side-wall profile measurements using the Koden ultrasonic technique showed very large overbreak to an extent that was considered unacceptable. The third trial pile was therefore constructed using the reverse-circulation drilling (RCD) technique with polymer fluid. According to Corbet *et al.* (2005), another 22 test piles with lengths ranging between 25 and 72 m were later constructed using the same method. The report authors did not explain why drilling with the RCD technique produced better results than those with the rotary auger. However, given that the ground consisted mainly of sand and gravel, the soil on the side walls of the second test pile was probably disturbed by the auger during its repeated entry into and exit from the pile bores.

All the 22 test piles were tested for axial and lateral load capacities at not less than 28 days from concreting. The test results showed that the unit shaft resistance in clay was higher than the original design value, but no improvement in the shaft resistance was found for the sand and gravel layers. A weaker than expected unit base resistance (end bearing) was also concluded from the back-analyses, and this prompted a revision of the design equations and more careful base cleaning for the working piles. The original and revised design equations, which were based on SPT results, are provided by Corbet *et al.* (2005). According to the piling contractor, 8200 working piles of 2.0 m diameter and up to 60 m deep were later constructed using polymer fluid. This case study again shows that polymer fluids can be used to stabilise excavations in gravelly soils. Proper base cleaning is also shown to be particularly important when polymer fluid is used.

Duann *et al.* (2004) and Chin and Chen (2006) reported the pile load test results for another section of the Taiwan High Speed Rail project, delivered under Contact 291. The ground conditions at the test site consisted of alluvial deposits of interbedded sand and silty clay with mudstone at greater depth. A total of six instrumented bored piles were constructed by the RCD method and tested in compression using a loading frame. Duann *et al.* mentioned that only four piles (TP1–TP4) were included in the original test programme. However, due to the unsatisfactory performance of these piles, two additional test piles (TP-1B and TP-3B) were constructed using improved methods, including (*a*) use of polymer fluid to reduce the effect of filter cake on the shaft resistance, (*b*) shorter construction times and (*c*) introduction of multi-stage grout injections to the pile bases. Table 7.4 provides some basic information on the piles. From the load test results, it was found that piles TP-1B and TP-3B gave a much stiffer, near-linear settlement response with the applied load. The maximum achievable loads with these piles were also higher (see Table 7.4). Although the better pile response must be partly due to the multistage toe-grouting operation, higher shaft resistance was also found to be a contributing factor, and this is illustrated in Figure 7.6. The improvement in the unit shaft resistance seen in piles TP-1B and TP-3B was attributed to the use of polymer fluid and a shorter construction time.

Table 7.4 Summary of test pile information for Taiwan High Speed Rail Contract 291

Pile designation	Diameter: m	Length: m	Use of polymer[a]	Toe grouting	Total construction time: h	Maximum load: tons
TP-1	1.8	65.0	No	Single stage	68.5	2200
TP-1B	2.0	64.0	Yes	Multi-stage	24.0	4220
TP-2	2.0	62.0	No	Single stage	72.0	2800
TP-3	2.0	56.5	No	Single stage	36.0	2800
TP-3B	1.8	60.0	Yes	Multi-stage	18.0	4160
TP-4	1.8	54.5	No	Single stage	38.0	4090

Data after Duann *et al.* (2004).

[a] Duann *et al.* (2004) did not specify the stabilisation method for those piles constructed not using polymer fluid, but it is believed that bentonite slurry was used.

Figure 7.6 Measured unit shaft resistance plotted against SPT *N* values for Taiwan High Speed Rail Contract 291: (a) clay and (b) sand (adapted from Duann *et al.* 2004)

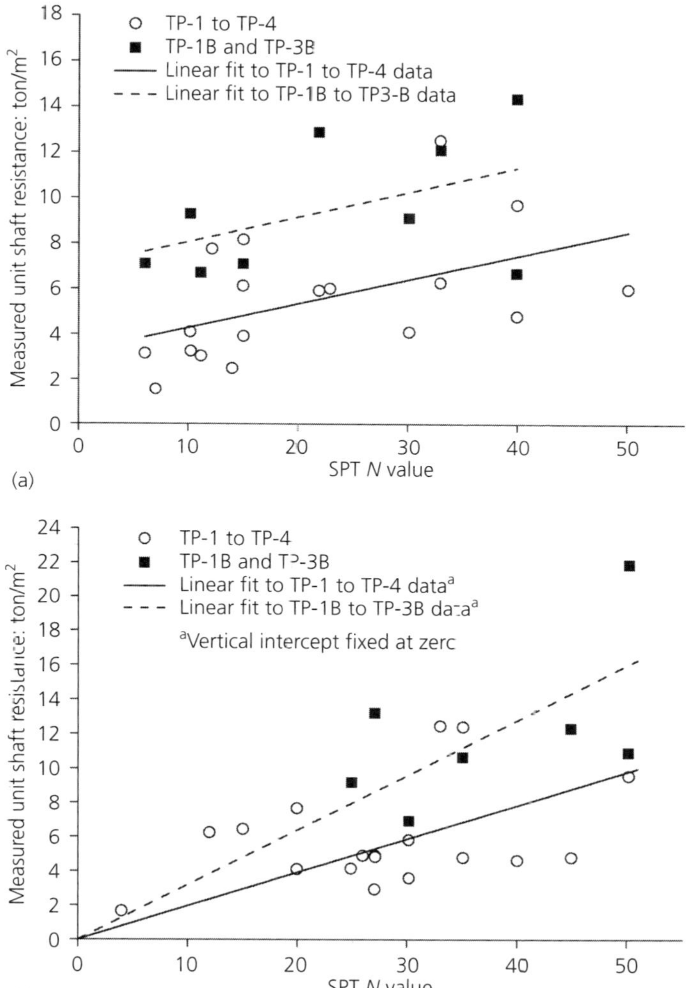

BORED PILES IN TAIPEI CITY

Lin *et al.* (2007) presented the load test results for six drilled shafts (bored piles) constructed using polymer fluid in Taipei City, Taiwan. The site was underlain by about 50 m of low-plasticity clay and silt followed by about 15 m of sandy gravel and then sandstone bedrock. Detailed ground investigation results including SPT *N* values and undrained shear strength for the clay/silt layer were provided by Lin *et al.* The aim of the trial was to assess the shaft resistance of the piles through the gravel layer and also the effect of tip (toe) grouting. To this end, three of the test piles were toe grouted. All the piles were drilled using the reverse-circulation method using polymer fluid. It was noted that the solid-laden fluid was removed from the excavations to sedimentation tanks for self-cleaning; the use of mechanical fluid-cleaning equipment is not mentioned. The polymer used for the trial was not described by the original report authors, but Lin (personal communication, 2016) gave its trade name as A-303. According to the manufacturer (Pacific Orb Trading), this product is an anionic polyacrylamide (i.e. a PHPA) supplied in powder form, and its method of use is very similar to that for other PHPA polymers. No construction problems were reported, and all the test piles were constructed successfully. From the load test results, the original report authors derived the unit shaft resistance (f_s) values for the test piles, and presented design equations for use in that area.

DIAPHRAGM WALLS IN TAIPEI CITY

Hsieh and Ho (2011) discussed the experience gained from a diaphragm walling project in Taipei City. Diaphragm walls were constructed for Dingpu Station, which forms part of the Taipei MRT (Mass Rapid Transit) system. The ground conditions at the station site were challenging in that there were about 20 m of highly permeable gravel sandwiched between the topsoil and the sandstone bedrock. The average particle size for the gravel was about 30–40 mm, and the largest boulder found in this layer had a size of 1.2 m. The groundwater table was 12 m below the surface. The trenches were excavated using a hydraulic grab and supported by the polymer fluid AIDDRIL 300, which is believed to be a granular PHPA. Hsieh and Ho noted that 15 kg of polymer materials were mixed with 50 m^3 of mix water, and adjustment was made depending on the actual site conditions. Due to the gravel layer, the contractor stockpiled some clay clods on site to use as a fluid-loss-control material. The contractor also employed an observational approach to monitor the amount of fluid loss when excavation reached 10, 15, 20 and 25 m below the ground level. At these depths, digging was halted and the support fluid level was topped up to about 35–40 cm below the ground level. The support fluid level was then recorded 10 and 15 min later, and the rate of fluid loss calculated using the difference in the fluid level. The degree of fluid loss was classified into three categories, as follows:

- Minor fluid loss: flow rate less than 1.0 m^3/h. Remedial action can be maintaining a stable supply of support fluid or increasing the viscosity of the fluid.
- Moderate fluid loss: flow rate between 1.0 and 2.5 m^3/h. Remedial action can be adding clay slurry into the trench. The clay slurry is prepared by mixing the stockpiled clay with water.
- Serious fluid loss: flow rate higher than 2.5 m^3/h. Remedial action can be adding clay clods to the trench or backfilling the trench with sand. The backfilled trench will be excavated following ground improvement works.

At the station site, the recorded fluid loss was typically classified as moderate to serious, and it was found that adding clay slurry/clods to the trench was an effective means to reduce the fluid loss to minor or negligible levels.

7.4.2 Hong Kong

TUEN MUN STATION

In Hong Kong, polymer support fluid was used for the construction of the West Rail Tuen Mun Station between 2001 and 2002. This case study was documented by Lam and

Jefferis (2014). In this project, the piling contractor proposed to use polymer fluid as an alternative to bentonite slurry, and cited the Taiwan High Speed Rail project as a successful case study, which was designed and supervised by the same consultant company. The project involved the construction of nine rock-socketed bored piles, ranging from 1.5 to 1.8 m in diameter and up to 51 m in length, to support a new railway viaduct crossing the Tuen Mun River. Part of the river was diverted so that during construction the piling rigs operated on a platform built on top of the original riverbed; this removed the need for a barge. The ground conditions consisted of soft alluvial clay overlying residual soil overlying granite bedrock. Prior to the start of the excavation, a 10 m-long temporary casing was inserted into the ground at each pile location to support the alluvial clay at the top. Polymer fluid mixed at a concentration of at least 0.7 kg/m^3 was then introduced into the pile bores during excavation below the casing level. The polymer used was SlurryPro CDP. At the end of the digging, the pile base was cleaned by air-lifting. The support fluid was tested before, during and after excavation for the conventional parameters (pH, viscosity, density and sand content) and checked against the project fluid specifications proposed by the contractor. Although the completed piles were not load tested, their structural integrity was checked by cross-hole ultrasonic logging. No problems were found with the completed piles. Although the boring for this project was not particularly technically challenging compared with some others, it is interesting to note that if the contractor had not proposed the use of a polymer fluid, the consultant would not have proposed it first and would by default have opted for the conventional options of either bentonite or full-length casing. Indeed, in most countries and situations it is typically the contractors who first propose use of polymer fluids as a way to produce a more competitive tender by reducing the construction costs

7.4.3 Thailand

BANGKOK ELEVATED ROAD AND TRAIN SYSTEM

Littlechild and Plumbridge (1998) reported one of the first documented uses of polymer fluid in Thailand, although little information about the polymer used was given. The project was the Bangkok Elevated Road and Train System (BERTS), which required 16 000 bored piles with a diameter of 1.5 m. Due to the large number of piles to be constructed, instrumented trial piles were first constructed using different techniques, including bentonite and polymer fluids and base grouting. Since Bangkok is located at the southern end of the Chao Phraya alluvial plain, the ground conditions in Bangkok comprise interbedded layers of medium-dense to dense sand (SPT $N = 50$ to >100 blows/300 mm) and firm to very stiff clay ($S_u = 50$–200 kPa) to a significant depth. The top 12–18 m of soil is soft marine clay known as Bangkok Soft Clay ($S_u = 10$–25 kPa). Due to the ground conditions, during construction a temporary steel casing was required to support the upper Bangkok Soft Clay, and bentonite or polymer fluid was used for excavation below the casing.

An interesting feature of this project, and indeed of all the identified Bangkok case studies, is that the polymer fluid was not mixed with water but with dilute bentonite. In the case of the BERTS project, bentonite was added to the mix water for the polymer fluid because high fluid loss into the sand layers was observed in the initial polymer-supported trial bores, making the use of polymer fluid uneconomic according to Littlechild and

Plumbridge (1998). It was found that the fluid loss could be significantly reduced if the polymer fluid was mixed with water preloaded with bentonite at a concentration of 5 kg/m^3, compared with the typical concentration of 30–60 kg/m^3 for bentonite fluid. It is interesting to note that Tallard (1992) has also discussed the effect of bentonite addition on the filtration behaviour of a biodegradable polymer fluid. Tallard (1992) reported that adding a small amount of bentonite clay particles to the polymer fluid can significantly reduce its filtration volume through the American Petroleum Institute (API) cell. The beneficial effect was attributed to the clay particles allowing the polymer molecules to bridge over the voids on the filter, to form a strong and thin cake.

Before the BERTS project, the piling contractor had already carried out two successful pile loading tests on other projects (Asoke and Worajack) using polymer–bentonite fluid. However, since the construction time of these trial piles was only between 14 and 16 h, two additional trial piles (C21 and C22) were carried out on the BERTS project site. Their purpose was to assess the effect of construction time and for the design consultant to set the polymer fluid requirements. A summary of the construction details and the loading test results is given in Table 7.5. The results of five bentonite trial piles (C5, C8, C10, SRT1 and SRT3) at the same site have also been included for comparison. The ratio of the actual to the calculated shaft resistances for each pile has been plotted against the construction time and fluid viscosity, as shown in Figures 7.7 and 7.8, respectively.

Regarding the effect of construction time, Figure 7.7 shows that for those piles constructed using mixed polymer–bentonite fluid, increasing the construction time to up to 42 h did not appear to have a negative effect on the performance of the pile. However, for those constructed using bentonite fluid, a significant reduction in the shaft resistance was found with increasing construction time. Similar findings were also reported by Hosoi and Matsushita (2016). Figure 7.8 shows the effect of fluid viscosity increasing the fluid viscosity appears to have a positive effect for the piles constructed using polymer–bentonite fluid, but has a clear negative effect on piles constructed using bentonite.

Despite the apparent advantages of using mixed polymer–bentonite fluid in Bangkok, the authors have reservations about this practice, as even though the bentonite dosage is small (5 kg/m^3), filter cakes can still form on the excavation side-walls and reduce the shaft resistance. In fact, the adverse effect of adding bentonite to polymer fluid has been noted in a laboratory study by Thasnanipan *et al.* (2003). In this study, a series of laboratory interface shear box tests were carried out. The results showed that the interface resistance between the cement mortar (simulating the pile concrete but crucially not using full-size aggregate particles – see Lam *et al.*, 2014) and the sand was reduced by both pure bentonite and polymer–bentonite fluids. To overcome problems of high fluid loss in areas such as Bangkok, the authors prefer the use of the non-cake-forming additives for fluid loss control (e.g. superabsorbent polymers), as they can plug the pores of the side walls without forming a filter cake at the surface. Whether pore blockers were considered for the BERTS project has not been stated. However, it is worth noting that polymer–bentonite fluids are commonly used in the petroleum industry, but for very different reasons. For oil well drilling, it is often found that drilling fluids with a high solids content cause reductions in the penetration rate, and hence part of the bentonite

Table 7.5 Construction details and load test results for trial piles constructed using polymer and bentonite fluids in Bangkok, Thailand

Support fluid type	Pile	Pile diameter: m	Pile length: m	Marsh funnel time: s	Construction time: h	Actual shaft resistance: MN	Calculated shaft resistance: MN	Ratio of the actual to the calculated resistance
Polymer–bentonite	Asoke	1.0	55	32–38	14	11.50	10.90	1.06
	Worajack	1.5	56	48–55	16	25.00	19.70	1.27
	C21	1.5	50	35–45	41	21.00	17.70	1.19
	C22	1.5	50	45–55	41	22.00	16.80	1.31
Bentonite	C5	1.5	55	50	42	6.35	20.45	0.31
	C8	1.5	55	38	30	12.00	20.80	0.58
	C10	1.5	55	33	29	13.90	19.50	0.71
	SRT1	1.5	53	32	16	23.00	18.54	1.24
	SRT3	1.5	55.7	37	21	19.80	20.48	0.97

Data after Littlechild and Plumbridge (1998).

Figure 7.7 Pile performance plotted against the construction time in Bangkok. (Data after Littlechild and Plumbridge, 1998)

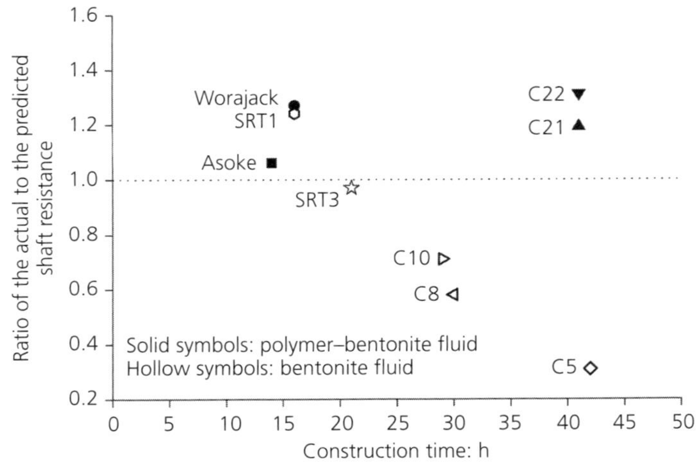

Figure 7.8 Pile performance plotted against the polymer–bentonite fluid viscosity. (Data after Littlechild and Plumbridge, 1998)

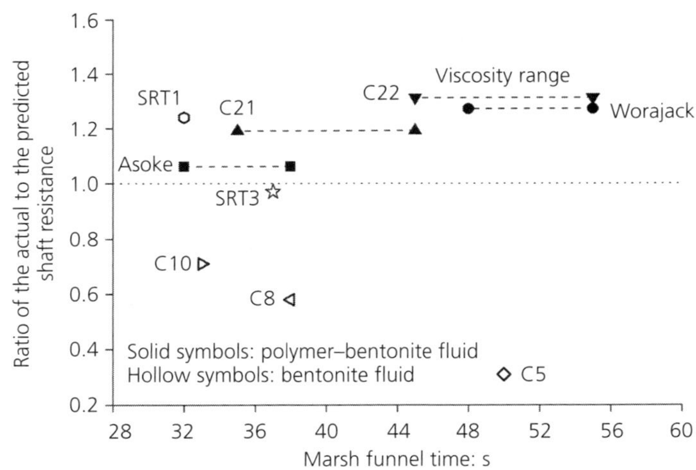

is replaced by polymers (usually polyanionic cellulose or xanthan gum) in order to reduce the amount of solids in the fluid but maintain the cuttings-carrying capacity. The polymers in the fluid can also help to reduce the swelling of water-sensitive formations. These systems are commonly known as a 'low-solids mud' by petroleum engineers (Darley and Gray, 1988). Mixed-polymer (typically CMC) and bentonite fluids are also commonly used in horizontal directional drilling (HDD) for their good filtration control and high solids-carrying capacities at the low flow velocities of large diameter

HDD bores. It must be borne in mind that, for HDD, the formation of a moderate filter cake on the side wall is not a concern.

OTHER PROJECTS IN BANGKOK

Probably due to the experience gained in the BERTS project, the other two published case studies in Bangkok also reported the use of dilute bentonite slurry as the mix 'water' for the polymer fluid. Thasnanipan *et al.* (2002) noted that the reason for using the mixed-fluid system was to reduce the amount of fluid loss that was observed from trial boreholes drilled using pure polymer fluid. However, they also commented that, depending on the thickness of the sand layers at the piling locations, the addition of bentonite may not always be necessary. The results of 11 trial piles were reported, and they were all found to show lower settlement than predicted. Three of the trial piles were instrumented with strain gauges, and the measured shaft resistance was found to be significantly higher than the predicted values, giving a measured-to-predicted shaft resistance ratio of between 1.41 and 1.58. These numbers are slightly higher than those given by Littlechild and Plumbridge (1998; see Figures 7.7 and 7.8), but this may be due to the predictions made by Thasnanipan *et al.* (2002) being more conservative. Based on back-calculated results of instrumented pile load tests, Thasnanipan *et al.* (2006) suggested design β values for bored piles in Bangkok constructed using polymer (strictly, a polymer–bentonite mix) and bentonite fluids ($\beta = K_s \tan \delta$, where K_s is the coefficient of the lateral earth pressure and δ is the pile–soil friction angle). Their suggested β values for polymers are 0.1–0.15 greater than those for bentonite, but the difference will depend on the internal friction angle of the sand.

7.4.4 Singapore

TAMPINES

In Singapore, three case studies of polymer use have been reported by Ho and Tan (1996) and Ho *et al.* (1999). Ho and Tan (1996) discussed the performance of a test barrette constructed for a two-level basement in Tampines, Singapore. The site is underlain by the Old Alluvium, which is a medium-dense to dense sand also with a significant amount of gravel, silt and clay. At the test location, the SPT N values increased from less than 10 near the surface to over 100 blows/300 mm at a depth of 17 m. The dimensions of the test barrette were 0.8 m (width) × 2.8 m (length) × 17 m (depth), and it was excavated by a hydraulic grab. Strain gauges and rod extensometers were installed at different levels down the barrette. The type of polymer used was not stated, although the properties of used fluid samples taken from the trench were presented. In particular, the Marsh funnel time of the fluid was reported to be in the range 28–32 s – this is a very low value by any standards, as the viscosity of water is 26 s. Surprisingly, despite the low fluid viscosity, no instability was reported for the barrette.

The completed barrette was load tested by the maintained load method 26 days after casting. For each stratum, the measured unit skin friction (f_s) was normalised by the corresponding SPT N value, and the ratios compared with those obtained previously from bored piles constructed using full-length casings at other sites. It was found that there was little difference between the two construction methods. However, for the barrette, the base resistance did not pick up until a toe displacement of 4.8 mm had been reached. Thereafter, the observed base response was found to represent the consolidation of weak debris under the test load. The reason given by Ho and Tan (1996) for the poor base response is that, because of the long waiting period (4.5 h) between completion of the desanding process and casting, soft sediment was able to form at the base of the trench due the settlement of soil particles from the fluid. It is important to note that although the sand content of the used fluid was only between 0.2% and 1.0%, the test barrette still had a 'soft toe'. That so relatively little sand could give a soft toe is surprising, but this was probably due to the viscosity of the fluid being so low that it was incapable of limiting the settlement of the particles. This was exacerbated by a long trench open time between cleaning and concreting, thus allowing the time for considerable sedimentation process to complete. Because of the issue of base quality, the working barrettes, which were constructed in the same manner as the test barrette, were designed without relying on the end resistance (i.e. friction only). In the authors' opinion, cost savings would have been achieved if the foundations had been constructed using higher-viscosity fluid, and if after excavation the used fluid had been properly cleaned (e.g. by allowing the particles to settle in the excavation overnight). This would have allowed the end resistance to be considered in the final design.

CHOA CHU KANG

Ho *et al.* (1999) presented the load test results of an instrumented bore pile constructed using polymer fluid for a housing project in Choa Chu Kang, Singapore. The ground conditions consisted of 25 m of residual soil overlying 5 m of weathered granite followed by bedrock. The SPT N value was less than 10 blows/300 mm in the residual soil, and increased to between 50 and 100 blows/300 mm in the weathered granite. A 5 m-long temporary casing was used, and excavation below the casing was carried out under a polymer fluid, for which no information is given. The pile was instrumented with strain gauges and rod extensometers at several levels, and was tested by the maintained-load method. From the load test results, it was found that the end resistance showed a sudden drop when the toe displacement reached 49 mm. A check on the installation records revealed that the depth of the pile bore was reduced by 200 mm during the period between the end of boring and just before lowering the reinforcing steel cage. This indicated accumulation of soft sediment at the bottom of the bore. The soft sediment was suspected to have mixed with the fresh concrete during casting, and the sudden drop in the end resistance was believed to be the result of overstressing and crushing of the weakened concrete. The peak unit shaft resistance was expressed as the ratio f_s/N and compared with data obtained from piles constructed using bentonite. No significant differences were found between the two construction methods, although the data appear to be quite scattered.

To summarise, the two Singaporean case studies show remarkable similarity despite the difference in the local geology (Old Alluvium and granitic residual soil) and the form of the foundation elements (barrette and pile). Both case studies reported good shaft resistance but poor base response due to the accumulation of sediment during the waiting period after base cleaning but before concreting. However, it should be noted that in Singapore the problem of 'soft toes' was not confined to piles constructed using polymer fluids, as Ho (2002) found that piles constructed under bentonite also suffered from 'soft toes'.

7.4.5 Vietnam

HO CHI MINH CITY

Nguyen and Fellenius (2014) reported the results of static load tests on two bored piles constructed for the Binh Loi Bridge, which crosses the Saigon River at Ho Chi Minh City in Vietnam. Both piles were drilled using the reverse-circulation technique to 73.0 m depth. One of the piles had a diameter of 1.5 m and was drilled using bentonite fluid (pile D1500), and the other one had a diameter of 2.0 m and was drilled using polymer fluid (pile D2000). A temporary casing was used to support the excavation down to 10 m depth. The type of polymer used was not mentioned, nor the properties of the support fluid. The site was underlain by about 15 m of soft clay and silt followed by dense to very dense silty sand. At the location of pile D2000 there was also a layer of clayey silt of between 30 and 40 m in depth. Ground investigation results, including those from SPTs, CPTs and basic index tests, were given in the original article.

Nguyen and Fellenius (2014) noted two construction issues for pile D2000 that are worth mentioning. The first was that, after the tremie pipe had been lowered to the base of the pile bore, it was found to be blocked by sediment (debris) that had accumulated at the base. The pipe was thus extracted and the sediment removed. The second one was that, after construction, the pile was cored, and debris and slurry were found at the pile toe. These issues were probably due to the polymer fluid becoming heavily loaded with silt and sand particles at the end of the excavation. Also, due to the significant depth of the excavation (75 m), considerable sedimentation and sediment build-up probably occurred after the end of excavation. As discussed in Section 5.7, cleaning of polymer fluids prior to concreting is very important. Deep bores like this one are likely to benefit from the use of chemical additives to accelerate the sedimentation process. After construction, both piles were load tested using single-level bi-directional cells. The results showed little difference between the two piles in terms of their unit shaft resistances. Nguyen and Fellenius postulated that both piles were affected by the filter cake between the concrete and the soil. Polymer fluids, if heavily laden with fine soil particles, will also form a filter cake on the exposed soil surfaces, although the thickness will be less than that of a bentonite filter cake.

7.4.6 India

CHENNAI

Khare and Ramanathan (2013) described the collapse of a diaphragm wall trench supported by polymer fluid. The location of the site was not revealed by these authors, but was possibly in Chennai, India. The geology of the site comprised 2.5 m of made ground followed by about 22 m of silty sand and clay and then weathered rock of 3 m thickness. Below the weathered rock was fresh bedrock. The SPT N_{60} values increased from less than 10 for the top 5 m of soil to around 50 at a depth of 20 m. Diaphragm walls 5 m long, 1 m thick and 25.2 m deep were required for a metro station built using the cut-and-cover construction method. The wall trenches were excavated using a hydraulic grab. The excavation of the collapsed trench was delayed by heavy rain and a public holiday. When work resumed on the sixth day after the start of the excavation, the trench was found to have partially collapsed. Side-wall profiling using the ultrasonic technique revealed that the collapse occurred between 1 and 9.5 m below the ground level. Khare and Ramanathan noted that the heavy rains caused a rise in the groundwater table to a depth of 2.5 m below the ground level. This resulted in a reduction in the net positive fluid head (see Figure 1.1) and thus the factor of safety for the top 10 m of loose sandy soil. Khare and Ramanathan conducted back-analyses of the stability of the collapsed trench. They concluded that increasing the unit weight (density) of the support fluid alone led only to a marginal increase in the factor of safety, and that the support fluid head (level) played a major role in trench stability. The report authors also noted another case study of diaphragm wall trench collapse, but bentonite fluid was used in this second case. The collapse was again due to rise of the groundwater table after heavy rain. These two case studies show that, regardless of the type of support fluid used, it is very important to maintain a sufficient stabilising fluid pressure in the excavation, especially during the rainy season when the groundwater table can rise very rapidly.

DELHI

Trindade (2014) reported the use of polymer fluid for the construction of the Mandi House and Janpath Road metro stations in Delhi, India. The ground conditions at both station sites were quite similar, comprising about 3 m of made ground, followed by a thick layer of sandy and silty soil. The groundwater table was found at 9 m below the ground level. Diaphragm walls and bored piles were required to form the main structures of both stations. The diaphragm wall panels were excavated using hydraulic grabs, and the pile bores were excavated using drilling buckets and augers. The polymer used was PolyMud (see Section 4.3.10). The polymer was mixed using pH 12 make-up water, to provide fluid with a Marsh funnel time of between 55 and 65 s. The use of additives was not required at these sites. Trindade noted that during construction the support fluid level was always maintained at no less than 1.5 m below the guide wall (for the diaphragm wall trenches) and 3 m below the surface casing (for the pile bores). Probably for this reason, no collapses were encountered, and some excavations remained open for up to 7 days. The average concrete overbreak was 11% for the diaphragm wall panels and 3% for the bored piles.

7.5. The Americas

7.5.1 USA

EARLY EXPERIENCES

As polymer support fluids were first developed and used in the USA, a relatively large number of US case studies have been identified. O'Neill (1991) presents early experiences with polymer fluids in the USA, and touches upon a wide range of site issues such as borehole stability, fluid loss and fluid contamination by silt particles. O'Neill also mentioned two early comparative studies that showed that drilled shafts (bored piles) constructed with polymers were not in any way inferior to piles constructed by other methods. Despite these promising initial trials, many problems were encountered during the early years of polymer use, and these are described by Berkovitz and Long (1995). They identified that many state departments of transportation (DOTs) reported side-wall sloughing, excavation collapse and bottom sediments when polymer fluids were used. Experience showed that the sediment build-up could occur even after bottom cleaning, thus reducing the bearing capacity. It was also noted that most of the problems were not detected by standard field fluid tests, sand content measurements and shaft-bottom probing. Some were identified only during or after concrete placement. As many questions raised by the DOTs remained unanswered at that time, Berkovitz and Long recommended that until there was further research and specifications to facilitate consistent construction performance, bentonite fluids should be used for the routine production of piles, and any use of polymer fluids in federal or state DOT projects should be designated as 'experimental' under the Federal Highway Association (FHWA) system, and special testing and reporting requirements would therefore have to be followed. Berkovitz and Long asserted that extensive full-scale trials conducted under controlled conditions were necessary before any routine use of polymer fluids. Other recommendations included development of specifications for polymer fluids, resolution of all environmental issues, development of an expert system, etc.

In a general discussion of the advantages of using biodegradable polymer fluids for construction, Tallard (1992) mentioned their use for the construction of some structural diaphragm wall panels at the World Bank Headquarters in Washington, DC. According to Tallard, this was the first time biodegradable polymer fluid was used in North America for diaphragm walls. The polymer used was described as a low-molecular-weight cellulosic polymer. Details of the particular project were not given, as the focus of the article was the advantages of biodegradable polymer fluids in general. By cross-referencing with promotional literature from Liquid Earth Support, it was ascertained that the polymer used for the World Bank Headquarters project was PREMIUM. The properties of this product have been discussed in Chapter 4.

OHIO RIVER, KENTUCKY

Holmes (1993) reported one of the first uses of SlurryPro CDP fluid in the USA for the construction of drilled shafts (bored piles) for a new bridge over the Ohio River in Kentucky in 1992–1993. To form the two piers for the main span of the bridge, 88 bored piles with an original design depth of up to 140 ft (43 m) and a socketed length of 22–55 ft (7–17 m) were required. The ground conditions consisted of sand and gravel overlying shale (mudstone). Laboratory tests showed that the shale disintegrated immediately upon contact with water or mineral slurry. However, it was not possible to construct the pile bores in the dry due to the presence of alluvial sediments (sand and gravel) with a water head of about 90 ft (27 m) at the top level of the shale. Holmes reported that the original design was driven piles but that this was later changed to bored piles, as the polymer fluid was proven to be able to preserve the integrity of the water-sensitive shale by laboratory immersion tests. To verify the design assumptions, a test pile was constructed under the conditions of actual construction at each of the two main pier locations. Construction started with the installation of a 74 in (1.9 m) internal-diameter casing through about 30 ft (9 m) of sand and gravel to the top of the shale formation. The casing was filled with polymer fluid, and the pile bore was drilled. River water was used as the make-up water for the granular polymer, which was initially mixed in situ at the pile bore but later pre-mixed in tanks due to the scale of the operation. The fluid was run at a Marsh funnel time ranging from 40 to 65 s. The pH of the fluid typically remained within the range 6–7. This pH range is low by the current standard, as an acidic environment is unfavourable for polymer dissociation and viscosity development (see Section 3.5.2). The sand content was reported to be about 0.25% prior to concrete placement. After excavation, side-wall profiling using a caliper showed no pile bore enlargement or necking-in. At the base of each test pile, an Osterberg cell (O-cell) with a capacity of 3000 short tons (26.5 MN) was fitted for a bi-directional load test. Although the load test results are not given by the report author, two related articles (Goodwin, 1993a, 1993b) present the measured load–movement curves for both the upward and downward directions as well as the equivalent surface load–settlement curve (as if the pile was loaded from the surface) for one of the tests. A total of 6000 tons of soil resistance (3000 tons from the end bearing plus 3000 tons from the pile shaft) was achieved for each of test piles. In light of the load test results, it was noted that the original design could have used fewer piles and made significant cost savings; it is unclear whether this was actually done.

CHRISTOPHER BOND BRIDGE, KANSAS CITY

It is interesting to note that the Ohio River case study is not the only one that reports the preserving effect of polymer fluids in shales. Axtell *et al.* (2009) and Brown and Axtell (2010) describe the use of the PHPA polymer POLY-BORE for the construction of the foundations for the Christopher Bond Bridge, which spans the Missouri River in Kansas City. The ground conditions comprised approximately 55 ft (17 m) of alluvium (sand and gravel) overlying bedrock. The majority of the bedrock encountered was shale, with lesser amounts of limestone at depth. The shale at the top was found to be highly weathered and relatively soft. There were concerns that the shale would deteriorate in the presence of support fluids, so slake durability tests were performed on core samples in accordance with ASTM D4644 (ASTM, 2004). In essence, this test measures the amount of physical breakdown of a rock specimen after tumbling in a rotating drum over two wet–dry cycles. The results are expressed as the slake durability index (second cycle), $I_d(2)$ – the higher the value (maximum 100%), the more durable the specimen under the slaking fluid. Both Missouri River water and POLY-BORE polymer fluid were used as slaking fluids in the tests. The former gave an $I_d(2)$ value of 72.2%, and the latter 98.2%. This showed that POLY-BORE fluid was able to preserve the integrity of the shale better than the river water. The polymer fluid was therefore used to drill the rock socket of a test pile, which was later tested by an O-cell. From the results of the load test, it was found that, even though the pile bore was deliberately left open for a period of 4 days, the mobilised unit shaft resistance was still similar to that developed by piles in other western Missouri shale. However, it should be noted that a wire brush tool known as a 'back-scratcher' was used to scarify the side wall of the rock socket, and this may have increased the roughness of the side wall and thus the shaft resistance.

The performance of piles constructed under polymer fluids in other soil types is reported in several other studies, including those by Brown *et al.* (2002) (silty sand), Brown (2002) (silty and clayey sand), Camp *et al.* (2002) (sand and clay), Frizzi *et al.* (2004) (sand and sandstone), and Ata and O'Neill (1998) (sand and clay). Each of these studies will be briefly discussed below.

CAPE FEAR RIVER, NORTH CAROLINA

Brown *et al.* (2002) reported the use of polymer fluids for the foundation piles of a new bridge over the Cape Fear River near the North Carolina coast. The project called for over 500 drilled shafts (bored piles) with diameters of between 1.2 and 2.4 m and lengths of between 22 and 30 m. The soil conditions at the test pile locations were dense to very dense silty fine sands of the Pee Dee Formation. The sands were calcareous with lightly cemented layers and phosphate particles, and SPT N values ranged from 50 to 100 blows/300 mm. The piling contractor proposed to use polymer fluid to construct the piles, but the North Carolina DOT provided specifications only for bentonite slurry. To gain acceptance from the authority for the alternative fluid, two trial piles with a diameter of 1.22 m and a length of 24 m were constructed, one with polymer fluid and the other with bentonite fluid. The trial piles were constructed with care to achieve identical conditions for the two support fluids. The polymer used was provided by KB Technologies (now KB International), but the name of the product was not given. The polymer fluid had the following properties: a density of 1.01 kg/m^3, a Marsh funnel time of 65 s, a pH of 10.5 and a sand content of 1%. The bentonite used was Aquagel (manufactured by Baroid), and the slurry had a density of 1.05 kg/m^3, a Marsh funnel time of 35 s, a pH of 9 and a sand content of 0.5%. These values are all within typical ranges, although it is unusual to see the sand content of a bentonite fluid (0.5%) lower than that of a polymer (1%) – the opposite is usually the case due to the thixotropic (gelling) property of bentonite providing a greater particle-holding capacity. After construction, the trial piles were tested by a Statnamic rapid load-testing device capable of applying a downward load of up to 18 MN. The test results showed that a maximum static capacity of 13.8 MN was mobilised at a displacement of 24 mm in the pile constructed under polymer. For bentonite, a maximum resistance of only 8.4 MN was mobilised at a displacement of 58 mm. Furthermore, the pile appeared to plunge at a displacement of around 25 mm. As the trial piles were instrumented with strain gauges, the mobilised shaft and base resistances for each pile could be determined. It was found that the unit shaft resistance of the pile constructed under polymer fluid was approximately three times higher than that of the pile constructed under bentonite slurry. The difference was thought to be due to improved bond at the shaft–soil interface. The base resistance was also higher for the pile constructed with polymer fluid, although the difference appears to be greater than that which can be logically attributed to a cleaner pile base.

AUBURN, ALABAMA

Brown (2002) reported a more comprehensive comparative field study in silty soil. The study was conducted at the Auburn University National Geotechnical Experimentation Site at Spring Villa, Alabama. The site geology comprises Piedmont residual soils, which can be classified as micaceous sandy or clayey silt. Eleven piles were constructed using a variety of drilling techniques to assess their effects. Two piles (1B and 24B) were constructed using bentonite slurry, four (1DP, 1LP, 24DP and 24LP) were constructed using polymer fluid, four were constructed using temporary casings advanced ahead of shaft excavation (1C, 1CDef, 24C and 24CDef) and one was constructed with a continuous-flight auger. The numbers in the pile designations indicate the duration of the soil–fluid or soil–casing exposure time in hours. For the piles constructed under support fluids, 'B', 'DP' and 'LP' denote bentonite, dry polymer and liquid polymer, respectively. For those piles constructed using a temporary casing, 'C' denotes 'casing' and 'Def' indicates 'artificial defects' (sandbags) in the pile concrete. All the support fluid materials came from the same manufacturer (Baroid). Although the names of the

Figure 7.9 Load–settlement curves for bored piles constructed using different installation techniques and with different exposure times: (a) bentonite slurry, (b) polymer fluid and (c) temporary casing. (Reproduced with permission from ASCE International)

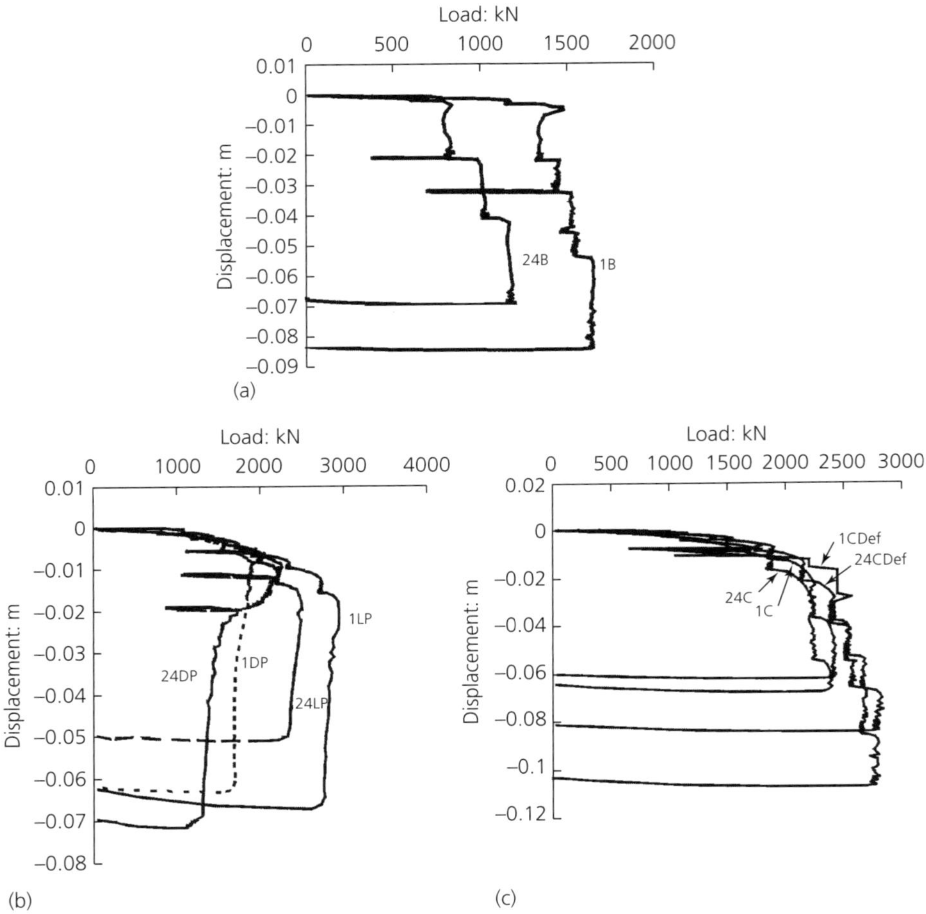

products used were not given, it was mentioned that the bentonite was a high-grade Wyoming bentonite and that the polymers were PHPAs. The load–settlement curves for the ten bored piles are reproduced in Figure 7.9, and the properties of the support fluids used are summarised in Table 7.6.

As can be seen in Figure 7.9, the piles constructed using bentonite slurry (1B and 24B) had the lowest capacities, and longer exposure times had an adverse effect on the capacity. Among the four piles constructed using polymer fluids, those constructed with liquid polymer (1LP and 24 LP) had better performance than those constructed using dry polymer (1DP and 24DP). Longer exposure times again led to a lower capacity, but note that this finding differs from the findings in several other case studies. Lastly, the

Table 7.6 Support fluid properties for six trial piles constructed at Spring Villa, Alabama

Pile designation	Support fluid type	pH	Marsh funnel time: s	Density g/cm^3	Sand content: %
1B	Bentonite	10	52	1.04	2
24B		10	52	1.06	1.5
1DP	Dry polymer	9	57	1.05	1
24DP		10	44	1.01	1
1LP	Liquid polymer	10	46	1.01	0.25
24LP		10	47	Not available	Not available

Data after Brown (2002).

piles constructed using a temporary casing (1C, 1CDef, 24C and 24CDef) all behaved rather similarly, although two of them had artificial soil inclusions built into the pile bodies. Subsequent inspection of pile 24B revealed a thin seam of easily identifiable bentonite between the soil and the pile shaft, whereas no distinct film was found at the interface of pile 24DP. These findings confirm that it is the presence of a layer of filter cake left at the concrete–soil interface that contributes to the lower capacities commonly found for piles constructed using bentonite fluid. Brown (2002) also found the pronounced capacity reduction surprising, even for pile B1, which had a very limited exposure time. Research conducted by the authors reveals that the interface shear strength between concrete and coarse soil is reduced approximately linearly with the square root of the cake formation (filtration) period, and that the rate of reduction is a function of the soil type and the construction method (Lam *et al.*, 2014).

CHARLESTON, SOUTH CAROLINA

Camp *et al.* (2002) reported one of the few US case studies on the effect of polymer fluids on pile performance in clay. A total of 12 trial bored piles of diameter 1.8–2.4 m and length 30–46 m were constructed at three different test sites along the route of the US Highway 17 Replacement Bridge. The names and designations of the three sites were Charleston (C), Mt Pleasant (MP) and Drum Island (DI). The bearing stratum for all piles was Cooper Marl, which is a stiff calcareous marine clay with an average SPT N value of 15 blows/300 mm. For all the piles, the loose sands and/or soft clays above the Cooper Marl were supported by a casing that could be temporary or permanent. The Cooper Marl was excavated either dry (pile designations MP4, MP6, C4, C5, DI1 and DI2) or with one of three types of support fluid: bentonite (MP1), polymer (MP2, MP3, C2 and C3) or water (C1). The polymer used was SlurryPro CDP (manufactured by KB International), and the bentonite used was Aquagel (manufactured by Baroid). The properties of the support fluids and information on additive usage are listed in Table 7.7. Given that the Cooper Marl could be excavated dry, it is surprising to see that a wide range of additives was used with the polymer fluids. Some of the additives used, such as InstaFreeze, WeightIt, SandSeal and KobbleBlok, are typically only required to help prevent excessive fluid loss and to stabilise collapsing side walls. Unfortunately,

Table 7.7 Properties of support fluids for six trial bored piles constructed for the US Highway 17 Replacement Bridge project in South Carolina

Pile designation	Support fluid type	pH	Marsh funnel time[a]: s	Density: g/cm^3	Sand content: %	Additives
MP1	Bentonite	8–8.5	57–59 (32–42 from supply)	1.03–1.04	<1	None
MP2	Polymer	11–11.5	55–75	1.01–1.04		LA-1, InstaFreeze, WeightIt, SandSeal, KobbleBlok, SlurryPro MPA
MP3	Polymer	11.5	51–58	1.02–1.03		LA-1, InstaFreeze, WeightIt, SandSeal, KobbleBlok, SlurryPro MPA
C1	Fresh water	n/a	28	1.00		None
C2	Polymer	10–12	53–75	1.02–1.04		InstaFreeze, WightIt, SlurryPro MPA
C3	Polymer	11–12	50–68	1.01		InstaFreeze, WightIt, SlurryPro MPA

Data after Camp *et al.* (2002).
[a] From the bottom of the pile bore unless stated otherwise.

Camp *et al.* did not provide any explanations as to why these additives were used. An interesting feature of this case study is that the thickness of soft soil or sediment at the pile base was measured by a miniature drilled-shaft inspection device (mini-SID). At the end of excavations, the device was lowered to the base of the completed pile bores, where it was allowed to penetrate into the sediment or disturbed soil under its own weight. Table 7.8 shows the mini-SID results. It can be seen that, as expected, the pile constructed using water (C1) had the largest amount of sediment or disturbed soft soil (37 mm). This was most probably due to the lack of particle-holding capacity for water. The piles constructed using polymer and bentonite fluids had an average value of about 20 mm, although the individual average values for the piles ranged from 16 to 30 mm. It is interesting to note that the piles bored dry still had an average soft soil or sediment thickness of 15 mm, which was believed to be due to the disturbed soil at the pile base resulting from the cleaning action. After construction, the trial piles were tested by O-cells, and some were also tested by the Statnamic method for axial and lateral pile capacities. From the axial load test results, it was concluded that support fluids did not appear to have a significant effect on the shaft resistance in Cooper Marl. This was attributed to the fact that Cooper Marl is a relatively impermeable material, and hence no filter cake formed on the side-wall surface. Attempts were also made to correlate the mini-SID results with the observed end resistances of the piles, but no identifiable trend was found.

MIAMI, SOUTH FLORIDA

Frizzi *et al.* (2004) reported a case study in Miami, South Florida, where the ground conditions comprised layers of limestone, sandstone and cemented sands belonging to the Miami and Fort Thompson Formations. To aid the design of a new building, an unreinforced 'method' pile (designation MS) and three instrumented and reinforced piles (PS1, PS2 and BS1) with a diameter of 1.83 m and a length of 36.6 m were constructed and tested. The purpose of the tests was to evaluate the effects of using bentonite and polymer fluids – the latter had not been previously used in that area. The type of polymer used for piles PS1 and PS2 was not given and was only described as a proprietary product. In terms of polymer use, Frizzi *et al.* mentioned that premixed polymer fluid was added to the top of the excavation and that proprietary additives were also added to the polymer-supported bores to facilitate rapid sediment settlement. For pile BS1, a high-yield bentonite was used to prepare the slurry; a slurry additive was also used, although details were not given.

The properties of the support fluids are summarised in Table 7.9. It can be seen that the values of the various polymer parameters were within the typical ranges for common PHPA-based polymer systems. Table 7.9 also gives the amount of soft sediment measured at the bottom of the piles by using the mini-SID device. It can be seen that a large amount of sediment was found in the three piles (MS PS1 and PS2) constructed using polymer fluids, despite the fact that additives had been used to accelerate the sedimentation process. It is unclear from the original text whether the sediment was removed before concreting, although the use of a clean-out bucket was mentioned. Table 7.9 also lists the concrete overbreak factor (the ratio of the actual to the theoretical volume) for the four trial piles. The high overbreak factor for pile MS (1.21) was attributed to a

Table 7.8 Sediment thicknesses at the bottom of 12 trial bored piles constructed for the US Highway 17 Replacement Bridge project in South Carolina

Pile designation	Construction technique	Cleaning method[a]	Sediment or soft soil thickness at various sampling locations: mm									Average: mm	Average for each construction technique: mm
			Centre	North-west	North	North-east	East	South-east	South	South-west	West		
C1	Water	1	36	52	25	30	35	34	40	45	34	37	37
MP1	Bentonite	1, 2	25	26	18	17	12	15	15	19	25	19	19
MP2	Polymer	1	25	27	26	35	25	11	15	13	20	22	21
MP3		1, 2	7	34	12	16	35	5	5	25	13	17	
C2		1, 2	20	40	30	33	30	20	35	30	35	30	
C3		1, 2	10	30	10	10	27	18	30	0	11	16	
MP4	Dry	1	9	38	19	11	15	25	25	17	15	19	15
MP6		1	8	?	?	10	14	16	10	30	27	16	
C4		1	5	10	26	15	5	7	10	10	13	11	
C5		1	15	10	11	11	6	6	15	25	10	12	
DI1		1	9	13	5	5	19	9	7	21	13	11	
DI2		1	6	13	15	10	41	10	30	16	8	16	

Data after Camp *et al*. (2002).

[a] 1, clean-out bucket; 2, down-hole hydraulic pump.

Table 7.9 Support fluid properties, sediment thickness and concrete overbreak factors for four trial bored piles constructed in Miami, Florida

Pile	Marsh funnel time: s	pH	Sand content: %	Density: lb/ft^3	Sediment thickness: mm	Concrete overbreak factor
Method polymer pile (MS)	50–70	10–12	1.5	63	30–90	1.21
Polymer pile 1 (PS1)	50–70	10–12	1.0	63	10–60 (average 40)	1.19
Polymer pile 2 (PS2)	50–70	11–12	1.8	63	70	1.09
Bentonite pile (BS1)	40	10	2.0	67	15–20 (average 17.5)	1.07

Data after Frizzi *et al.* (2004).

partial side-wall collapse that occurred after the bore had been left undisturbed for 38 h to evaluate hole stand-up time. The actual cause of the partial collapse was not given. The high overbreak factor for pile PS1 (1.19) was attributed to the volume change in the sandy soil zones above the rock socket due to casing vibration during installation and subsequent extraction. After construction, the three instrumented piles (PS1, PS2 and BS1) were tested using O-cells. No significant difference was found between the three piles in terms of the measured unit shaft resistance. However, the two piles constructed using polymers (PS1 and PS2) were found to show significantly lower end resistance than the one constructed using bentonite (BS1). This was probably due to the large amount of sediment collected at the base of these piles (see Table 7.9). The end resistance curves are shown in Figure 7.10.

HOUSTON, TEXAS

Ata and O'Neill (1998) report the performance of bored piles constructed using polymer support fluids at two different sites in Texas. The first site is a National Geotechnical Experimentation Site located at the University of Houston (NGES-UH). The soil at this site comprises a highly plastic clay (the Beaumont Formation) that extends from the ground surface down to a depth of about 8 m, which is followed by a very stiff to hard sandy clay (Montgomery formation) to a great depth. The shear strength of the soil in the top 10.7 m, as predicted by unconsolidated undrained triaxial compressions tests, ranges between 48 and 190 kPa. The piezometric groundwater table lies at approximately 2 m below the ground surface. The second site is an NGES located at the Texas A&M University (NGES-TAMU). The soil at this site consists of 12 m of sand overlying soft clay shale. The sand deposit can be described as medium dense, with SPT N values ranging from 7 to 25 blows per 0.3 m. The undrained shear strength of the clay shale is about 370 kPa. The piezometric groundwater table lies at approximately 5 m below the

Figure 7.10 Base resistance curves of three trial bored piles constructed in Miami, Florida. (Adapted from Frizzi *et al.*, 2004)

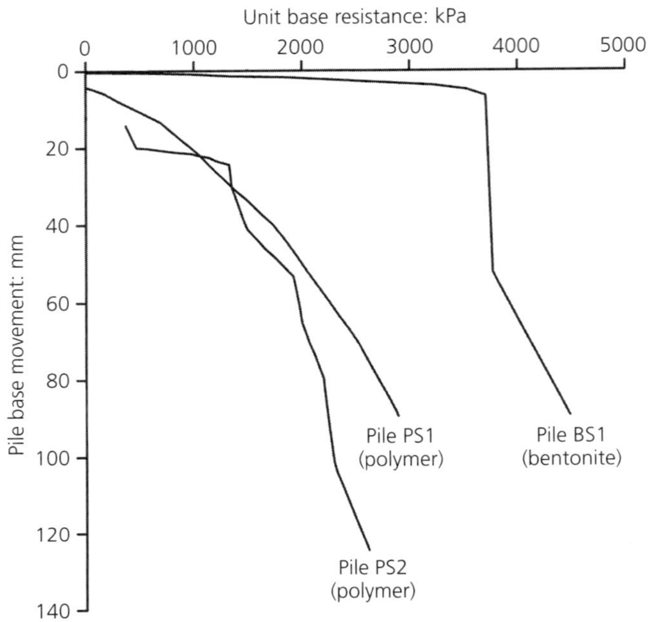

ground surface. The polymer used was a granular PHPA (POLY-BORE manufactured by Baroid; see Section 4.3.5). The polymer was mixed at concentrations of 0.85 and 1.28 g/l at NGES-UH and NGES-TAMU, respectively. Two 'technique' shafts, 0.914 m in diameter, were excavated at each site using polymer fluid. At both sites, during the excavation of the first shaft (Tech I) the introduction of polymer fluid was intentionally delayed until side-wall sloughing started to occur. In contrast, during construction of the second technique shaft (Tech II) the fluid was introduced before the excavation depth reached the piezometric groundwater table. At all times after introduction, the fluid level was maintained within 0.6 m of the ground surface. The results from the Tech I shafts at both sites showed that delay in introducing the support fluid after the borehole had penetrated below the groundwater table did not prevent sloughing and irremediable side-wall collapse. Results from the Tech II shafts also showed that polymer fluid, when introduced with the right dosage and prior to the borehole reaching the groundwater table, maintained the stability of open boreholes for more than 18 h at both sites. The Tech II shafts were later concreted and load tested using O-cells. At NGES-UH (the clay site), the values for the developed unit side resistance were comparable to or higher than the values reported previously for piles constructed with bentonite slurry at nearby sites. The back-calculated shaft adhesion (α) value was 0.74, which is considerably higher than the design value of 0.55 recommended by the FHWA at the time. At NGES-TAMU (the sand site), the values of the developed unit side resistance also compared favourably with those of piles constructed with bentonite slurry at the same site.

7.5.2 Canada

GOLDEN EARS BRIDGE

In Canada, polymer support fluids have been used, but apparently at a much smaller scale than the USA. At the time of writing, only two documented case studies have been identified. The first concerns the Golden Ears Bridge across the Fraser River in British Columbia (Amini *et al.*, 2008). The geology in that area comprises post-glacial normally consolidated to lightly overconsolidated soft to stiff silty clay. To evaluate the effects of construction methods on the shaft resistance of piles in clay, two 2.5 m-diameter bored piles and two 0.35 m-diameter driven concrete cylinder piles were installed. The first bored pile had a length of 32 m, and was excavated within a full-length temporary casing. The second bored pile was much longer at 74 m, and the top 21 m of soil was excavated with the support of a temporary casing. Below the casing level, polymer fluid was used to maintain the stability of the pile bores, and a fluid head of about 7.5 m was maintained above the groundwater table. The type and properties of the polymer fluids used were not mentioned, although the polymer is believed to have been a product from KB International according to information on the company's website. Based on the load test results, it was found that the α values back-calculated from the two bored piles were similar (i.e. no difference between the full-length casing option and polymer fluid support), and both were considerably greater than those recommended by the US FHWA for bored piles (O'Neill and Reese, 1999). The back-calculated values for the bored piles also matched those recommended by the API (2000) for driven steel pipe piles. This positive finding was attributed to the construction procedures. According to the piling contractor, polymer fluid was also used for 48 working piles supporting four bridge piers. The working piles had the same diameter as the trial piles (2.5 m) and lengths of between 70 and 80 m.

EDMONTON, ALBERTA

The second case study concerns the use of polymer fluid for drilling through oil sands containing methane gas (Hagy, 2013). The site was located north of Edmonton and Fort McMurray, Alberta. For the construction of a mechanically stabilised earth wall, drilled shafts (bored piles) with diameters of between 36 and 60 in (0.9 and 1.5 m) and lengths of between 125 and 160 ft (38 and 49 m) were required. During excavation, methane gas was detected to be migrating into the bores, and, as a result, drilling had to stop. The methane gas concentration was found to have a 100% lower explosive limit (LEL) – five times higher than the common upper limit of 20%. This was extremely dangerous, as the rotating tool could create a spark and so cause an explosion. To mitigate the risk, air blowers were inserted in the top of the bore to blow out the gas. However, the gas was found to be migrating into the bores faster than the air blowers could disperse it, and the methane concentration remained at 100% LEL. Polymer fluid was therefore used to suppress the entry of the gas into the bores and also to act as a spark arrestor. The fluid would also provide support to the excavation side walls. The product used was CETCO's Shore Pac, but details about the fluid properties were not given. It was reported that the methane concentration returned to 0% LEL with the use of the polymer fluid, and construction was completed without further issues.

7.5.3 Brazil

SÃO PAULO

Paschoalin Filho *et al.* (2013) compared the costs of using polymer and bentonite fluids for the construction of diaphragm walls for a project in São Paulo. The total excavation volume for the project concerned was 3344 m^3. The cost comparison showed that if bentonite fluid was used, then 25% of the total excavated soil would be considered as contaminated, whereas if polymer fluid was used, no soil would become contaminated. In addition, the unit cost of disposal of contaminated soil would be about 40% higher than that of non-contaminated soil. When both the volume and the unit cost of soil disposal were considered, it was found that, if polymer fluid were to be used, the total disposal cost would be 28% lower than the total cost of disposal if bentonite fluid were to be used. However, the cost of applying polymer fluid would be considerably higher than the cost of applying bentonite fluid, and this made the estimated total cost of construction about 6% higher if polymer fluid were to be used. Nonetheless, Paschoalin Filho *et al.* concluded that polymer fluid was a feasible alternative because of the environmental benefits it offered.

REFERENCES

Amini A, Fellenius BH, Sabbagh M, Naesgaard E and Buehler M (2008) Pile loading tests at Golden Ears Bridge. *Proceedings of the 61st Canadian Geotechnical Conference, Edmonton*. Canadian Geotechnical Society, Richmond, BC, Canada.

API (American Petroleum Institute) (2000) API RP 2A-WSD. Recommended practice for planning, designing and constructing fixed offshore platforms – working stress design, 21st edition. API, Washington, DC, USA.

ASTM (American Society for Testing and Materials) (2004) D4644-04. Standard test method for slake durability of shales and similar weak rocks. ASTM International, West Conshohocken, PA, USA.

Ata A and O'Neill M (1998) Side-wall stability and side-shear resistance in bored piles constructed with high-molecular-weight polymer slurry. *Proceedings of the 3rd International Geotechnical Seminar on Deep Foundations on Bored and Auger Piles, Ghent*. Balkema, Rotterdam, the Netherlands, pp. 111–117.

Axtell PJ, Thompson WR, III and Brown DA (2009) Drilled shaft foundations for the kcICON Missouri River Bridge. *Proceedings of the 34th Annual Conference on Deep Foundations, Kansas City*. Deep Foundations Institute, Hawthorne, NJ, USA, pp. 3–12.

Bachy Soletanche (2010) Marine works. Large diameter bored piling and pre-cast concrete deck. ExCel, Victoria Dock, London, UK. http://www.bacsol.co.uk (accessed 25/09/2016).

Beresford JJ, Cashman PM and Hollamby RG (1989) The merits of polymeric fluids as support slurries. *Proceedings of the 3rd International Conference on Deep Foundations*, London (Burland J and Mitchell J (eds)). Balkema, Rotterdam, the Netherlands, vol. 1, pp. 3–10.

Berkovitz BC and Long CS (1995) Use of polymer slurries for drilled shaft construction. *Proceedings of the 31st Symposium on Engineering Geology and Geotechnical Engineering, Logan, Utah*. Idaho State University, Pocatello, ID, USA, pp. 17–23.

Brown D (2002) Effect of construction on axial capacity of drilled foundations in Piedmont soils. *Journal of Geotechnical and Geoenvironmental Engineering* **128(12)**: 967–973.

Brown D and Axtell PJ (2010) Design and construction challenges at the kcICON Bridge. *Deep Foundations: The Magazine of Deep Foundations Institute*, spring, pp. 73–76.

Brown D, Muchard M and Khouri B (2002) The effect of drilling fluid on axial capacity, Cape Fear River, NC. *Proceedings of the 27th Annual Conference on Deep Foundations, San Diego*. Deep Foundations Institute, Hawthorne, NJ, USA.

Bustamante M and Boato R (2005) Polymer slurry in large diameter pile drilling – case histories. *Proceedings of the 16th International Conference on Soil Mechanics and Geotechnical Engineering, Osaka*. Millpress, Amsterdam, the Netherlands, pp. 2083–2086. (In French.)

Bustamante M, Gianeselli L, Boato R and Conedera A (1998) Performance of polymer slurries in large diameter bored pile. *Proceedings of the 3rd International Seminar on Deep Foundations on Bored and Auger Piles (BAP III), Ghent*. Balkema, Rotterdam, the Netherlands, pp. 119–127.

Camp WM, Brown DA and Mayne PW (2002) Construction method effects on axial drilled shaft performance. *Proceedings of International Deep Foundations Congress 2002, Orlando*. American Society of Civil Engineers, Reston, VA, USA, pp. 193–208.

Chin CT and Chen JR (2006) Foundation engineering practice in Taiwan – High Speed Rail Experiences. *Proceedings of a Seminar on the State-of-the-Practice of Geotechnical Engineering in Taiwan and Hong Kong, Hong Kong*. Hong Kong Institution of Engineers, Hong Kong SAR, Hong Kong, pp. 28–51.

Corbet SP, Culley DS, Sherwood DE and Cockroft JEM (1991) Testing and analysis of preliminary test piles in very weak chalk. *Proceedings of the 4th International Conference on Piling and Deep Foundations, Stresa*. Balkema, Rotterdam, the Netherlands, pp. 57–63.

Corbet SP, Peters M, Huppert F and Seitz J (2005) Taiwan High Speed Railway – contracts 260 & 270 – geotechnical aspects and pile testing. *Proceedings of the 16th International Conference on Soil Mechanics and Geotechnical Engineering, Osaka*. Millpress, Amsterdam, the Netherlands, pp. 2095–2098.

Darley HCH and Gray GR (1988) *Composition and Properties of Drilling and Completion Fluids*, 5th edn. Gulf Publishing, Houston, TX, USA.

Duann SW, Chen MS, Seah TH and Fujita M (2004) Optimization of pile foundation design through full-scale pile load tests in Taiwan High Speed Rail project. *Proceedings of the 15th Southeast Asian Geotechnical Society Conference, Bangkok*. Southeast Asian Geotechnical Society, Pathumthani, Thailand, pp. 223–230.

Frizzi RP, Meyer ME and Zhou L (2004) Full scale field performance of drilled shafts constructed utilizing bentonite and polymer slurries. *Proceedings of the GeoSupport Conference 2004, Orlando, Florida*. American Society of Civil Engineers, Reston, VA, USA, pp. 573–586.

Goodwin JW (1993a) Bi-directional load testing of shafts to 6000 tons. *Proceedings of Sessions of the 1993 ASCE Annual Conference and Exposition: Design and Performance of Deep Foundations: Piles and Pies in Soil and Soft Rock, Dallas, Texas*. American Society of Civil Engineers, Reston, VA, USA, pp. 204–217.

Goodwin JW (1993b) Osterberg cell sets record in Kentucky. *Foundation Drilling* **32(5)**: 21–22.

Guadagnini G (2001) L'organizzazione di prove di carico sup ali in alveo – L'esempio del ponte Vasco de Gama a Lisbona. *INARCOS* **622**: 469–477. (In Italian.)

Hagy, M (2013) Polymer slurry used in methane gas suppression. *Foundation Drilling* **34(1)**: 45, 48.

Hardy B (2009) ExCel-lent adventure. *European Foundations* **43(summer)**: 26.

Ho CE (2002) Design strategy for deep slurry stabilized bored piles in compression. *Proceedings of the 9th International Conference on Piling and Deep Foundations, Nice, France*. Deep Foundations Institute, Hawthorne, NJ, USA, pp. 505–512.

Ho CE and Tan CG (1996) Barrette foundation constructed under polymer slurry support in Old Alluvium. *Proceedings of the 12th Southeast Asian Geotechnical Conference, Kuala Lumpur*. Southeast Asian Geotechnical Society, Pathumthani, Thailand, pp. 379–384.

Ho CE, Tan CG and Lim CH (1999) An instrumented polymer slurry bored pile in weathered granite. *Proceedings of the 52nd Canadian Geotechnical Conference, Regina, Saskatchewan*. Canadian Geotechnical Society, Richmond, BC, Canada, pp. 211–218.

Holmes MM (1993) Drilled shafts and new slurry technology used instead of driven piles: sets record on the Ohio River. *Foundation Drilling* **32(5)**: 13–15.

Hosoi T and Matsushita S (2016) Recent diaphragm wall technologies and future challenges. *Geotechnical Engineering Journal of the SEAGS and AGSSEA* **47(2)**: 112–125.

Hsieh YH and Ho CC (2011) Diaphragm wall construction on gravel stratum – taking Taipei MRT Contract CD271A as an example. *Rapid Transit Systems and Technology* **44**: 171–178. (In Chinese.)

Khare MG and Ramanathan M (2013) Investigation of collapse of slurry trench in underground metro construction. *Forensic Engineering: Informing the Future with Lessons from the Past: Proceedings of the 5th International Conference on Forensic Engineering, London*. ICE, London, UK, pp. 205–214.

Lam C and Jefferis SA (2014) The use of polymer solutions for deep excavations: lessons from Far Eastern experience. *HKIE Transactions*, **21(4)**: 262–271.

Lam C, Troughton V, Jefferis S and Suckling T (2010) Effect of support fluids on pile performance – a field trial in east London. *Ground Engineering* **43(10)**: 28–31.

Lam, C, Jefferis SA and Martin CM (2014) Effects of polymer and bentonite support fluids on concrete-sand interface shear strength. *Géotechnique* **64(1)**: 28–29.

Lam C, Jefferis SA and Suckling TP (2015) Construction techniques for bored piling using polymer fluids. *Proceedings of the Institution of Civil Engineers – Geotechnical Engineering* **167(6)**: 565–573.

Lam C and Jefferis SA (2016) Performance of bored piles constructed using polymer fluids: lessons from European experience. *Journal of Performance of Constructed Facilities* **30(2)**: 04015024.

Lennon DJ, Ritchie D, Parry GO and Suckling TP (2006) Piling projects constructed with vinyl polymer in Glasgow, Scotland. *Proceedings of the 10th International Conference on Piling and Deep Foundations, Amsterdam*. Deep Foundations Institute, Hawthorne, NJ, USA, pp. 499–506.

Lesemann H (2010) *Anwwndung polymerer Stützflüssigkeiten bei der Herstellung von Bohrpfählen und Schlitzwänden*. PhD thesis, Technical University Munich, Munich, Germany. (In German.)

Lesemann H and Vogt N (2012) Untersuchungen zur hydraulischen Stützung mit Polymerlösungen. *Geotechnik* **35(1)**: 11–21. (In German.)

Lin SS, Wang KJ, Hsieh HS, Chang YH and Huang CS (2007) Field testing of axially loaded drilled shafts in clay/gravel layer. *Journal of GeoEngineering* **2(3)**: 123–128.

Littlechild B and Plumbridge G (1998) Effects of construction technique on the behaviour of plain bored cast in situ piles constructed under drilling slurry. *Proceedings of the 7th International Conference and Exhibition on Piling and Deep Foundations, Vienna*. Deep Foundations Institute, Hawthorne, NJ, USA, pp. 1.6.1–1.6.8.

Manuel Correia R and Sêco e Pinto PS (1999) The Vasco de Gama Bridge over the river Tagus in Lisbon – Main geotechnical aspects. *Proceedings of the 12th European Conference on Soil Mechanics and Geotechnical Engineering, Amsterdam*. Balkema, Rotterdam, the Netherlands, pp. 447–458.

Nguyen MH and Fellenius BH (2014) Bidirectional-cell tests on two 70 m long bored piles in Vietnam. *Proceedings of Sessions of Geo-Congress 2014, Atlanta, Georgia*. American Society of Civil Engineers, Reston, VA, USA, pp. 482–496.

O'Neill MW (1991) Effects of mineral and polymer slurries on drilled shafts. *Proceedings of the 16th DFI Annual Members' Conference, Chicago*. Deep Foundations Institute, Hawthorne, NJ, USA, pp. 141–154.

O'Neill MW and Reese LC (1999) *Drilled Shafts: Construction Procedures and Design Methods*. US Federal Highway Administration, Department of Transportation, Washington DC, USA, Report FHWA-IF-99-025.

Paschoalin Filho JA, Guerner Dias AJ and Brum de Lima Duarte E (2013) The use of polymer slurry as an economic and sustainable way to build diaphragm walls – a case study of a construction work in São Paulo/Brazil. *Exacta* **11(3)**: 299–306.

Schwarz J and Lange U (2004) Brückengründung mit 70 m tiefen flüssigkeitsgestützten gebohrten Pfählen in Benin/Afrika. *Proceedings of the 19th Christian Veder Kolloquium, Graz*. Graz University of Technology, Graz, Austria, pp. 73–89. (In German.)

Sêco e Pinto PS and Oliveira R (1998) A recent difficult foundation problem: the case of the new Tagus Bridge. *Proceedings of the 4th International Conference on Case Histories in Geotechnical Engineering, St. Louis, Missouri*. Missouri University of Science and Technology, Rolla, MO, USA, pp. 188–209.

Tallard GR (1992) New trenching method using synthetic bio-polymers. In *Slurry Walls: Design, Construction, and Quality Control* (Paul DB, Davidson RR and Cavalli NJ (eds)). American Society for Testing and Materials, Philadelphia, PA, USA, pp. 86–102.

Thasnanipan N, Baskaran G and Anwar MA (1998) Effect of construction time and bentonite viscosity on shaft capacity of bored piles. *Proceedings of the 3rd International Geotechnical Seminar on Deep Foundations on Bored and Auger Piles (BAP III), Ghent*. Balkema, Rotterdam, the Netherlands, pp. 171–177.

Thasnanipan N, Aye ZZ, Submaneewong C and Teparaksa W (2002) Performance of wet-process bored piles constructed with polymer-based slurry in Bangkok subsoil. *Proceedings of International Deep Foundations Congress 2002, Orlando*. American Society for Testing and Materials, Reston, VA, USA, pp. 143–157.

Thasnanipan N, Aye ZZ and Boonyarak T (2003) Behaviour of polymer-based slurry for deep-seated bored piles in multi-layered soil of Bangkok. *Proceedings of the 4th International Geotechnical Seminar on Deep Foundations on Bored and Auger Piles (BAP IV), Ghent*. Millpress, Rotterdam, the Netherlands, pp. 267–274.

Thasnanipan N, Aye ZZ, Boonyarak T, Arayasiri M and Intraratchaiyakit C (2006) The contest on the prediction of capacity of a bored pile constructed under polymer-based slurry in Bangkok subsoil. *Proceedings of the 10th International Conference on Deep Foundations, Amsterdam*. Deep Foundations Institute, Hawthorne, NJ, USA, pp. 387–396.

Trindade R (2014) Third-generation polymers. *GeoDrilling International*, May, pp. 17–19.
Troughton VM (1992) The design and performance of foundations for the Canary Wharf development in London Docklands. *Géotechnique* **42(3)**: 381–393.
Wheeler P (2003) Clear as mud. *New Civil Engineer*, 17 April, pp. 22–24.

Polymer Support Fluids in Civil Engineering
ISBN 978-0-7277-5786-9

http://dx.doi.org/10.1680/psfce.57869.187

Chapter 8
Detailed analysis of case studies

8.1. Introduction

In the previous chapter, selected case studies from around the world were reviewed to show current experience with polymer fluid technology in different countries. In this chapter, two case studies, one in Glasgow and one in London, are discussed to further illustrate some site-specific and polymer-specific issues relating to construction and fluid testing.

8.2. Glasgow

8.2.1 Introduction

Glasgow is one of the first cities in Scotland where synthetic polymer fluids were used for the construction of bored piles. The first use is believed to have been for a project on Glassford Street in 1999. Lennon *et al.* (2006) presented a brief introduction to the polymer technique and the experience gained from two projects in Glasgow, one for piles founded in dense alluvial sand and one for piles founded in bedrock. No problems were reported, although it was noted that the use of some additives required careful control. Static load test results from both sites showed near-linear load–settlement relationships up to the proof loads and small permanent settlements upon unloading. Another Glasgow case study is presented here to highlight some unexpected behaviour of polymer fluids and also the importance of adopting suitable equipment and construction techniques. This project took place in the summer of 2008, and was discussed by Lam *et al.* (2010a, 2014).

8.2.2 Site and ground conditions

The construction site was located at 1 Waterloo Street, which is about 500 m north-west of the Argyle Street station and about the same distance north of the River Clyde in central Glasgow. Typical for that area, the ground conditions consist of a layer of made ground underlain by a thick layer of alluvial silty fine sand and then bedrock consisting of bands of mudstone, siltstone and sandstone. Thin layers of coal (lignite) are also commonly found above the bedrock in the area. Table 8.1 summarises the depths and engineering descriptions of the soil layers encountered at the site.

Figure 8.1 shows an overview of the site during construction. This was a very small urban site measuring just 24 m wide by 40 m long and requiring 62 bored piles with a diameter of 750 mm to support the proposed development; that is, an average pile spacing of 4 m. The piles were 18–20 m deep, and were all socketed into the bedrock. Prior to the start of the piling work, excavation was carried out to reduce the ground level from the street level (about 13 mOD (metres above the Ordnance Datum)) to 8 mOD, which was the

Table 8.1 Summary of the soil layers and their description for a site in central Glasgow, Scotland

Depth below the ground level:[a] m	Soil description
0–3	Made ground: concrete and tarmac surface underlain by brick, ash fill and sandstone fragments
3–5	Firm brown silty clay and loose sand (alluvium)
5–22	Medium-dense becoming dense reddish brown silty sand (alluvium)
22–23	Coal and sandstone
23 to >30	Bedrock: mudstone/siltstone/sandstone

Lam *et al.* (2014).
[a] The piling platform level was 4 m below the original ground (street) level.

piling platform level. The groundwater table was found to lie on average at about 5 m below the piling platform level during construction; that is, at about 3 mOD. However, due to the close proximity of the site to the River Clyde, the groundwater table at the site is believed to have been tidal. Although no groundwater monitoring work was

Figure 8.1 A small construction site in central Glasgow where polymer fluid was used. (From Lam *et al.*, 2014)

Figure 8.2 Particle size distribution curve of the alluvial sand in Glasgow. (From Lam *et al.*, 2014)

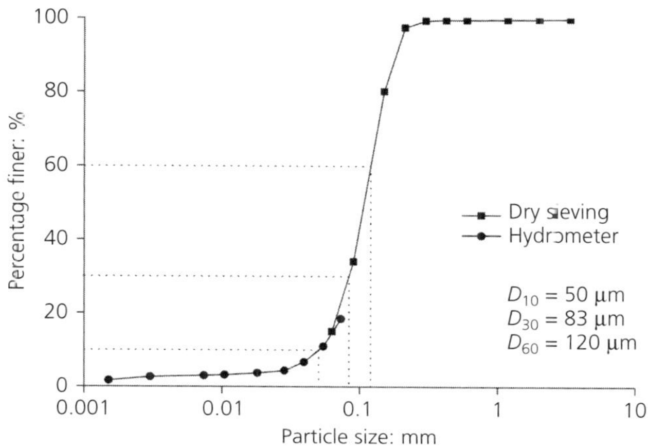

conducted for this project, the nearby tidal gauge showed a daily fluctuation of about 4 m between high and low tides, so the groundwater table should also fluctuate, although probably by a smaller amount.

As for other sites in central Glasgow, bored piling using polymer fluid was the preferred method of construction because this does not require the bulky ancillary plant for cleaning or considerable storage space for the full-length casing option. It is for this reason that polymer fluids have been favoured over bentonite slurry for piling projects in central Glasgow in recent years. Continuous-flight auger piles were a possible option, but flighting has occurred in loose granular deposits in the past, resulting, on occasion, in building damage. They are therefore only used in low-risk situations (Lennon *et al.*, 2006).

Figure 8.2 shows the particle size distribution curve of the alluvial sand collected from the site. The sand consists of about 2% clay, 13% silt and 85% sand-sized particles, with D_{10}, D_{30} and D_{60} particle sizes of 50, 83 and 120 μm, respectively. From these figures, the coefficient of uniformity (C_U) and the coefficient of curvature (C_Z) are 1.45 and 1.15, respectively, indicating poorly graded uniform sand (classification SPu). The water content of the sand below the groundwater table was about 30% and the specific gravity of the soil grains was 2.66, as measured in accordance with BS 1377 (BSI, 1990). The saturated bulk unit weight of the material was 19 kN/m^3, calculated from the moisture content, assuming full saturation. The angle of shearing resistance ranged from 34° to 36° for the medium-dense becoming very dense sand at this site.

8.2.3 Polymer fluid behaviour

8.2.3.1 Basic information and preparation

The polymer fluid used for this project was the SlurryPro CDP system, manufactured by KB International in the USA. This polymer system consists of the granular base polymer SlurryPro CDP and a range of additives for mix water treatment, fluid cleaning, fluid loss control and fluid weighting. This polymer system was used to construct two of the four piles supporting the tower crane and all the working piles for the main building. Another synthetic polymer (Shore Pac) was used to construct the other two tower crane piles, so that the contractor could evaluate the difference between the two polymer systems in the same environment. Both polymers are understood to contain high-molecular-weight partially hydrolysed polyacrylamides (PHPAs).

The mixing of the polymers was carried out in a 20 m^3 open top tank by aeration. Potable water with an initial electrical conductivity of 58 μS/cm and a hardness equivalent of 7.5 mg/l of calcium carbonate was used for mixing. These results show that the water was soft. To prepare the SlurryPro CDP fluid, the water was first treated with a small amount of the additive Protek (potassium hydroxide) to increase the pH to 11. The SlurryPro CDP base polymer was then sprinkled into flowing water to produce a concentration of about 0.9 kg/m^3. To prepare the Shore Pac fluid, the mix water was first treated with soda ash (sodium carbonate) to a pH of about 10. The granular polymer was then added to the water at a concentration of 0.6 kg/m^3. This concentration was a third lower than that used for mixing the SlurryPro CDP fluid, and was chosen following the advice from the manufacturer's representative who attended the site. This dosage was also in line with the general recommendations of the manufacturer (CETCO). To promote dissolution of the polymer granules, each polymer was mixed for at least 60 min by aeration using bubble bars in the mix tanks.

8.2.3.2 Viscosity change before use

Immediately after mixing of each of the two polymers, the air compressor was turned off and the viscosity of the fluid measured by a Marsh funnel. The fluid was then left to sit in the storage tank overnight. On the next day when excavation started, a centrifugal pump with a maximum capacity of 320 m^3/h was used to transport the fluid from the mixing and storage tank to the pile locations. The pump formed part of a fluid circuit, which is shown schematically in Figure 8.3. The pipework was laid out so that the fluid was recirculated back to the storage tank when the ball valve in the feed line to the excavation was shut. This meant that the pump did not need to be turned repeatedly on and off during excavation – a useful feature, as the pump and the storage tank were at some distance from the excavation. During the construction of the tower crane piles,

Figure 8.3 Schematic drawing of the plant set-up. (Modified after Lam *et al.*, 2010a)

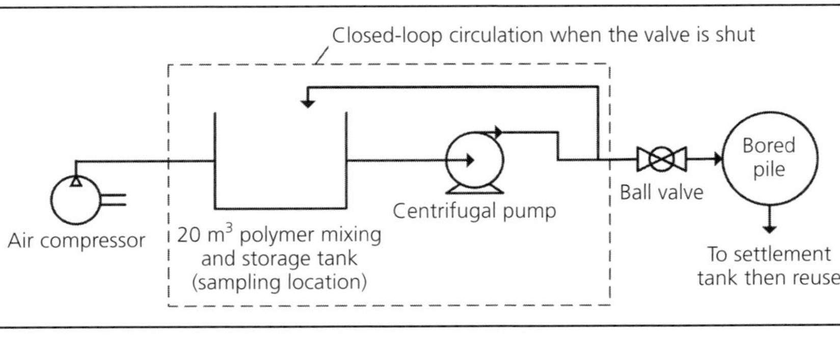

Figure 8.4 Change in the Marsh funnel time before and during construction. (Modified after Lam *et al.*, 2010a)

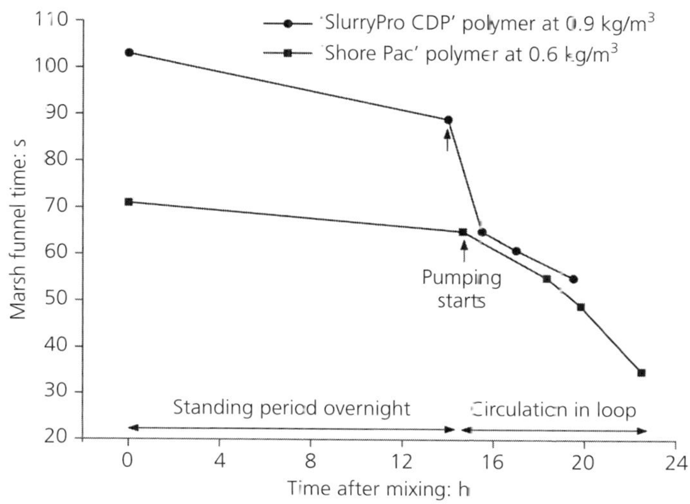

specimens of the two polymer fluids were taken from the storage tank and tested with a Marsh funnel to assess the effect of pumping.

Figure 8.4 shows the Marsh funnel time of the two polymer fluids during the monitoring period. It can be seen that during the initial standing period between the end of mixing and the start of pumping on the following day, the viscosity dropped from 103 to 89 s for SlurryPro CDP and from 71 to 65 s for Shore Pac, which were equivalent to reductions of 14% and 9%, respectively. These results were contrary to expectations, because it was thought that the extra time would have allowed any undissolved polymer granules to break up and further disentanglement of polymer molecules to occur – both processes being expected to contribute to viscosity development. The storage tank was open at the top, but as the weather during the overnight period was fine, the fluids could not have been diluted with rainwater. Since intermediate measurements during the overnight period were not practicable, it was not known how the viscosity decreased with time and, more importantly, why this occurred. However, laboratory measurements later showed that the reduction was caused at least in part by the escape of air bubbles introduced to the fluid during mixing by aeration. The bubbles affected the flow of the fluid through the Marsh funnel, and their dissipation during the standing period thus resulted in lower viscosity values.

8.2.3.3 Viscosity change during construction of the tower crane piles

Excavation of the tower crane piles started about 14 h after the mixing of the fluid required to stabilise the bores. Although gravity feed from the tank was possible, it was considered too slow, so the fluid was sent to the pile bores by a centrifugal pump. As digging progressed, the valve on the feed line was used by site operatives to adjust the volume of the flow. When the fluid was not being drawn off for use, the centrifugal pump continued to run, and the fluid was recirculated to the storage tank (see Figure 8.3). The consequence of this arrangement was a significant reduction in the fluid viscosity and this is shown in Figure 8.4. It can be seen that for the SlurryPro CDP fluid the viscosity dropped from 89 to 55 s over a period of 6 h, which is equivalent to a reduction of 38% in the Marsh funnel time. The situation with the Shore Pac fluid was similar: the viscosity dropped from 65 to 35 s over 8 h, equivalent to a 46% reduction. Both polymers are understood to be based on high-molecular-weight PHPAs, and the reduction was believed to be due to damage to the long-chain polymer molecules caused by the strong mechanical action of the pump.

Figure 8.4 shows that, towards the end of the monitoring period, the rate of viscosity reduction accelerated. This was because each batch of polymer fluid had an initial volume of about 20 m^3, but a significant portion (approximately 9 m^3) was drawn off to the pile bores during excavation. This resulted in an increase in the rate of recirculation for the remaining fluid, and so the rate of fluid damage accelerated. Near the end of the monitoring period, the viscosity was approaching such a low level that concerns were raised about the stability of the excavations. As a remedial measure, the viscosity was boosted by adding extra polymer directly to the bores. The tower crane piles were later successfully completed without any problems. The working piles for the proposed building were later constructed using only SlurryPro CDP fluid, with the pipework rearranged to avoid recirculation.

8.2.4 Excavation and base-cleaning techniques

8.2.4.1 The potential problem of 'soft toe' and concrete contamination

Since the ground conditions at the Glasgow site consisted of silty fine sand, piles constructed using polymer fluids actually carry a greater risk of having 'soft toes' and concrete contamination. This is because, unlike a gelling bentonite slurry that can typically suspend soil particles up to the size of coarse sand, polymer fluids have a considerably lower particle-holding ability due to their different rheological properties. Under typical polymer fluid conditions, particles smaller than coarse silt (<0.02 mm) will tend to remain in suspension, as their settling rate will be too low for them to reach the base of the excavation in significant quantity. However, coarser particles (coarse silt size and above) are more likely to accumulate at the base of a polymer-fluid-supported excavation than for an equivalent bentonite-slurry-supported excavation. The potential amount of sediment will depend on the concentration of soil in the fluid, the period from the end of excavation to concreting, the viscosity of the fluid, and the height of the fluid column through which the soil must settle.

The potential problems of 'soft toe' and concrete contamination in piles constructed using polymer fluids have been reported widely. For example, Brown (2004) reported that there have been numerous cases of projects using polymer fluids where removal of silts and fine sands from the fluid has been difficult. On one project near the Atlantic coast in the USA, the contractor had to overpour the concrete by over 2 m in order to remove concrete contaminated with silt. On another project in Florida with very fine silty sands, the base of the pile bores were inspected using a down-hole camera, which revealed the bottom 0.6 m of fluid to be heavily laden with silt so that it appeared as a gel. Based on these and other anecdotal experiences with polymer fluids in silty fine sands, Brown (2004) urges engineers and contractors to exercise caution in the use of polymers in these ground conditions.

To avoid the potential issues of 'soft toe' and concrete contamination in the working piles, a simple yet effective fluid-testing regime and adjustment to the construction techniques was used in the Glasgow project. These together with the construction techniques are described below.

8.2.4.2 Excavation technique

The construction process started with the insertion of a 6 m-long steel casing and removal of the soils within the casing under dry conditions. The support fluid was then introduced into the pile bore to allow the excavation to be continued until the required founding level. During the excavation process, the support fluid level was kept at 1 m below the piling platform level, to counterbalance the groundwater pressure and also to exert a net stabilising pressure on the side walls. Since the amount of fluid loss to the ground was not significant ($\approx$0.5 m^3 over a 12 h period in a 20 m-deep completed bore) and the pile bores were generally stable, no additives were required, and only occasional top up of the support fluid was necessary.

To minimise soil disturbance, a twin-flight piling auger and a flat-sided cleaning bucket were used for the excavation and cleaning, respectively. Both tools allow fluid passage during insertion into and retrieval from the pile bore, so minimising pressure changes in the support fluid and in the adjacent soil pore water as the tools are moved in the bore. A flat-sided cleaning bucket can be seen in Figure 5.17. The twin-flight auger is shown in Figure 5.18 immediately after retrieval from one of the pile bores. Note the ability of the polymer to encapsulate the cuttings (saturated sand) so that they retain their shape on the flights.

8.2.4.3 Polymer fluid: test procedures and properties after excavation

Immediately after reaching the desired founding level, fluid samples (used fluid) were taken from the pile bores at three different levels: top (2 m), middle (10 m) and bottom (18 m). Three types of test were conducted on each sample: density, sand content and Marsh funnel time. For the density measurements, since the resolution of the mud balance (0.01 g/cm^3) is too low for polymer fluid management, a portable density meter (see Figure 5.9) with a resolution of 0.0001 g/cm^3 was used. The portable density meter could be used on this site because the particle size of the alluvial sand was no greater than 0.2 mm (see the particle size distribution curve in Figure 8.2). The density meter would not have been suitable if the maximum particle size had been greater than, say, 0.5 mm because of the risk of blocking the capillary tube (which is the sensing element of the instrument) and because their contribution to the fluid density would not have been registered. The procedures for the density, sand content and Marsh funnel time tests can be found in Chapter 6 and are not repeated here.

Before each fluid test, the sample was stirred vigorously to ensure uniform dispersion of all the soil particles – a step that is especially important for the density and sand content

Figure 8.5 Fluid density, sand content and Marsh funnel time profiles for a polymer-supported pile bore. (From Lam *et al.*, 2014)

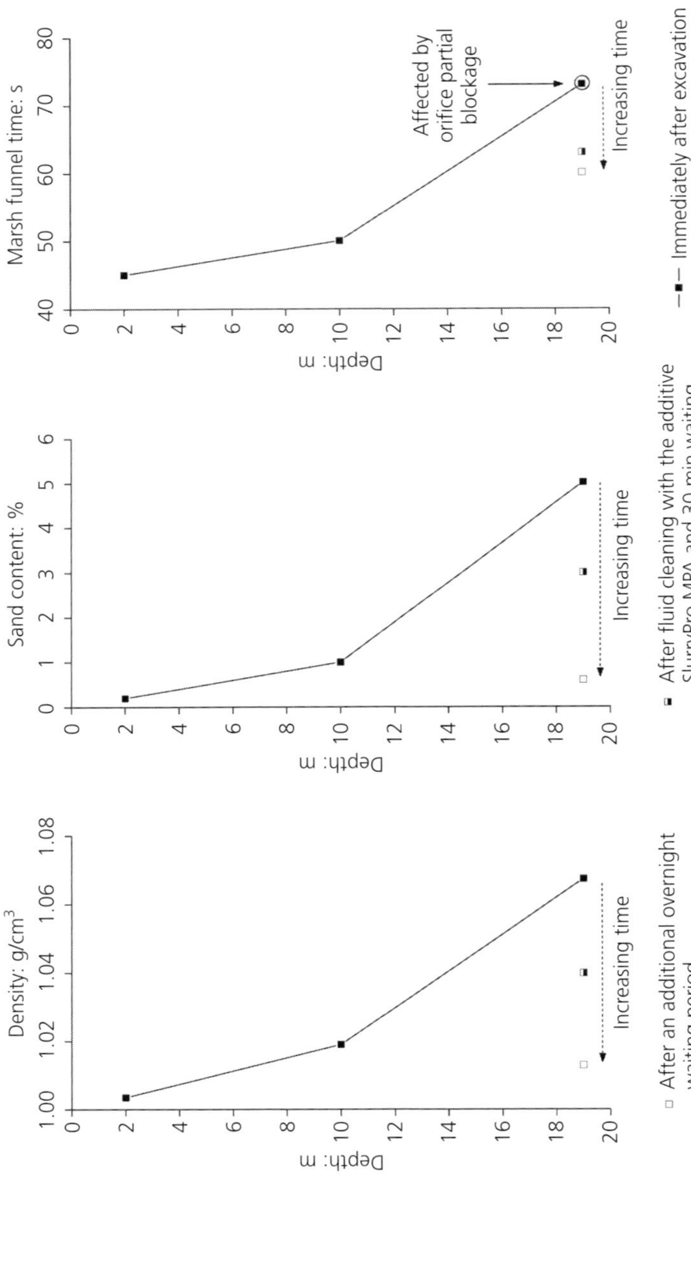

tests. At this site, although the alluvial silty sand did not contain any gravel-sized particles, the thin coal (lignite) layer above the bedrock was broken into small gravel-sized fragments by the piling auger, and some of them were found in suspension in the support fluid. This made it necessary to pre-screen the fluid before the Marsh funnel tests using the screen built into the funnel. Pre-screening was not necessary for the density tests, as the syringe used to fill the density meter had a small-diameter tip that functioned as a screen for coarse particles, and so their effect was not registered in the density reading.

Figure 8.5 plots the density, sand content and Marsh funnel test results against depth for one of the working piles for three stages: (*a*) immediately after excavation, (*b*) after fluid cleaning with the chemical additive SlurryPro MPA and a 30 min wait and (*c*) after an additional overnight waiting period.

The results for the first stage are discussed here. It can be seen from the plots that all three measured parameters show an increasing trend with depth. This indicates that the suspended soil particles were settling in the pile bores at the time of sampling. It is also worth noting that some sand particles were found to have accumulated inside the Marsh funnel near the orifice after the test on the fluid sample taken from the bottom of the bore. As a result, the flow path was slightly obstructed by the accumulated particles. The corresponding viscosity result was probably affected, and this is annotated in Figure 8.5.

8.2.4.4 Fluid cleaning by the use of additives

As the test results showed that the support fluid was heavily laden with solids immediately after excavation, it was necessary either to clean the fluid using additives or to exchange it for fresh fluid. The additive used was SlurryPro MPA, which was supplied as a heavy viscous solution. The fluid-cleaning procedure involved dropping a weighted package containing 5–10 litres of 1% (of the as-delivered concentration) SlurryPro MPA solution to the base of the pile bore. The package was broken by rotating a cleaning bucket, and the contents dispersed by gently raising and lowering the bucket a few times over a distance of 3–5 m from the base of the excavation. A 30 min waiting period then followed, to allow the additive to coagulate the suspended particles and to allow the collected solids to settle to the base of the bore. As previously shown in Figure 5.19, a coagulated gel mass can be formed by adding SlurryPro MPA to used SlurryPro CDP fluid laden with soil. It should be noted that in this bucket demonstration, undiluted SlurryPro MPA solution was used so that the effect was more dramatic than would be expected in a pile bore, for which a 1% solution (×100 dilution) is recommended by the manufacturer.

After the 30 min waiting period, a further fluid sample was then taken from the bottom of the bore to assess the effects on the fluid properties. The results are also shown in Figure 8.5. It can be seen that all three test parameters had reduced as a result of the cleaning action and the 30 min wait. However, the values still exceeded the allowable upper limits in the project specifications, which required that the sand content should be less than 1% and the fluid density should be less than 1.02 g/cm^3. These tests suggested that fluid cleaning with additives alone was not sufficient for the silty fine sand ground conditions, and that either a complete fluid exchange or further cleaning was required to eliminate the risks of 'soft toe' and concrete contamination.

8.2.4.5 Fluid cleaning by further waiting

To reduce the amount of suspended soil particles further, it was decided to leave the pile bore open overnight, to give the particles extra time to settle out of suspension. In this regard, the pile bore was used as a sedimentation 'tank' to allow the support fluid to self-clean – an option available for polymer fluids but not bentonite slurries (see Section 5.7). After the overnight waiting period, a sample taken from the bottom of the pile bore showed that all the three parameters had been further reduced to within the upper limits for the sand content and density (see Figure 8.5). It is worth noting that Tallard (1992) also proposed a similar approach to allow the polymer fluid to self-clean in a diaphragm wall trench. Tallard noted that allowing a full day between the end of excavation and the time of concreting produced a satisfactory result.

8.2.4.6 Sediment build-up during the waiting periods

To obtain data on the sediment accumulation at the base of the pile bore during the waiting periods, the depth of the bore was also checked regularly with a weighted measuring tape. Although the accuracy of this measurement method is rather low (±50 mm approximately), an interesting pattern was revealed, which is shown in Figure 8.6. It can be seen that as soon as the excavation was completed, sediment started to accumulate at the pile base and continued throughout the waiting periods. All the sediment was removed by a cleaning bucket at the end of the test period, and so the thickness was reduced to zero prior to concreting. The results of the depth measurements confirmed that there was no side-wall collapse during the waiting periods. Had such a collapse occurred, the sediment depth would have shown a sharp increase between depth readings. A non-linear function has also been fitted to the data points, as a possible

Figure 8.6 Sediment build-up at the base of a polymer-supported pile bore. (From Lam *et al.*, 2014)

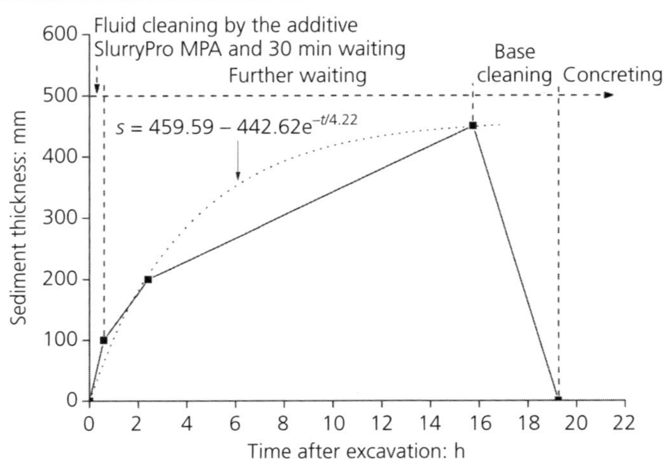

simulation of the build-up of the sediment thickness (s) over time (t). The fitted equation shows that after 16 h the sedimentation process was almost complete. This suggests that further waiting would not have brought extra benefits at this site. For other sites, the sediment build-up characteristics could be different, depending on the soil grading, the fluid properties at the end of excavation and the depth of the excavation.

The important message from the fieldwork is that although there may be some soil settlement and removal in the surface tanks, the main mechanism for soil removal from polymer fluids is settlement in the bore. Sufficient time must be allowed for this settlement and, if necessary, treatment chemicals must be used.

8.2.4.7 Concreting

To avoid intermixing of the wet concrete and the support fluid, the piles were concreted using a 150 mm-diameter tremie pipe. The hopper and the tremie pipe sections used can be seen on the right-hand side of Figure 8.1. A vermiculite plug was used to separate the first pour of concrete from the support fluid in the tremie. During casting, the tremie pipe was lifted (and the pipe reduced in length as necessary) as the concrete level rose, to maintain a pipe embedment depth of between 3 and 6 m. The concrete at the top of the concrete column appeared clean, confirming the effectiveness of the cleaning methods employed, although a minor degree of intermixing was still observed, as shown in Figure 8.7. A small amount (approximately 0.2 m^3) of overpouring was therefore

Figure 8.7 Fresh concrete as it first appeared at the top of a polymer-supported bore during tremie concreting. The boundary between fresh concrete (lighter shade) and used polymer fluid (darker shade) has been highlighted in white. (From Lam *et al.*, 2014)

necessary to ensure the quality of concrete at the pile top. A much larger amount of over-pouring would have been necessary if the fluid had not been cleaned prior to concreting.

8.2.5 Pile integrity and loading test results

To confirm the structural integrity of the completed working piles, they were tested by the low-strain pulse-echo method (55 piles) and ultrasonic cross-hole logging (four piles). A specialist subcontractor was employed to conduct this work about 3–4 weeks after construction. Descriptions of the test procedures can be found in manuals such those by Turner (1997) and the Federation of Piling Specialists (FPS, 2006), so they are not repeated here. None of the piles tested was found to show any sign of defects or integrity problems such as inclusions or necking. Figure 8.8 is a typical cross-hole

Figure 8.8 Typical cross-hole ultrasonic logging profiles for piles constructed using the polymer fluids in Glasgow. (From Lam *et al.*, 2014)

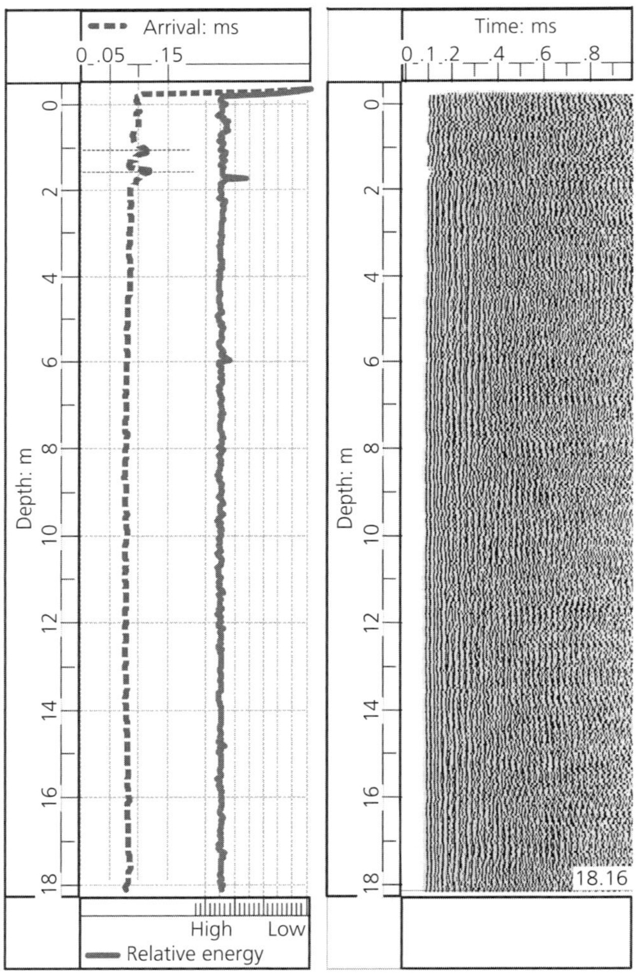

ultrasonic profile result for one of the piles. The traces on the left-hand side show the first arrival time and relative energy (dimensionless) against depth. The right-hand traces show the full sonic profile against time – the positive part of the waveform is shown light and the negative portion is shown dark. It can be seen that the pile has uniform first arrival time and energy profiles over the entire length of the sonic tubes, suggesting that the pile material is of sound and uniform quality. Concrete containing inhomogeneities, such as soil inclusions, wash-out and pockets of support fluids would show localised lower arrival times and attenuated signals.

To confirm that the piles as constructed would perform satisfactorily under the design working loads, two of the piles were tested by the high-strain dynamic method about a month after construction. The dynamic tests involved dropping a 3.5-tonne hammer from a height of between 800 and 1300 mm onto the pile head. The pile strain and acceleration were measured by strain transducers and accelerometers placed on the two opposite faces of the shaft about two pile diameters below the pile head. The data collected were analysed using the Case Pile Wave Analysis Program (CAPWAP) signal-matching technique, to estimate the mobilised shaft and toe resistances and also the likely immediate settlement under static loads. During the analyses, soil parameters were adjusted until a good match was obtained between the measured and simulated response. Table 8.2 summarises the test pile information and the CAPWAP results. The results indicated that both piles had mobilised total resistances considerably greater than the working loads and that the predicted pile head settlements under the working loads were small. As both piles only settled by about 1 mm per blow (5–6 mm per 5 blows), the total resistance values given in the table are likely to be considerably smaller than the ultimate resistance, which typically requires a minimum of 2–3 mm set per blow to mobilise during the test. Overall, the results showed that the completed piles were sound, both structurally and geotechnically.

8.2.6 Lessons learned

The lessons learned from this case study are summarised as follows:

- Polymer fluids mixed by the aeration method using materials from two different manufacturers showed a significant reduction in Marsh funnel times after standing overnight. This is believed to be due to the gradual escape of entrained air bubbles. When preparing fresh polymer fluids or boosting used fluids, it may be necessary to make allowance for this 'bubble' effect by raising the control value.
- The viscosity of PHPA-based polymer fluids can be severely reduced if they are circulated through a centrifugal pump in a loop system. It is recommended that for the delivery of the polymer fluid to an excavation, centrifugal pumps should be avoided and, for preference, diaphragm pumps used. If diaphragm pumps are not available, circulation of the fluid should be minimised. It is recognised that a centrifugal pump may be necessary to pump fluid from the excavation during concreting, as the pump must be submersible and high flow rates are necessary.
- At the Glasgow site where the ground conditions consisted of silty fine sand, fluid testing at the end of excavation showed that the polymer fluid was heavily laden

Table 8.2 Summary of high-strain dynamic loading test results for two bored piles constructed using polymer fluids

Pile data					CAPWAP results										
					Test set: mm/blow	Maximum compressive stress: MPa	Maximum tensile stress: MPa	Maximum transferred energy: kN m	Case damping factor		Mobilised soil resistance: kN			Predicted pile head settlement: mm	
Pile No.	Length: m	Rock socket length: m	Pile diameter: m	Specified working load: kN					Shaft	Toe	Shaft	Toe	Total	At working load	At 150% working load
P1	21.5	2.5	0.75	3000	5/5	16.48	–5.59	16.3	0.378	0.504	2599	1948	4548	5.7	9.8
P2	19.9	1.0	0.75	2550	6/5	15.46	–2.29	14.2	0.387	0.436	2012	1981	3994	5.5	8.8

Lam *et al.* (2014).

with temporarily suspended solids. The use of chemical additives alone was found to be insufficient to remove the solid particles within a 30 min period. A further overnight waiting period was thus required for the density and sand content values to fall below their specified maximum values. Depth checking in one of the pile bores also showed that sediment started to accumulate at the pile base immediately after excavation and continued throughout the 16 h waiting period. If the pile bore were concreted without the further waiting period, a 'soft toe' would have developed during insertion of the reinforcing steel case and build-up of tremie pipes. The fresh concrete would also have been contaminated by the settling soil particles, as described in the US case studies by Brown (2004).

- Structural integrity testing conducted at this site showed that the completed piles were of sound quality without any sign of defects. High-strain dynamic load tests conducted on two completed piles showed high total resistances despite the overnight waiting period.
- Allowing (and indeed on this Glasgow site requiring) the pile bores to be left open overnight allows the piling contractor to improve efficiency and productivity, as work can be sequenced more flexibly with less standing time for the workforce and plant. For example, digging of pile bores can be started one day and finished the next – a schedule that can be undesirable for piling under bentonite slurry as excessive build-up of filter cakes on the excavation side walls may occur. This advantage has been noted by Wheeler (2003).

8.3. Stratford, London

8.3.1 Introduction

Although polymer fluids have been used in the UK since the 1990s, at the time of writing many British engineers and their clients are still sceptical about the performance of polymer fluids due to a lack of previous research and industry guidance. Even when polymer fluids are used, many still prefer to use the existing design rules, which were largely developed based on historic case studies with bentonite slurries. For example, during the construction of a bored piling project in the Elephant and Castle area in London, where the ground conditions consisted of mainly stiff clay, the piling contractor had to reduce the design shaft adhesion (α) factor from the typical 0.5 to 0.35 for any polymer-supported piles that were left open overnight. This effectively increased the lengths of those piles. To address this issue and to provide further evidence of the effect of polymer fluids, a field trial comprising three bored piles was undertaken at a site in Stratford, east London. The specific objectives of the trial were to

- assess the difference between piles constructed using polymer and bentonite fluids
- assess whether a polymer-supported bore can be open for an extended period without compromising the performance of the completed pile
- assess the effect of polymer and bentonite support fluids on the quality of concrete.

Some parts of this project have been discussed by Lam *et al*. (2010b, 2015).

8.3.2 Site and ground conditions

8.3.2.1 Site location

Figure 8.9 shows an aerial photograph of the Stratford site taken in 2008. As can be seen, the test site is located to the south-east of the Stratford International station box in London. The land surrounding the station was raised by 7 m using excavated material from the Channel Tunnel Rail Link tunnel and from the station box (Dyson and Blight, 2007). Since the site was being developed for the Stratford City project, a comprehensive ground investigation had been carried out prior to the trial. The results of the investigation were used to identify the soil profiles, groundwater conditions and geotechnical design parameters. These were all required for the back-analysis of the load test results, and are described individually as follows.

8.3.2.2 Soil profiles

From the borehole logs and by examination of the soil cuttings during the excavation of the test piles, the site was found to be underlain by the made ground/land raise, the Lambeth Group, Thanet Sand and then chalk. The Lambeth Group is a stratigraphic group consisting of the Woolwich, Reading and Upnor Formations. At this site, the Woolwich Formation consists of two members, namely the Laminated Beds and the Lower Shelly Clay. The names, depths and geotechnical descriptions of these soil layers according to the British Standards Institution (BSI, 1999) are listed in Table 8.3. For the purpose of back-analysis, the made ground and the River Terrace Gravel layer are considered as one unit, and the two members of the Woolwich Formation are also treated as a single unit. Figure 8.10 shows the resulting idealised soil profiles together with some details of the test piles.

Figure 8.9 An aerial photograph of the test site and the nearby structures. (From Google Earth)

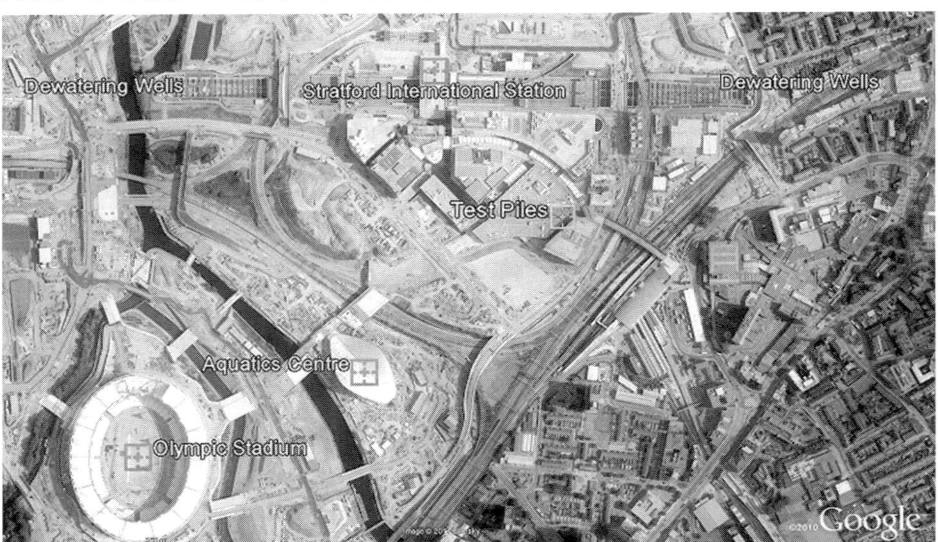

Table 8.3 Soil profiles at the Stratford test site, East London

Idealised unit for back-analysis	Soil layer	Depth: mOD	Depth: m bgl	Description
Made ground	Made ground (land raise)	5.5 to −0.5	0–6.0	*Made ground*: dark-brown gravelly fine to coarse sand with occasional shell fragments and pockets of stiff clay. Gravel-sized fragments are angular to subrounded fine to coarse flint, chalk, brick, concrete and wood
	River Terrace Gravel	−0.5 to −1.0	6.0–6.5	Brown fine to coarse *sand* and rounded to angular fine to coarse *gravel* of flint
Woolwich and Reading Formations	Laminated Beds, Woolwich Formation, Lambeth Group	−1.0 to −4.0	6.5–9.5	Firm greyish brown slightly sandy thinly laminated *clay* with closely spaced thin to thick laminations of silt and occasional shell fragments
	Lower Shelly Clay, Woolwich Formation, Lambeth Group	−4.0 to −5.5	9.5–11.0	Stiff dark greyish green sandy *clay* with many shells and shell fragments
	Lower Mottled Clay, Reading Formation, Lambeth Group	−5.5 to −8.0	11.0–13.5	Stiff to very stiff green very sandy slightly gravelly *clay* with occasional shell fragments. Gravel is subrounded to rounded fine to coarse flint. From 12.5 to 13.5 m bgl, light green limestone hardpan (calcrete duricrust)
Upnor Formation	Upnor Formation, Lambeth Group	−8.0 to −14.0	13.5–19.5	Very stiff dark-green very sandy gravelly *clay* with occasional shell fragments at top. Gravel is rounded to subrounded flint
Thanet Sand	Thanet Sand	–14.0 to –27.5	19.5–33.0	Very dense grey slightly silty fine to medium *sand*
Chalk (not required)	Chalk	–27.5 to –29.5 (end of boreholes)	33.0–35.0	Structureless *chalk* (grade Dm, Mundford grade VI)

m bgl, metres below ground level.

Figure 8.10 Schematic soil profiles and test pile details. (From Lam *et al.*, 2015)

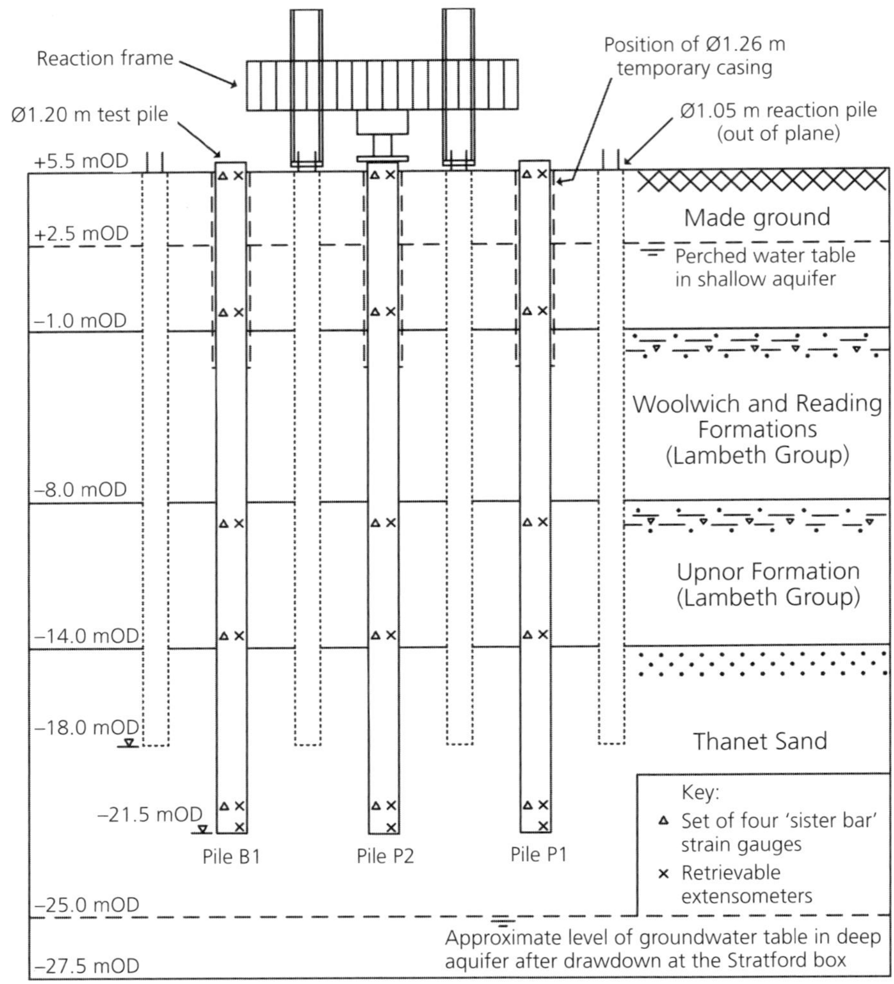

8.3.2.3 Groundwater conditions

There are two aquifers at the test site: a shallow aquifer in the made ground and River Terrace Gravel, and a deep aquifer in the Thanet Sand and the underlying chalk. These two aquifers are separated by an aquiclude, which is the cohesive Lambeth Group at the test site and the London Clay in other parts of London. According to Whitaker (2004), the groundwater level in the deep aquifer near the Stratford station box has been rising steadily since the late 1950s, and in 1997 the piezometric surface had reached about −11 mOD. However, for the construction and operation of the Channel Tunnel Rail Link Stratford box, a permanent dewatering scheme was required to depress the piezometric level back to a minimum of 10 m below the base slab level of the box (the slab level is approximately −10 mOD at the west end and −15 mOD at the east end). As a

Figure 8.11 Piezometric data and the idealised pore pressure profile. (From Lam *et al.*, 2015)

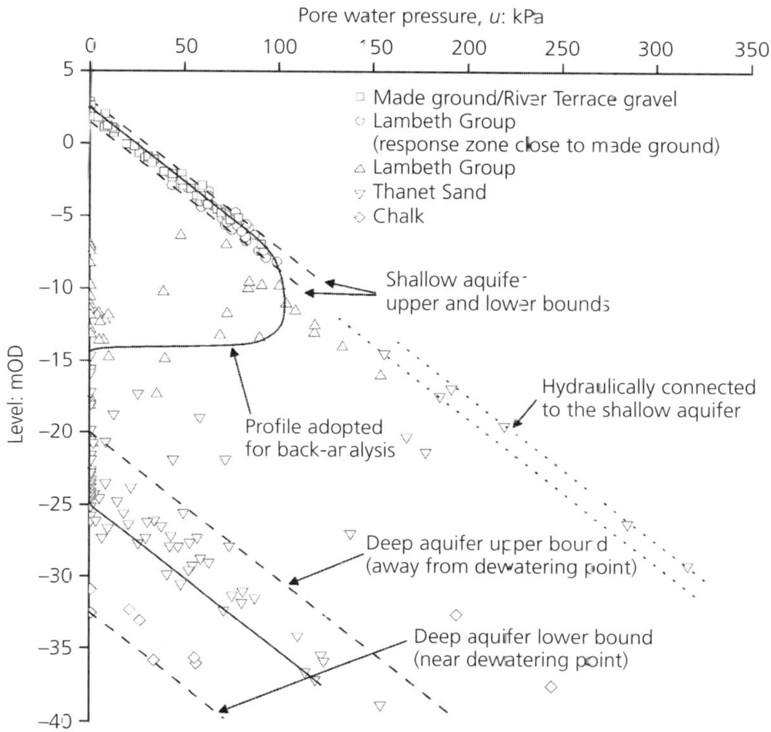

result, dewatering wells were installed at both ends of the box, and they have been pumping continuously since late 2001.

To study the pore pressure profile with depth, piezometric measurements were taken during the ground investigation. Figure 8.11 shows the measurement results and an idealised profile. It can be seen that the shallow and deep groundwater tables stood at 2.5 mOD and 25 mOD, respectively. Because of the under-drainage, the pore pressure in the Lambeth Group was below hydrostatic. This is typical of many other London sites.

8.3.2.4 Shear strength parameters

The undrained shear strength (S_u) of the Lambeth Group was investigated using data from UU (unconsolidated undrained) triaxial tests, standard penetration tests (SPTs) and cone penetration tests (CPTs). Figure 8.12 shows the combined plot with some idealised profiles. In this plot, the S_u values from the SPTs were obtained by multiplying the N values by an empirical factor of 4.5, and the S_u values from the CPTs were obtained by dividing the cone resistance (q_c) by a factor of 25. It can be seen that the interpreted S_u values from the UU tests and the SPTs correspond reasonably well, but those from the CPTs were significantly higher than the others. This was probably due to the high sand

Figure 8.12 Undrained shear strength values derived from laboratory and in situ tests. (From Lam *et al.*, 2015)

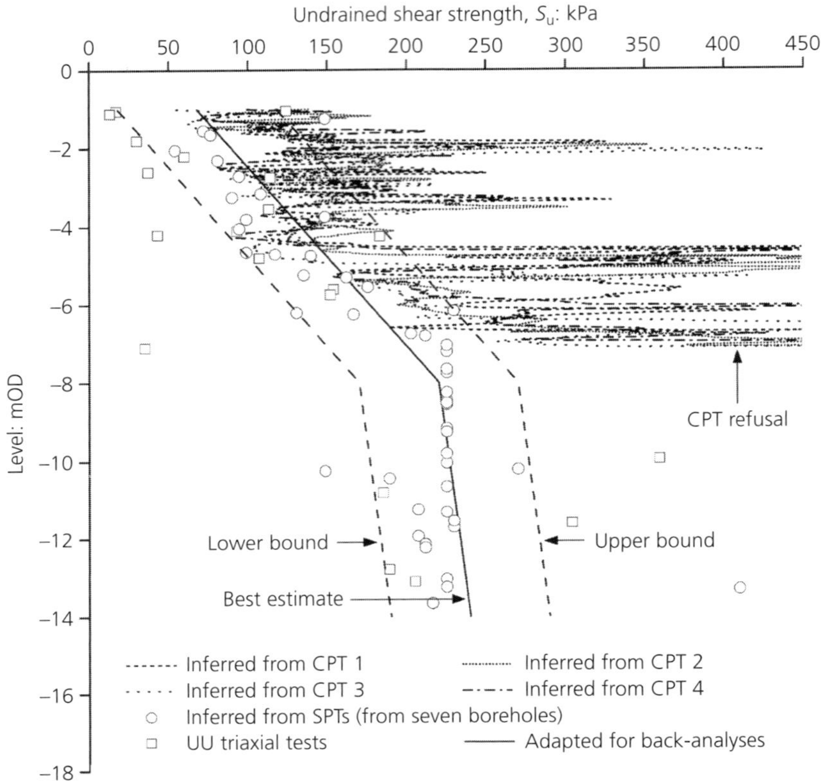

and gravel contents of the clay causing an increase in the value of q_c. The profile labelled 'best estimate' was used for the back-analyses.

For the underlying Thanet Sand, the peak angle of the shearing resistance (ϕ') is typically taken as 36° for pile design, as this is the value commonly obtained from direct shear box tests on recompacted samples (Troughton, 1992; Troughton and Platis, 1989). The critical state angle of Thanet Sand is about 32° (Ventouras, 2005).

8.3.3 Construction details

8.3.3.1 Test pile and instrumentation details

To achieve the three objectives of the pile tests, three full-scale bored piles were constructed side by side at the test site, as previously shown in Figure 8.10. The piles were constructed as follows:

- pile P1 – excavated under polymer support fluid (SlurryPro CDP) and concreted on the same day

- pile P2 – excavated under polymer support fluid (SlurryPro CDP) and concreted on the next day
- pile B1 – excavated under bentonite slurry (Berkbent 163) and concreted on the same day.

As shown in Figure 8.10, the test piles had an embedded length of 27.0 m and a stick-up length of 0.5 m. The nominal diameter was 1.20 m, but the actual outside diameter for the top 8 m was 1.26 m due to the use of a temporary casing during excavation. The piles were reinforced axially with 12 ribbed steel bars (32 mm diameter) in the upper half of the pile and four bars in the lower half. To create a pile head suitable for the load test, the upper 1.2 m was cast in a permanent steel casing 1.2 m long. The outside diameter of this casing was 1.26 m and the inside diameter was 1.23 m Eight continuous-flight auger piles were installed to provide the reaction force required for the load tests. The continuous-flight auger piles had a diameter of 1.05 m and an embedded length of 23.5 m.

'Sister bar' strain gauges (model VWS-4000, MGS Geosense) and retrievable extensometers (model A-9, Geokon) were installed near the boundaries between the different soil layers to measure the strain distribution and compression of the piles, respectively. The strain gauges were installed 0.75, 6 2, 14.9, 19.4 and 26.5 m from the pile top. The retrievable extensometers were also installed at these levels, and at the pile base at 27.5 m.

8.3.3.2 Mix water and support fluid: basic information

The mix water used for preparing the support fluid was potable water delivered by a tanker. The pH was about 7.5, the electrical conductivity was 670 μS/cm and the total hardness was about 250 mg/l (ppm). These properties are typical of fresh water in London.

The polymer used was the SlurryPro CDP system supplied by KB International. As discussed in Chapter 4, this polymer system consists of the base polymer SlurryPro CDP and several additives for enhanced fluid performance. Since the soil at the test site consisted of mainly stiff clay and dense sand, the excavations were largely stable, so only a few additives were needed. The products used, their dosages and their key functions were as follows:

- ProTek 100: concentrated potassium hydroxide solution. Prior to adding the base polymer, ProTek 100 was added to the mix water, to raise the pH to about 11.0. The key function of this additive was to treat the water hardness and to create a suitable alkaline environment for the base polymer to generate viscosity.
- SlurryPro CDP: base PHPA polymer supplied as dry granules. The manufacturer states that it is a 'vinyl' polymer. It was mixed by aeration at a concentration of 1.0 kg/m^3 for the trial. The key function of this product was to generate fluid viscosity and be the carrier of other additives. Figure 4.9 shows a 25 kg bag of SlurryPro CDP and a container of ProTek 100 on site.
- LA-1-D: additive supplied as a fine powder. The chemistry of this product is not published although it is probably an associative thickener. It was added at a

concentration of 0.08 kg/m^3 for the trial. According to the manufacturer, this additive helps the formation of a fluid membrane on the side wall.

- SlurryPro MPA: additive supplied in liquid form. The chemistry of this product is not published but the molecules are understood to carry a positive charge and to have a hydrophobic tail. SlurryPro MPA reacts with the PHPA, initially producing a mass of fine hair-like stands to which suspended soil sticks. Over a period of a few minutes the hairs coalesce to form larger and tighter masses that settle more rapidly. According to the manufacturer, the key function of this additive is to accelerate the settlement of suspended soil particles so they can be removed prior to concreting. As soon as the excavation was complete, 10 litres of 1% solution (in water) was dropped to the bottom of the pile bore in a plastic bag, which was weighted with a brick or stone. On site, this process is known as dropping a SlurryPro MPA 'bomb', and was shown in Figure 4.11. If added during the excavation (this was not done during the trial), SlurryPro MPA can reduce fluid loss in coarse soils, as the coagulated mass formed from its reaction with the base polymer can plug the pores in the side walls. Figure 5.19 shows the cleaning and gel-forming effects of this additive in a bucket trial, but note that the effect shown in the photograph was greatly exaggerated because undiluted SlurryPro MPA solution was used.

To mix the polymer, the SlurryPro CDP base polymer and the additive LA-1-D were sprinkled into the treated water in a mixing tank. After mixing, the fluid had a slightly milky appearance, as the LA-1-D additive did not dissolve in the water but stayed as fine colloidal system.

The bentonite used for the trial was Berkbent 163, manufactured by Tolsa Minerals (formerly Steetley Bentonite and Absorbents). This product is a sodium carbonate (soda ash) activated calcium bentonite, and is commonly used in the UK for the construction of bored piles, cut-off walls and diaphragm walls. According to the manufacturer, the moisture content of the bentonite powder ranges from 10% to 14%, and the cation exchange capacity (CEC) is about 85 mEq/100 g. The methylene blue index (MBI) is about 89 mEq/100 g. Methylene blue is a cationic dye that is adsorbed onto the negatively charged surfaces of clay particles. The value of the MBI is therefore a direct indication of the negative charge on the clay. There is also often a strong correlation between the MBI and other clay properties such as the CEC, specific surface area and swell potential.

Depending on the ground conditions, the slurry concentration recommended by the bentonite manufacturer ranges from 30 to 80 kg/m^3 (i.e. 3–8% w/w). For the trial, the bentonite slurry was prepared at 40 kg/m^3, as the soil at the test site was relatively stable. Although treatment of the mix water with soda ash was recommended by the manufacturer in hard-water areas such as London, a preliminary trial showed that the residual soda ash in the product was capable of removing 90% of the water hardness; that is, to reduce it from about 250 to 25 mg/l, with the remaining 10% believed to be caused by magnesium ions, which are not precipitated by soda ash. As a result, no soda ash treatment was deemed necessary, as it would have made little difference. After mixing, the slurry was left in the storage tank to hydrate, and at the time of use the fluid was about 3 days old. Figures 2.3 and 2.4 show the equipment used for mixing the bentonite slurry and its storage. No bentonite cleaning equipment was required as, for the trial, only a single pile was being constructed under bentonite slurry.

8.3.3.3 Excavation and base-cleaning procedures

Excavation of the test piles generally followed the common procedures for that area. The initial steps included inserting a steel casing through the made ground down to the top of the underlying clay, as shown schematically in Figure 8.10. Since the made ground was underlain by stiff overconsolidated clay (the Lambeth Group), after the installation of the casing, the made ground and the majority of the Lambeth Group were excavated dry without any support fluid. To maintain the stability of the pile bores, the support fluid was introduced into the bores, by centrifugal pump, when the excavations reached about 17 m bgl (−11.5 mOD). As discussed in the Glasgow case study (see Section 8.2), centrifugal pumps can cause severe damage to the polymer fluid if it is allowed to circulate in a loop system. A centrifugal pump was used because at the time of the trial its adverse effect on polymers had not yet been proven.

Once the support fluid was in the pile bores, the excavation followed predetermined guidelines deploying the right tools and procedures to prevent suction occurring in the fluid column and in the surrounding soil during raising of the excavation tool – an issue not always appreciated by foundation contractors. To prevent suction, either a twin-flight auger or a vented digging bucket was used, as shown in Figures 8.13 and 8.14. These tools allow the fluid to flow through the tool during extraction.

Figure 8.13 A twin-flight piling auger used for the Stratford field trial

Figure 8.14 (a) A vented digging bucket used for the Stratford field trial and (b) a close-up of the vent

(a)

(b)

Figure 8.15 The air-lifting operation to exchange used bentonite fluid

For pile B1, at the end of the excavation the used bentonite fluid was exchanged for fresh fluid, as the sand content was above the specified limit; further details will be given in the following section. Figure 8.15 shows the fluid exchange operation by air lifting.

For the two piles constructed under polymer fluid, the bores were excavated down to about 0.5 m above the final base level, and the solids-laden column of fluid was then cleaned with the additive SlurryPro MPA as follows. A sealed bag of a dilute solution of SlurryPro MPA was dropped to the base of the bore. The bag was then broken with a cleaning bucket, and the contents dispersed by gently raising and lowering the tool several times. A waiting period of about 30 min was then allowed for the additive to react with the base polymer and for the coagulated soil to fall out of suspension. The effect of this procedure was checked by comparing the sand content measured before and after the treatment. The results will be discussed in the next section. It is worth noting that O'Neill (1991) also recommended contractors to let the used polymer fluids to stand in the pile bores without agitation for a minimum of 30 min.

After the exchange (bentonite slurry) or the in situ cleaning of the support fluid (polymer fluid), a flat-sided cleaning bucket was used to excavate the final 0.5 m of soil and to prepare the soil surface for a base hardness/cleanliness test. This test was made with a 100 mm × 100 mm × 20 mm (thickness) metal plate suspended on a tape measure. The impact as felt by the operator when the plate hit the pile base was graded on a scale of 1 (hard) to 5 (very soft). Details of this test are given by Berry (2009). All three test piles received a grade 3, interpreted as detection of a distinct base with slight embedding of the weight into the soil. These results confirmed that the three test piles had a similar base quality.

8.3.3.4 Support fluid: field test results

Immediately after the excavation was completed, fluid samples were taken from the middle and bottom of the pile bores for testing. The properties tested included pH, sand content, density, Marsh funnel time and viscosity (measured over a range of shear rates in a Fann coaxial cylinder viscometer). In the viscometer, the shear stress was measured over a shear rate range of 5–1022 s^{-1} so that a flow curve could be developed. Table 8.4 summarises the test results, which are discussed below.

PH

The pH of the fluid was measured using a hand-held pH meter. The clean polymer fluid had a pH of about 11 due to the addition of the highly alkaline Protek 100. It can be seen that the used fluids also had the same pH. For the bentonite slurry, although the mix water was not pretreated, the pH of the fresh fluid was still 10.5 as a result of the residual soda ash in the bentonite powder. Again, there was no change in the pH for the used bentonite slurry.

SAND CONTENT

The sand content of the fluid, which is defined as the fraction by bulk volume of particles larger than 75 μm, was measured using the standard sand content kit developed in the oil industry. It can be seen that, immediately after excavation, the sand content of the polymer fluid was between 1.3% and 1.4% near the bottom of the pile bores. However, after adding the dilute SlurryPro MPA solution and a waiting period of about 30 min, the sand content of the polymer fluid in the piles dropped to about 0.1%. This confirmed the effective removal of the suspended coarse particles with SlurryPro MPA. In contrast, the sand content of the bentonite fluid was 6.8% in the middle of the bore and 10% near the bottom of the bore. Since these values exceed the upper limit of 4% as specified by the Federation of Piling Specialists (FPS, 2006), the used bentonite fluid was exchanged (as shown in Figure 8.15) to prevent the formation of a 'soft toe'. It may be noted that the settlement of fines can be less of an issue for bentonite slurries because they can have a gel that holds solids in suspension, and hence the higher sand content limit of 4% for bentonite slurry as compared with the 1% figure adopted for polymer fluid.

DENSITY

The density of the fluid was measured using a hand-held density meter (model DMA 35N, Anton Paar) as previously shown in Figure 6.9. As can be seen from Table 8.4, the density of the used polymer fluid was between 1.02 and 1.03 g/ml as a result of suspended soil particles. Note that pure polymer fluid will have a density of 1.0 g/ml. As expected from the sand content results, there was significantly more soil in the bentonite slurry, which had a density of 1.10 g/ml in the middle of the bore and 1.15 g/ml near the base of the bore. As mentioned previously, the used bentonite fluid was exchanged because it did not satisfy the current industry specification for concreting.

MARSH FUNNEL TIME

The Marsh funnel time for the fresh polymer fluid was about 70 s at the mixing tank, and 56 and 50 s in the bores of piles P1 and P2, respectively. Such a drop is not unexpected when the aeration mixing method and a centrifugal pump are used, although some of the

Table 8.4 Properties of polymer and bentonite fluids used for the Stratford field trial

Pile No.	Fluid details and exposure time				Fluid test results[a,b]			
	Product name (manufacturer)	Dosage: kg/m^3	Total fluid–soil exposure time: h	Time of fluid in the pile bore during sampling: h	pH	Sand content: %	Density: g/cm^3	Marsh funnel time: s
P1	SlurryPro CDP	1.0 for CDP;	7.5	5	11.2/11.2/11.2	0/0.2/0.2 (1.3)[d]	1.00/1.02/1.03	70/56/56
P2	(KB International)	0.08 for LA-1-D Protek 100 and MPA[c]	26	23	11.0/11.0/11.0	0/0.1/0.1 (1.4)[d]	1.00/1.02/1.02	69/50/50
B1	Berkbent 163 (Tolsa)	40.0	7.5	3	10.5/10.3/10.3	0/6.8/10	1.02/1.10/1.15	34/40/40

Data after Lam *et al.* (2010b).

[a] See Figure 8.16 for viscosity curves.

[b] Fluid test results are given in this format: fresh fluid in the tank/used fluid from the middle of the pile bore/used fluid from the bottom of the pile bore.

[c] Protek 100 was used to raise the pH of water to 11; 10 litres of 1% SlurryPro MPA solution were used to clean the polymer fluid in situ in the pile bores.

[d] Sand contents in parentheses were measured on samples taken near the bottom of the pile bore immediately after excavation but before the addition of 1% SlurryPro MPA solution.

Figure 8.16 Viscosity curves of clean polymer and bentonite fluids used in the Stratford field trial

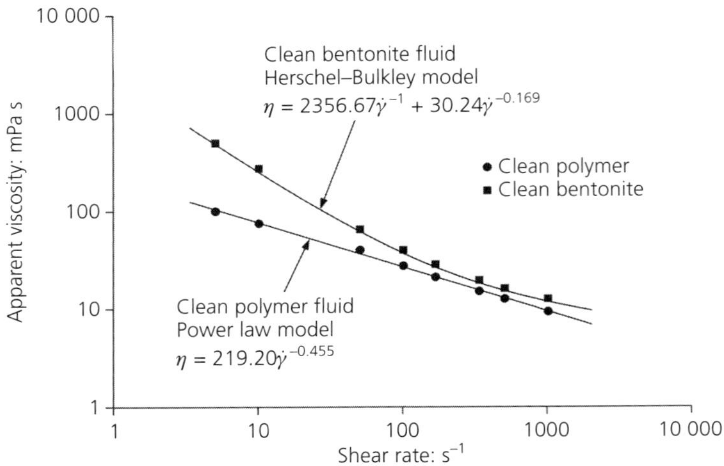

drop will have been the result of sorption of the polymer molecules on the soil during digging. A requirement to boost the active polymer fluid concentration after repeated reuse is thus normal. In contrast to the polymer fluids, the Marsh funnel time for the bentonite fluid rose from 34 s in the storage tank to about 40 s in the pile bore. This was caused by the suspension of soil particles, especially by clay-sized material, in the digging slurry.

SHEAR STRESS–SHEAR RATE (VISCOSITY) CURVE

Figure 8.16 shows the Fann viscometer test results. Power law and HerschelBulkley (a power law model with an additional yield stress term) models have been fitted to the data for the polymer and bentonite fluids, respectively. A discussion of these rheological models can be found in Chapter 9. In summary, the results show that the viscosity of both fluids decreases with increasing shear rate. The results also explain why the particle-holding ability of the bentonite fluid was so much greater than that of the polymer fluid. For particle settlement, low-shear rate viscosity is significant, and, for example, at a shear rate of 10 s^{-1} the apparent viscosity of the bentonite fluid was about four times higher than that of the polymer fluid. The difference is expected to increase as the shear rate decreases and the yield stress of the bentonite slurry dominates.

8.3.3.5 Concreting procedures and quality

Self-compacting concrete with the designated class C28/35 20 mm CIIB-V S4 DC-3 was used to form the piles. These designations have the following meanings: (*a*) the characteristic cylinder and cube compressive strengths at 28 days are 28 and 35 MPa, respectively; (*b*) the maximum aggregate size is 20 mm; (*c*) the cement type is Portland cement with 21–35% fly ash; (*d*) the target slump is between 160 and 210 mm; and (*e*) Building Research Establishment design chemical class 3 (Aggressive Chemical Environment for Concrete Class) (BRE, 2005).

Figure 8.17 Sampling from the top of the concrete column in a polymer-supported pile bore

Following standard bored piling practice, the concrete was delivered through a tremie pipe into the bores, to minimise the degree of intermixing between the fresh concrete and the support fluid. Initially, when the tremie pipe was rested on the pile base, a plug having the same internal diameter as the pipe was installed, and then the pipe was loaded with concrete. When the pipe was almost full, it was lifted slightly, and this allowed the concrete to flow out very quickly due to the large hydrostatic head inside the pipe. This was done to ensure any remaining soft sediment accumulated at the base could be flushed aside to ensure good contact between the concrete and the soil. The discharge rate of the first truck load of concrete was 1.2 m^3/min, and thereafter the pile bore was concreted at a lower rate of 0.5 m^3/min.

To assess the effects of the polymer fluids on the quality of the concrete, during casting of pile B1 (bentonite) and pile P2 (polymer), samples were taken from the chutes of the delivery trucks and from the top of the concrete column, as shown in Figure 8.17. The samples were then tested for their compressive strength at 3, 7 and 28 days as well as their stiffness at 28 days. Figure 8.18 shows the test results It can be seen that the strength of the concrete at the top of the concrete column (i.e. the concrete–support fluid interface) was clearly affected when compared with the strength of samples taken from the delivery truck. The overall strength reduction was 26% and 29% for the bentonite and polymer fluids, respectively, and a similar degree of reduction can also be seen in the elastic

Figure 8.18 Effect of bentonite and polymer fluids on the strength and stiffness of hardened concrete

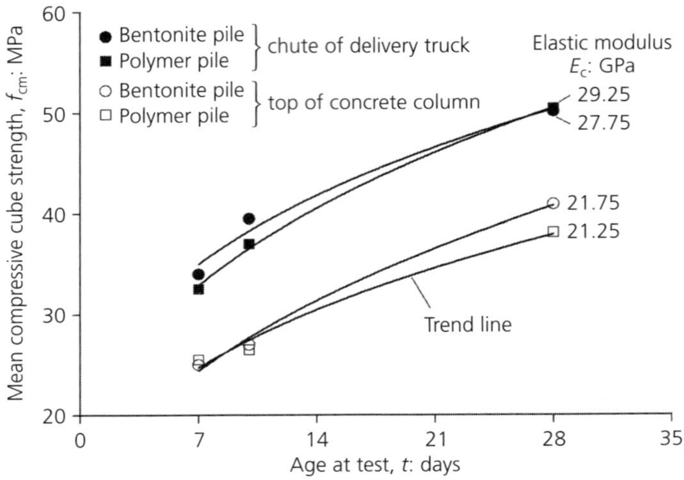

modulus results. This reduction was caused by the intermixing of fresh concrete and support fluid at their interface during casting. There does not seem to be any correlation with the support fluid type. The reduction shown in Figure 8.18 will not be representative of concrete deeper in the piles, the majority of which will not have had contact with the support fluid.

8.3.3.6 Side-wall profiles

Although no measurement of the sidewall profile was carried out after excavation, during casting the successive rises in concrete level were measured after each truck load of concrete was discharged. This information has been converted to an equivalent radius along the pile, and is shown in Figure 8.19. It is interesting to see that the profiles of the three piles were quite similar, suggesting comparable conditions of the pile bores before concreting. This finding also appears to be consistent with the results of Ata and O'Neill (1998) – that the side-wall profiles of their polymer-supported bores, one in sand and one in clay, remained almost the same for up to 19 h. However, the inferred radii given in Figure 8.19 should be treated with caution, as there are a number of factors that can affect their accuracy, including variations in concrete volumes between deliveries and material wastage.

8.3.3.7 Pile integrity

The structural integrity of the piles was assessed by two non-destructive methods: cross-hole sonic logging (CSL) and pile integrity testing (PIT). The operating principles of these tests are given by Turner (1997). In summary, the CSL results showed no anomalies in pile B1 but indicated a flaw in piles P1 and P2. Figure 8.20 shows the profiles of the first arrival time and attenuation of the signal energy. It can be seen that the signal energy reduced (attenuation increased) by 10–15 dB at around 18.0 m in pile P1 and

Figure 8.19 Equivalent radii of test piles at Stratford from concreting records

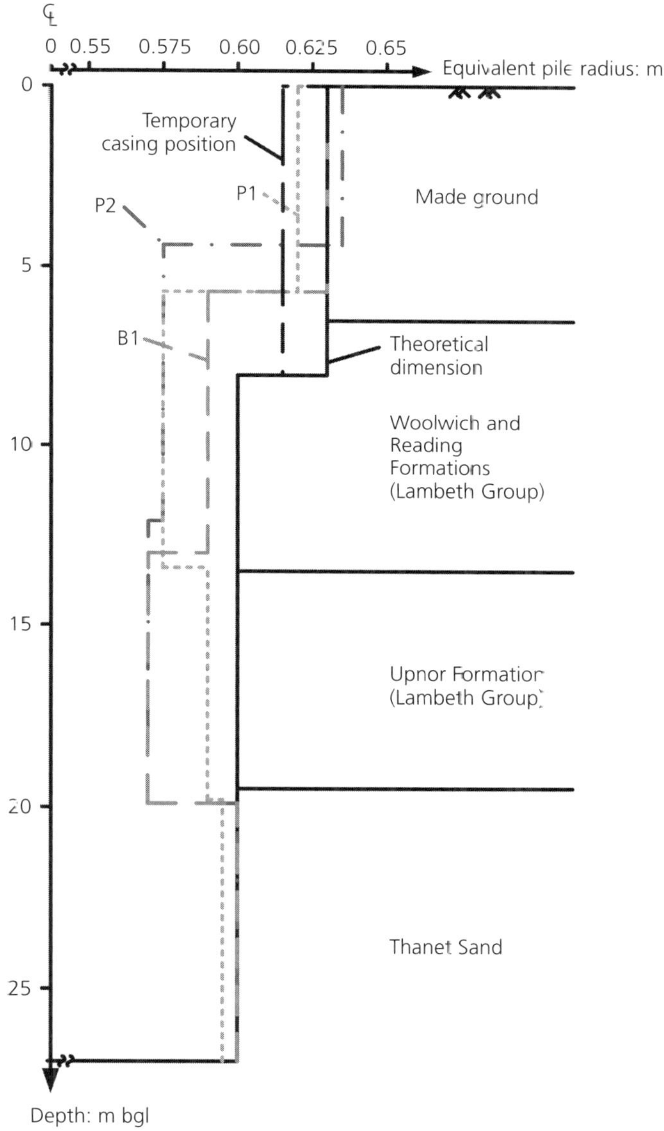

at 22.5 m in pile P2. This suggests a flaw or defect in the pile concrete. However, since the first arrival time increased only very slightly at these locations, the defects were likely to be minor. The reason why piles P1 and P2 had local defects is not known, but they were possibly caused by a damaged access tube, explained as follows. When CSL was being carried out on pile P1, it was noted that water was draining out of one of the tubes. To establish what level the water would settle at, this tube was filled with water again,

Figure 8.20 CSL results for the test piles constructed using polymer fluid at Stratford: (a) pile P1 and (b) pile P2

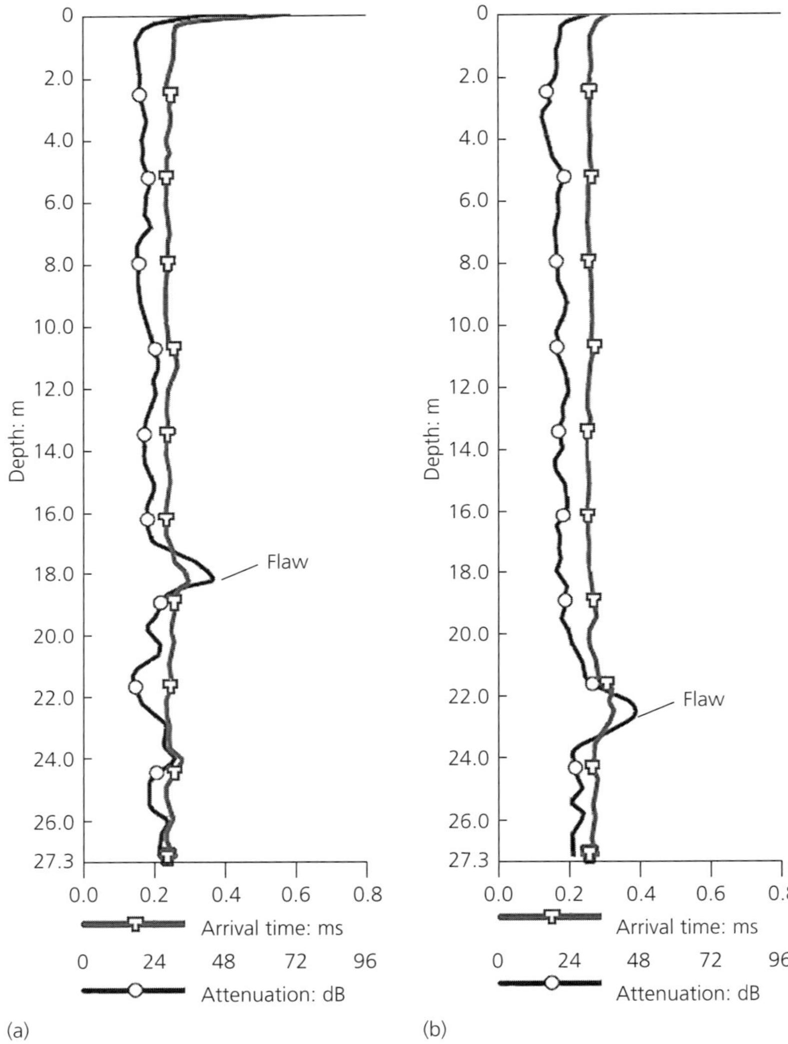

and the level checked on the next day. It was discovered that the water found its natural level at about 23 m; that is, 5 m below the location of the flaw. This means that there was one or possibly several leaks in the tube. Since the tube was filled with water prior to concreting, the water may have eroded the concrete locally around the leaking points.

Figure 8.21 shows the PIT results in the form of velocity–depth profiles. As no sudden change in the velocity can be seen in any of the plots, the cross-sections of the piles were likely to be uniform with depth; that is, no necking or bulging. This finding corresponds

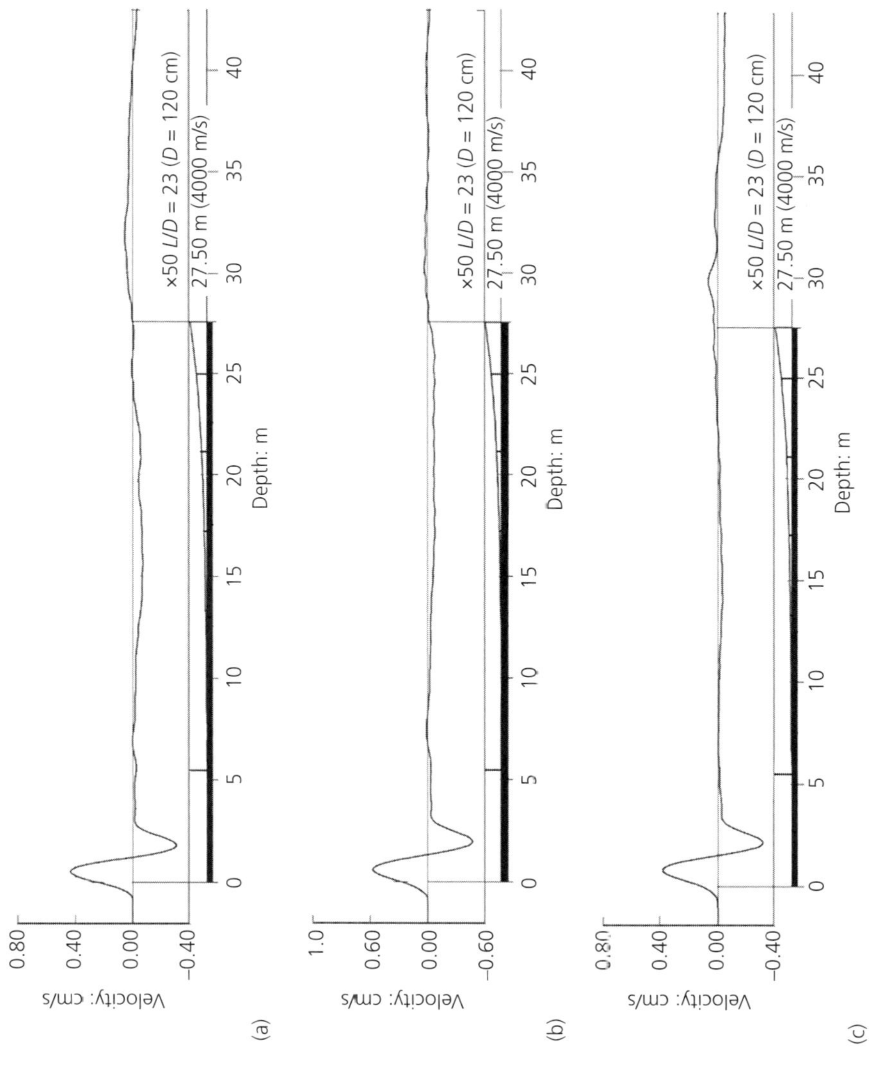

Figure 8.21 PIT results for the test piles at Stratford: (a) pile P1, (b) pile P2 and (c) pile B1

with the back-analysed side-wall profiles shown in Figure 8.19. Finally, it may be noted that a clear reflection of the wave at the pile base level (27.5 m from top of pile) was not detected in any of the piles. This is likely to have been due to the high soil resistance along the shaft, which considerably reduced the energy of the stress wave as it travelled down the pile shafts.

8.3.3.8 Load test method

Compressive axial load tests were carried out on the test piles when the concrete was about 28 days old. The piles were loaded following the 'slow' maintained load method specified by the *ICE Specification for Piling and Embedded Retaining Walls* (ICE, 2007). Figure 8.22 shows the reaction frame used for one of the tests. The frame was still under construction when the photograph was taken. Following each application of an increment of load, the load was maintained until the rate of pile head settlement satisfied one of three criteria as set out in the specification. For each load increment, the duration also could not be less than a specified minimum period. In addition to the requirement for load-holding periods, a single load cycle was also included in the test to assess the pile behaviour during unloading and reloading.

Although the specification requires only that the applied maximum load should be 100% of the design verification load (DVL) plus 50% of the specified working load (SWL), it

Figure 8.22 Reaction frame used for load testing the piles at Stratford

was decided to raise the maximum load to 100% DVL + 100% SWL, as it was felt that the original maximum load would not cause sufficient settlement for a detailed back-analysis. Based on the preliminary design calculations, the values of DVL and SWL of the test piles were 9.1 and 9.0 MN, respectively. The maximum (proof) load was therefore 18.1 MN.

8.3.4 Load test results

8.3.4.1 Load–settlement behaviour

Figure 8.23 shows the applied load and head settlement plotted against time for the three test piles. It can be seen that each of the three tests lasted for about 34 h and that the load-holding periods were only marginally different between the tests. It can also be seen from the figure that the two piles constructed under polymer fluid (P1 and P2), despite their different soil–support fluid exposure periods, behaved very similarly in terms of their settlement behaviour. The pile constructed under bentonite (B1), on the other hand, showed significantly larger settlements at high loads. At the maximum load of 18.1 MN (100% DVL + 100% SWL), the settlement values for piles P1, P2 and B1 were 29, 24 and 51 mm, respectively – note that the settlement of pile B1 was almost twice that of piles P1 and P2.

The reason for the superior behaviour of piles P1 and P2 was probably due to their higher shaft resistance. It is likely that the polymer fluid gave the piles a cleaner concrete–soil interface, as clean polymer fluids do not form a filter cake, and, during digging, polymer fluids form only a thin cake because of their low solids loading compared with bentonite slurries. The vigorous base-cleaning procedure adopted for the construction also should have contributed to a clean pile base. To gain further insight into the behaviour of the piles, the load test results were analysed using the methods proposed by Chin (1970) and Fleming (1992). These two methods are commonly used by UK engineers to deduce the likely ultimate pile shaft and base capacities (ICE, 2017).

8.3.4.2 Back-analysis using Chin's method

As none of the three test piles was taken to failure during the load tests, their load–settlement curves were analysed using the method proposed by Chin (1970) to obtain an estimate of the ultimate pile resistance. This method is also known as the Chin–Kondner method, since it was R. L. Kondner who first proposed the underlying hyperbolic stress–strain (or load–settlement) principle. This method was used because the deduced ultimate pile capacity is not based on some arbitrary definition of failure (e.g. settlement equals 10% of pile diameter) but corresponds to a state when the pile is displaced indefinitely with negligible increase in resistance. To use this method, one needs to divide the settlement at each loading stage by the applied head load at that stage and to plot it against settlement. For an end-bearing pile, the data points will plot along a straight line, with the slope of the line being the ultimate base resistance. For a combined friction and end-bearing pile, two straight lines will result. The slopes of the first and second lines give estimates of the ultimate shaft resistance and the total resistance, respectively. The results of the Chin analyses for the three test piles are given in Table 8.5.

Table 8.5 Summary of deduced pile capacities by Chin's method

Pile No.	Shaft resistance: kN	Base resistance: kN	Total resistance: kN
P1	16 562	12 588	29 150
P2	17 751	14 874	32 625
B1	15 658	9 543	25 201

Figure 8.23 Load and settlement changes with time for (a) pile P1, (b) pile P2 and (c) pile B1

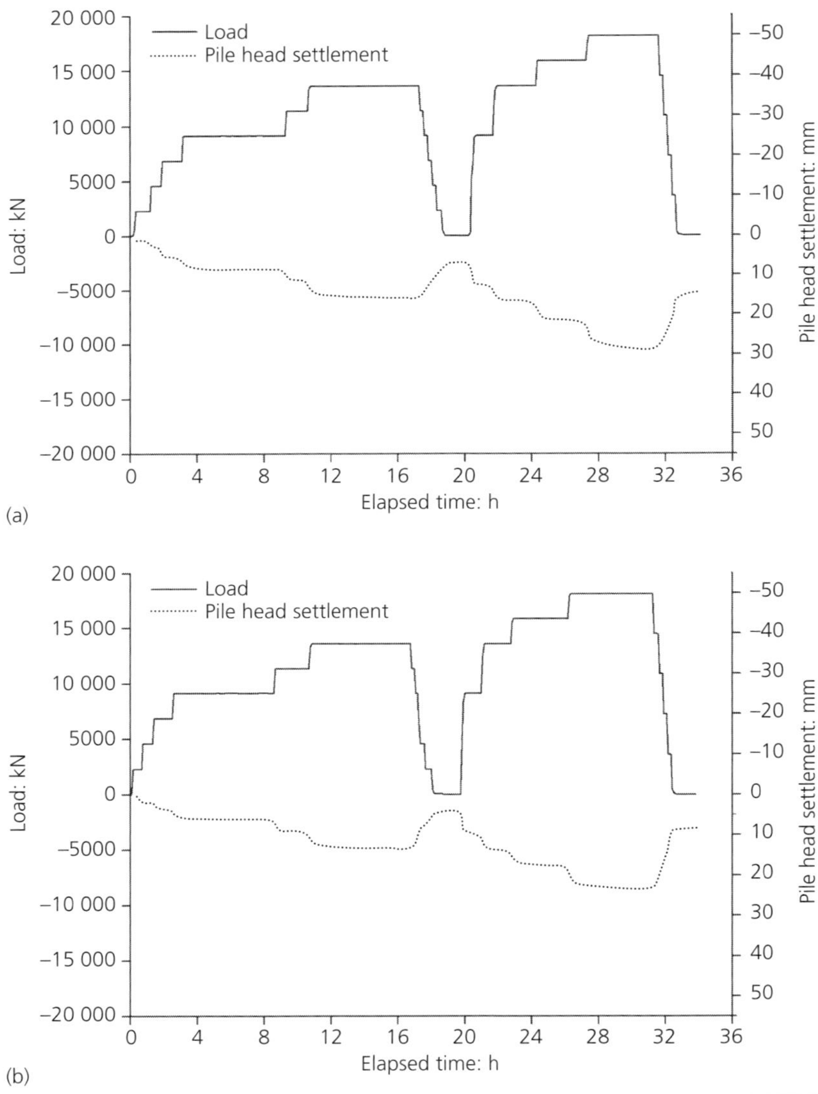

Figure 8.23 Continued

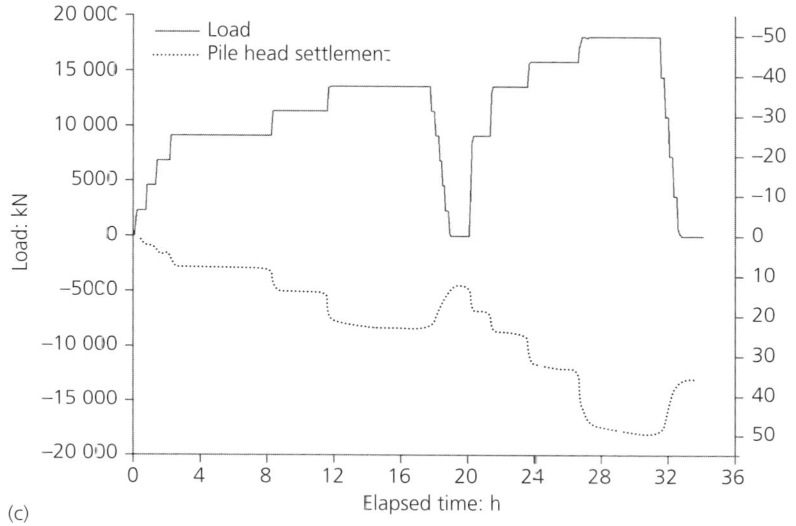

(c)

It can be seen that the deduced capacity of pile B1 was noticeably lower than those of piles P1 and P2. This is as expected, since pile B1 showed a much larger settlement at proof load (18.1 MN) than the other two (see Figure 8.23). It should be noted that, since a Chin analysis is essentially a curve-fitting exercise to a hyperbolic curve, a pile with a larger settlement will always be interpreted as having a lower ultimate resistance, and vice versa.

8.3.4.3 Back-analysis using Fleming's method

The computer program CEMSET is also commonly used in the UK to interpret pile load test results. Fleming (1992) proposed this method using parameters most geotechnical engineers would recognise and understand. Essentially, the method is based on the same hyperbolic function as used by Chin, but the overall load–settlement response is divided into three main parts, namely shaft resistance and settlement, base resistance and settlement, and elastic shortening. Since the underlying principles and the equations can be found in Fleming's original article, they are not repeated here, but the key parameters are described to aid discussion of the results:

- *Geometry*. There are five parameters that define the basic geometry and the load transfer length of the piles. These are the pile shaft diameter (D_s); the pile base diameter (D_b); the upper length of a pile carrying no load or low loads by the shaft resistance (L_0); the length of a pile transferring load to the soil by the shaft resistance (L_f); and the effective column length of the pile transferring the shaft resistance divided by L_f (K_e). For the test piles, D_s and D_b equal 1.2 m, and L_0 can be taken as the thickness of the made ground, which is 6 m. It follows that L_f is 21 m. The value of K_e was taken as 0.45, as this is the approximate value suggested by Fleming (1992) for piles in stiff overconsolidated clays.

Figure 8.24 Measured and simulated response of test piles using the method proposed by Fleming (1992): (a) pile P1, (b) pile P2 and (c) pile B1

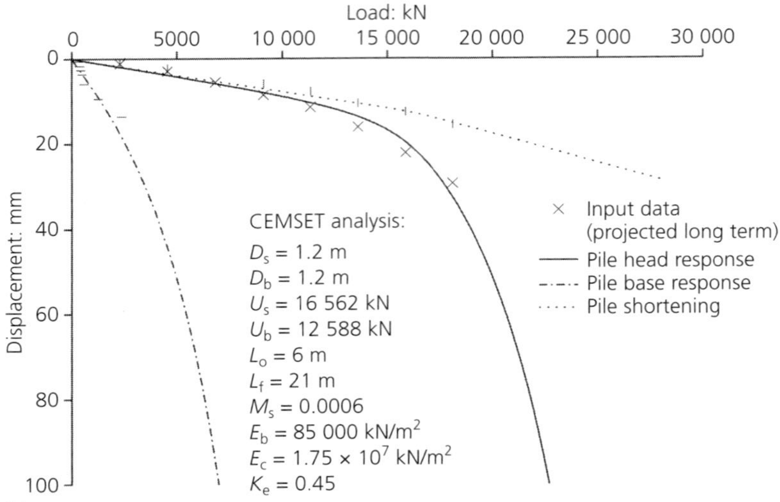

(a)

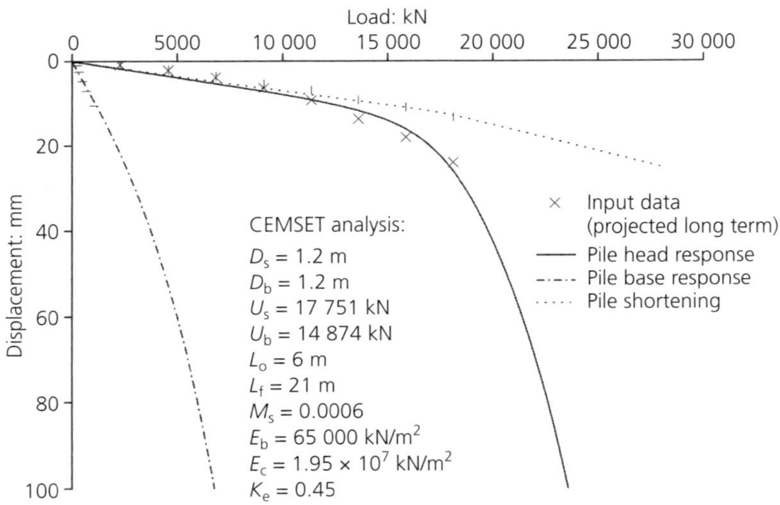

(b)

- *Resistance*. There are two resistance parameters in the formulation, namely the ultimate shaft resistance (U_s) and the ultimate base resistance (U_b). The values of these two parameters were taken directly from the previous Chin analyses, since the definition of the ultimate resistance is the same for both methods. The values corresponding to the long-term condition were used to remove the influence of time, as suggested by Fleming (1992).

Figure 8.24 Continued

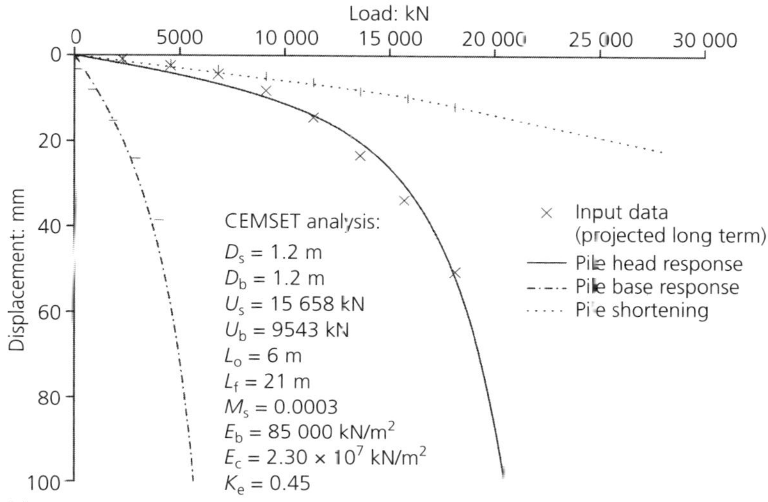

(c)

- *Stiffness*. The first stiffness parameter is the elastic modulus of pile material in compression (E_p). During the back-analysis, this parameter was adjusted until the computed values of elastic shortening at various loading stages matched the measured values as recorded by the extensometers that were installed in the sonic logging tubes before the load tests. The second stiffness parameter is the secant deformation modulus for soil beneath the pile base at 25% of the ultimate stress (E_b). This parameter was adjusted until the computed base settlement matched the actual response. The final stiffness parameter is a dimensionless flexibility factor that represents the movement of a pile relative to the soil when transferring load by friction (M_s). According to Fleming (1992), the value of this parameter typically ranges from approximately 0.0005 for very stiff soils or soft rocks to 0.004 for soft to firm or relatively loose soils. For stiff overconsolidated clays, it typically ranges between 0.001 and 0.002, but the actual value is known to be affected by construction practice and several other factors. During back-analyses, this parameter is the last to be adjusted since it is the least known.

Figure 8.24 shows the CEMSET analysis results of the three test piles. The values of the back-calculated parameters are also shown on the plots. Since the geometry and resistance parameters are either known or have already been determined, the following discussion is focused on the shaft flexibility factor M_s. It can be seen from the plots that the back-calculated values of M_s are 0.0006 for both piles P1 and P2 but 0.003 for pile B1. This means that, in terms of the shaft response, the behaviour of the two piles constructed using polymer fluids (P1 and P2) was similar to piles embedded in very stiff soils or soft rocks, whereas the behaviour of the pile constructed using bentonite (B1) was typical of those in soft to firm soil. In this case, the back-calculated M_s values can be used to indicate the effect of pile installation. For soil that experiences little disturbance, the value of M_s will be lower, and vice versa. To conclude, the CEMSET back-analysis results confirm the superior performance of piles constructed using polymer fluid.

8.3.5 Lessons learned

The lessons learned from this case study can be summarised as follows:

- Bentonite and polymer fluids have a similar degree of impact on the compressive strength and elastic modulus of hardened concrete. The effect should be limited to the interface region between the fresh concrete and the support fluid. The adverse effect is due to the increased amount of water in the mix in this interface region rather than to undesirable chemical reactions.
- By monitoring the properties of support fluids on site and rigorously applying defined fluid-cleaning procedures (for the polymer) and fluid exchange (for bentonite), the sand content values of the support fluids were reduced to appropriately low levels prior to concreting. This was achieved on a site where the soil strata were mainly stiff clay and dense sand. These procedures reduced the risk of soft sediment accumulating at the base of the excavations.
- The two piles constructed using polymer fluids (P1 and P2) showed a much stiffer load–settlement response than the pile constructed using bentonite (B1). The pile head settlements were 29, 24 and 51 mm for piles P1, P2 and B1, respectively, for the maximum applied load of 18.1 MN (100% DVL + 100% SWL). Back-analyses conducted using the method proposed by Fleming (1992) also revealed that the values of the flexibility factor M_s for piles P1 and P2 were similar to those commonly found in very stiff soils or soft rock, whereas the M_s value for pile B1 was typical of piles in soft to firm soils. These results clearly show the benefit of using polymer fluids when compared with bentonite at this site.
- Despite the different fluid–soil exposure times for the two piles excavated under polymer (7.5 versus 26 h), there was negligible difference between them in terms of their deduced side-wall profiles, their load–settlement response and the value of the back-analysed stiffness parameter M_s. This result confirms what was described in the news report by Wheeler (2003). The practical relevance of this finding is that a reduction in the design shaft resistance may not always be necessary for polymer-supported bores that are left open overnight. If this design practice were adopted, piling contractors would be able to organise site activities more efficiently.

REFERENCES

Ata AA and O'Neill MW (1998) Side-wall stability and side-shear resistance in bored piles constructed with high-molecular-weight polymer slurry. *Proceedings of the 3rd International Seminar on Deep Foundations on Bored and Auger Piles, Ghent*. Balkema, Rotterdam, the Netherlands, pp. 111–117.

Berry AG (2009) Method for cleaning and checking the base of diaphragm wall panels. *Ground Engineering* **42(11)**: 29–31.

BRE (Building Research Establishment) (2005) *Concrete in Aggressive Ground*, 3rd edn. BRE, Watford, UK.

Brown DA (2004) Zen and the art of drilled shaft construction: the pursuit of quality. *Proceedings of GeoSupport 2004 Conference, Orlando* (Turner JP and Wayne PW (eds)). American Society of Civil Engineers, Reston, VA, USA, pp. 19–33.

BSI (British Standards Institution) (1990) BS 1377-2:1990. Methods of test for soils for civil engineering purposes. Classification tests. BSI, London, UK.

BSI (1999) BS 5930:1999. Code of practice for site investigations. BSI, London, UK.

Chin FK (1970) Estimation of the ultimate load of piles from tests not carried to failure. *Proceedings of the 2nd Southeast Asian Conference on Soil Engineering, Singapore*. Asian Institute of Technology, Bangkok, Thailand, pp. 81–90.

Dyson S and Blight I (2007) Channel Tunnel Rail Link section 2: Stratford. *Proceedings of the Institution of Civil Engineers – Civil Engineering* **160(6)**: 29–32.

Fleming WGK (1992) A new method for single pile settlement prediction and analysis. *Géotechnique* **42(3)**: 411–425.

FPS (Federation of Piling Specialists) (2006) *Handbook on Pile Load Testing*. FPS, Beckenham, UK.

ICE (Institution of Civil Engineers) (2007) *ICE Specification for Piling and Embedded Retaining Walls*, 2nd edn. Thomas Telford, London, UK.

Lam C, Jefferis SA and Goodhue, Jr KG (2010a) Observations on viscosity reduction of PHPA polymer support fluids. *Proceedings of Sessions of the GeoShanghai 2010 International Conference, Shanghai*. American Society of Civil Engineers, Reston, VA, USA, pp. 184–191.

Lam C, Troughton V, Jefferis S and Suckling T (2010b) Effect of support fluids on pile performance – a field trial in east London. *Ground Engineering* **43(10)**: 28–31.

Lam C, Jefferis SA and Suckling TP (2014) Construction techniques for bored piling using polymer fluids. *Proceedings of the Institution of Civil Engineers – Geotechnical Engineering*, **467(GE6)**: 565–573.

Lam C, Jefferis SA, Suckling TP and Troughton VM (2015) Effect of polymer and bentonite support fluids on the performance of bored piles. *Soils and Foundations* **55(6)**: 1487–1500

Lennon DJ, Ritchie D, Parry GO and Suckling TP (2006) Piling projects constructed with vinyl polymer in Glasgow, Scotland. *Proceedings of the 10th International Conference on Piling and Deep Foundations, Amsterdam*. Deep Foundations Institute, Hawthorne, NJ, USA, pp. 499–506.

O'Neill MW (1991) Effects of mineral and polymer slurries on drilled shafts. *Proceedings of the 16th DFI Annual Members' Conference, Chicago, USA*. Deep Foundations Institute, Hawthorne, NJ, USA.

Tallard GR (1992) New trenching method using synthetic bio-polymers. In *Slurry Walls: Design, Construction, and Quality Control* (Paul DB, Davidson RR and Cavalli NJ (eds)). American Society for Testing and Materials, Philadelphia, PA, USA, pp. 86–102.

Troughton VM (1992) The design and performance of foundations for the Canary Wharf development in London Docklands. *Géotechnique* **42(3)**: 381–393.

Troughton VM and Platis A (1989) The effects of changes in effective stress on a base grouted pile in sand. *Piling and Deep Foundations: Proceedings of the International Conference on Piling and Deep Foundations, London*. Balkema, Rotterdam, the Netherlands, pp. 445–453.

Turner MJ (1997) *Integrity Testing in Piling Practice*. Construction Industry Research and Information Association, London, UK, CIRIA Report 144.

Ventouras K (2005) *Engineering behaviour of Thanet Sand*. PhD thesis, Imperial College London, UK.

Wheeler P (2003) Piles unlock polymer potential. *European Foundations,* summer, pp. 8–9.

Whitaker D (2004) Groundwater control for the Stratford CTRL station box. *Proceedings of the Institution of Civil Engineers Geotechnical Engineering* **157(4)**: 183–191.

Polymer Support Fluids in Civil Engineering
ISBN 978-0-7277-5786-9

http://dx.doi.org/10.1680/psfce.57869.229

Chapter 9
Rheological properties of polymer support fluids

9.1. Introduction

Rheology is the study of the deformation and flow of materials. As an example of flow behaviour, consider the concrete slump test, in which the amount of slump is used as an indicator of the workability of the concrete. If the slump process is considered in rheological terms, the final amount of slump and the rate of slumping are actually controlled by the yield stress and the plastic viscosity of the material, respectively, both of which are rheological parameters commonly used to model yield stress fluids (Ferraris and de Larrard, 1998). For grouts, a flow cone can be used for the rapid assessment of fluidity (apparent viscosity) in the field (Jefferis, 2003), or a viscometer can be used to characterise flow behaviour (e.g. see ASTM, 2014). For polymer support fluids, it is important to understand the rheological properties because it is the complex flow behaviour of these fluids that allows them to stabilise permeable soils. By way of example, Troughton (1992) noted that water could not stabilise a pile bore in the Thanet Sand at Canary Wharf, but the Stratford case history described in Section 8.3 has demonstrated that polymer fluids, which have the same density as water but higher viscosity (especially at low shear rates), could adequately support this formation. The rheological properties of support fluids also must be carefully considered when developing protocols for site work and quality control and to ensure satisfactory formation of the foundation elements as discussed in many of the previous chapters (e.g. Table 1.1). To understand how polymer fluids behave under different conditions, their steady state flow behaviour and transient viscoelastic properties have been characterised both in the laboratory and in the field. Before the results are discussed, some fundamental rheological concepts are introduced in the following section.

9.2. Fundamentals of non-Newtonian fluid flow

9.2.1 Definitions

The shear viscosity (η) of a fluid at any shear rate ($\dot{\gamma}$) is defined as

$$\eta = \frac{\tau}{\dot{\gamma}} \tag{9.1}$$

where τ is the shear stress. For Newtonian fluids, η is a constant, but for non-Newtonian fluids it will be a function of the prevailing shear rate and shear history. Therefore, when citing a viscosity value for a non-Newtonian fluid, the shear rate and conditions under

Figure 9.1 Non-Newtonian fluid behaviour

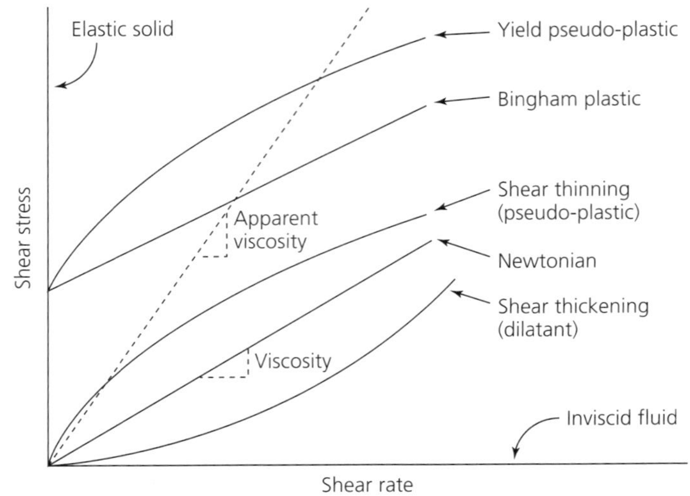

which it was measured should be stated. Water is a Newtonian fluid and has a viscosity of 1 cP at 20°C. Note that 1 cP (centipoise) is numerically equal to 1 mPa s (millipascal second). In this book, viscosity is taken to mean dynamic viscosity, and should not be confused with the related term 'kinematic viscosity' (ν), which is the ratio of dynamic viscosity (η) to the density (ρ) of the fluid. The common unit of kinematic viscosity is the stoke (St), and the SI unit is m^2/s. Water has a kinematic viscosity of 1.0 cSt (centistoke) or $1.0 \times 10^{-6}\ m^2/s$ at 20°C.

Figure 9.1 shows some different types of non-Newtonian behaviour as schematic plots of shear stress against shear rate. The types of fluid shown and some of their real-life examples are Newtonian (e.g. water and motor oil); shear thickening or dilatant (e.g. starch suspension); shear thinning or pseudo-plastic (e.g. polymer solutions and melts); Bingham plastic (e.g. clay slurries and cement grouts); yield pseudoplastic (e.g. paint and ketchup). For non-Newtonian fluids, the apparent viscosity can be defined at any shear rate as

$$\eta_a = \frac{\tau}{\dot{\gamma}} \tag{9.2}$$

Graphically, the apparent viscosity is the slope of a line joining the point (τ, $\dot{\gamma}$) to the origin on a shear stress–shear rate plot, as shown in Figure 9.1. It can be seen that the apparent viscosity line crosses the flow curves for three fluids. Each of these three very different fluids will have the same apparent viscosity at the points where the line crossses. In this book, η and η_a are used interchangeably unless otherwise specified.

9.2.2 Power law and Carreau models

Since polymer fluids commonly display shear-thinning behaviour – that is, a reduction in apparent viscosity with increasing shear rate – the rheological models suitable for this particular fluid type are discussed first. The simplest model for shear-thinning behaviour and indeed also for shear thickening is the power law model:

$$\tau = K\dot{\gamma}^n \tag{9.3}$$

where K is the flow consistency index (in Pa s^n) and n is the flow behaviour index (dimensionless). For $n < 1$ and $n > 1$, the model predicts shear-thinning and shear-thickening behaviour, respectively. For $n = 1$, Equation 9.3 reverts to Equation 9.1, and K becomes the Newtonian viscosity η. Substituting Equation 9.3 into Equation 9.1 gives the apparent viscosity of a power law fluid at shear rate $\dot{\gamma}$:

$$\eta = K\dot{\gamma}^{n-1} \tag{9.4}$$

When analysing shear stress–shear rate data, it is sometimes convenient to use a linear form of the flow model, and this can be achieved by taking logarithms on both sides of Equation 9.4:

$$\log \eta = \log K + (n - 1)\log \dot{\gamma} \tag{9.5}$$

Thus, the power law model plots as a straight line on a log η–log $\dot{\gamma}$ scale. This form of the model was used to fit the viscosity test results shown previously in Figures 6.6 and 6.8. However, as will be shown later, the power law model can represent the shear-thinning behaviour of partially hydrolysed polyacrylamide (PHPA) fluids only over a limited range of shear rates. At very high and low shear rates, the viscosity of PHPA fluids approaches limiting values, and thus the viscosity plot (also known as the flow curve, consistency curve or rheogram) deviates from a straight line on a log–log plot. For this situation, the Carreau model gives a better fit to the data:

$$\frac{\eta - \eta_\infty}{\eta_0 - \eta_\infty} = \left[1 + (\lambda\dot{\gamma})^2\right]^{(n-1)/2} \tag{9.6}$$

where η_∞ is the limiting viscosity at a very high shear rate (i.e. the infinite shear rate viscosity) and η_0 is the viscosity at a very low shear rate (i.e. the zero shear rate viscosity). The viscosity approaches constant values – being independent of the shear rate – at very high and very low shear rates, and these are known as Newtonian plateaus. In Equation 9.6, λ and n are curve-fitting parameters. For dimensional consistency, λ must have a unit inverse to that of $\dot{\gamma}$; that is, of time, and is thus sometimes called the time constant. In the Carreau model, λ sets the shear rate at which the transition occurs between the lower Newtonian plateau η_0 and the shear-thinning region at higher shear rates.

9.2.3 Bingham plastic and Herschel–Bulkley models

Bentonite slurries and cement grouts belong to another class of non-Newtonian fluids known as viscoplastic. This type of fluid will only flow when the applied shear stress is greater than a certain minimum shear stress – the yield stress. The simplest viscoplastic model is the two-term Bingham plastic model:

$$\tau = \tau_0 + \eta_p \quad \text{for} \quad \tau > \tau_0 \tag{9.7}$$

where τ_0 is the yield stress and η_p is the plastic viscosity. Plastic viscosity is defined as the slope of the shear stress–shear rate curve; that is, $\eta_p = d\tau/d\dot{\gamma}$. In the literature, τ_0 and η_p are commonly referred to as YP (yield point) and PV (plastic viscosity), respectively (Caenn *et al.*, 2011; FPS, 2006). For fluids that are known to exhibit Bingham plastic behaviour, the values of these parameters can be calculated directly from the results of a direct-indicating viscometer (see Section 6.5.3).

The three-term Herschel–Bulkley model may be used to simulate fluids with a non-linear τ–$\dot{\gamma}$ flow curve post-yield; that is, showing yield-pseudoplastic behaviour. Mathematically, the Herschel–Bulkley model is a combination of the Bingham and power law models:

$$\tau = \tau_0 + K\dot{\gamma}^n \quad \text{for} \quad \tau > \tau_0 \tag{9.8}$$

Schematic shear stress–shear rate plots for the Bingham and Herschel–Bulkley models are shown in Figure 9.1.

Both the two-parameter Bingham plastic and the three-parameter Herschel–Bulkley models have been used to represent the flow behaviour of bentonite fluids. The former is more commonly used due to its simplicity, although the latter can represent the flow data more accurately, given the extra parameter should the greater accuracy of fitting be required. By way of example, the viscosity test results for the bentonite fluid used for the Stratford field trial was fitted with the Herschel–Bulkley model, and is shown in Figure 8.16. It is important to recognise that all the models discussed so far can represent the τ–$\dot{\gamma}$ relationship only for steady state flow conditions. For fluids that display time-dependent behaviour such as thixotropy (as bentonite fluids do), the yield stress obtained by extrapolating Equations 9.7 and 9.8 to zero shear rate may be smaller than the shear stress required to mobilise the fluid after it is has been left undisturbed for a period of time. This latter shear stress is often known as the gel strength, and can be determined with the direct-indicating viscometer or a shearometer, although the latter is no longer in common use in civil engineering work (FPS, 2006) and can give results that are difficult to reproduce.

9.3. Laboratory techniques

9.3.1 Test instruments

The results presented in this chapter were obtained from laboratory tests using a number of instruments: the Ubbelohde capillary viscometer, the Marsh funnel, the direct-indicating (Fann) viscometer, the Brookfield viscometer and an advanced rheometer. With the exception of the advanced rheometer, descriptions of the instruments have been given in previous chapters and so are not repeated here. The advanced rheometer was used to carry out more sophisticated testing beyond the capabilities of the other instruments. A rheometer, as the name suggests, is an instrument that can be used to assess the rheological properties of a material. Depending on the particular model, the properties that can be measured by a rheometer commonly include shear stress–shear rate flow curves under steady state shear, viscoelastic properties during oscillatory shear, yield stress, creep and relaxation behaviour.

Figure 9.2 An advanced rheometer: the Bohlin Gemini

The rheometer used by the authors was a Bohlin Gemini rheometer as shown in Figure 9.2. Unlike the Fann and Brookfield viscometers, which can only measure the shear stress at fixed shear rates, the Bohlin rheometer can independently control both the shear stress and shear rate. Therefore, this rheometer can be used to perform a wide range of rheological investigations, including those listed above. The rheometer is fitted with a Peltier plate, which allows accurate temperature control from −30 to 200°C. For viscometry experiments, the cone-and-plate geometry is commonly used, as the shear rate is constant over the contact surface. The cone used had a diameter of 40 mm and an angle of 4°. For oscillatory experiments, a parallel-plate geometry is frequently used, as the gap between the plates (i.e. the thickness of the sample) can be varied to suit the needs of the test. Figure 9.3 shows schematic drawings of the cone-and-plate and parallel-plate geometries. Figure 9.4 shows a polymer fluid being tested in the Bohlin rheometer. Due to the use of an air bearing, the rheometer has a very wide torque range from 0.05 μN m to 200 mN m: this feature allows the viscosity of a fluid to be determined at very low shear rates, when the measured torque can be low.

Figure 9.3 Common measurement geometries for a rheometer

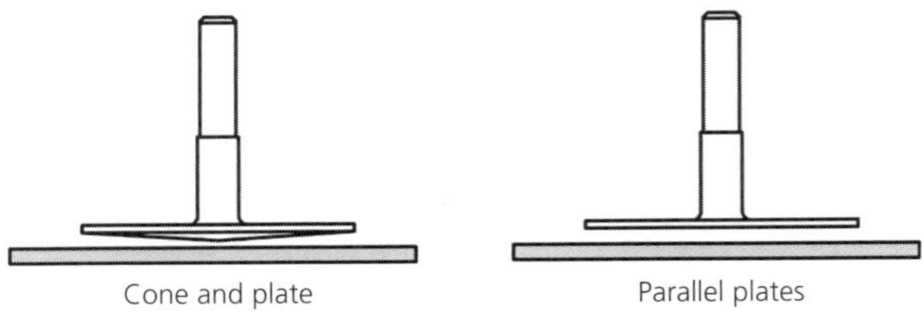

Figure 9.4 A fluid sample being tested in a rheometer

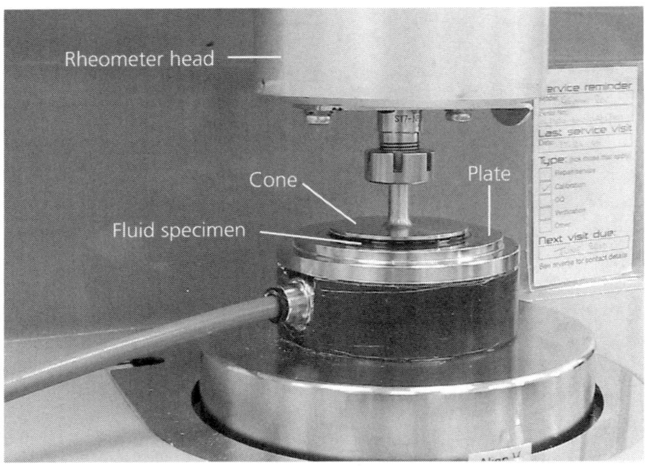

9.3.2 Preparation of polymer fluids

Special attention has to be given to the preparation of polymer fluid samples, as the quality of the mix water and the mixing method employed can have a significant effect on the results. Since the quality of potable water may vary considerably between different cities, and polymer fluids are sensitive to water quality, deionised water was used for all experiments discussed in this chapter to ensure reproducibility, unless stated otherwise. However, it is recognised that the mix water available at an actual job site will inevitably contain some dissolved salts, which could adversely affect the rheological properties of the fluids. The fluid properties presented in this chapter can therefore be regarded as the best possible for a given polymer under the stated test conditions.

To prepare polymer fluids for use on site, the aeration mixing method is commonly used. This involves sprinkling polymer powder into water filling into a tank and then bubbling air through the water in the tank. The method is easily scalable to mix large volumes of fluid. However, as will be shown later, entrained air bubbles that are slow to rise from the fluid can have a significant effect on the fluid properties. To eliminate this effect for laboratory testing, a mixer equipped with a low-shear axial-flow impeller was used for

Figure 9.5 A low-shear impeller for mixing polymer fluids in a laboratory

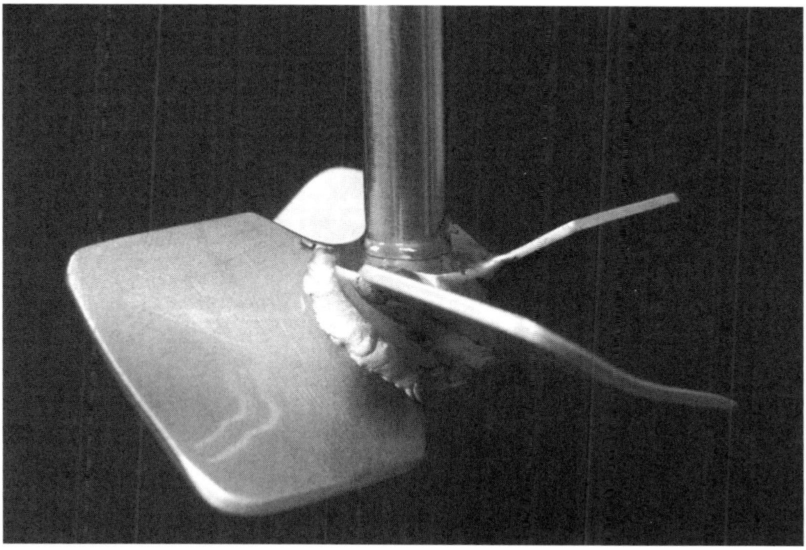

polymer fluid preparation. Figure 9.5 shows the impeller. It should be noted that synthetic polymers such as PHPAs are sensitive to shear degradation, so high-shear radial-flow impellers, which often have sharp edges to the blades, should not be used. Polymer granules have a tendency to stick together when wetted, so during the first stage of polymer preparation the mixer was set to rotate at about 500–600 rev/min, to promote wetting-out. Afterwards, the speed was turned down to about 200 rev/min for the dissolution stage. Low-speed stirring was typically continued for 40–50 min, although energy consumption readings on the mixer showed that a mixing time of around 30 min was probably sufficient, as there was no change in power consumption after this time. The longer mixing time was used to ensure complete dissolution of the granules and thus better consistency between test samples. These mixer settings were sufficient for mixing up to 5 litres of batches of fluid in a small bucket or a large beaker. It must be emphasised that high-speed mixing can cause significant damage to the polymer chains and thus reduce polymer performance and fluid viscosity (Nagashiro and Tsunoda, 1977; Nakano and Minoura, 1978). Therefore, as set out above, the mixing speed should be kept as low as possible while still imparting sufficient fluid movement to promote polymer dissolution and prevent particle clumping. As a precaution, all the fluids were used within 3 days from mixing, to eliminate possible effects of longer ageing, although noting that, on site, polymer fluid is often repeatedly used and then re-boosted with further polymer over many months without any fluid disposal.

9.4. Flow behaviour

9.4.1 Selection of rheological models – current practice

Before the viscosity of polymer fluids is discussed in detail, it is useful to reprise the way this parameter is currently measured and interpreted in civil engineering. As noted in previous chapters, viscosity is commonly assessed in the field using either the Marsh funnel or the direct-indicating viscometer. A review of the literature shows that the Marsh funnel has been used for the majority of field studies (e.g. Camp *et al.*, 2002; Ho and Tan, 1996; Holmes, 1993; Littlechild and Plumbridge, 1998; and many more). As these studies were field based, it is not surprising that the Marsh funnel was the preferred test instrument. The direct-indicating viscometer has been used in many investigations that were laboratory based or had a laboratory component (e.g. Ata and O'Neill, 1997; Howsam and Hollamby, 1990; Kheng *et al.*, 1991; Majano and O'Neill, 1993; and many more).

Among those studies that included measurements using the Fann viscometer, treatment of the test data was found to vary considerably. Some authors (e.g. Beresford *et al.*, 1989; Howsam and Hollamby, 1990) adopted the two-parameter power law model (Equation 9.3) as the most suitable for polymer fluids. In contrast, another research group took a broader view, and interpreted viscometer test data using both power law and Bingham plastic models (Ata and O'Neill, 1997; Majano and O'Neill, 1993). However, O'Neill's group preferred the Bingham plastic model in other research, as can be seen in Majano *et al.* (1994) and Ata and O'Neill (2000). Use of the Bingham model for data analysis seems to have been adopted for the last two versions of the Federal Highway Administration's drilled shaft manual (Brown *et al.*, 2010; O'Neill and Reese, 1999). As there is no consideration of the power law model in either edition of the manual, the reader could conclude that the Bingham plastic model is the correct model for the interpretation of viscometer data for polymer fluids (i.e. to derive plastic viscosity and yield stress) even though the example curve given in the manual clearly shows strong shear-thinning behaviour and hence a strongly variable plastic viscosity with a negligible yield stress.

9.4.2 Selection of rheological models – a simple comparison

Those familiar with polymer fluids might assume that a power-law-type model is generally recognised as more appropriate than the Bingham plastic model for such fluids. However, currently there appears to no consensus in the civil engineering literature as to the most appropriate model. To investigate the issues, a set of simple test data has been used to assess the relative applicability of the power law and Bingham plastic models to polymer fluids. The fluid tested was obtained from a construction site in the UK where the polymer SlurryPro CDP was used. The polymer concentration was 0.8 kg/m^3, and the fluid pH was 11. Figure 9.6(a) shows the Fann viscometer test results presented as a shear stress–strain rate (τ–$\dot{\gamma}$) plot. The viscometer readings were taken at 3, 6, 100, 200, 300 and 600 revs/min (multiply rev/min by 1.703 to obtain the corresponding shear rate in s^{-1}), as these are the speed settings on the common six-speed model. In addition to the power law model (Equation 9.3) and the Bingham plastic model (Equation 9.7), the Herschel–Bulkley model (Equation 9.8) was also included in the comparison, as it is effectively a combination of the two aforementioned models. It can be seen from Figure 9.6(a) that all three models are able to represent the test data reasonably well, with the power law and the Herschel–Bulkley curves effectively lying on top of each other, implying a near-zero yield stress. Given that the Bingham plastic line was drawn through the data points at 300 and 600 revs/min without applying a regression

Figure 9.6 Test results from a direct-indicating (Fann) viscometer using three rheological models: (a) fitted shear stress–shear rate curves and (b) fitted viscosity–shear rate curves

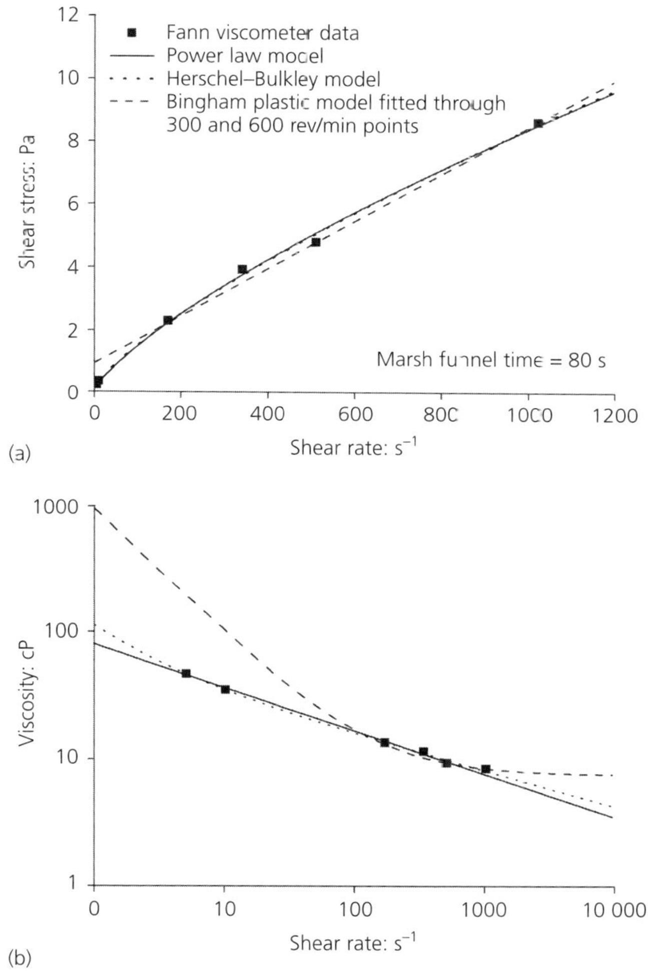

analysis (a practice inherited from the oil industry), this model may be preferred for simple curve fitting. However, if the models are evaluated on a viscosity–shear rate (η–$\dot{\gamma}$) plot, as shown in Figure 9.6(b), their fitting power is very different. It can be seen that the Bingham plastic model significantly overestimates the apparent viscosity of the fluid at low shear rates. For instance, at a shear rate of 5.1 s^{-1}, the measured apparent viscosity is 47 cP, while the Bingham plastic model gives 195 cP. Note that it is the apparent viscosity that is being considered here, not the plastic viscosity. Figure 9.6(b) also shows that the power law model gives a very good fit to the data over this range of shear rates. Although the data can also be fitted to the Herschel–Bulkley model, this is at the expense of using an extra parameter, yield stress. The choice between the

Herschel–Bulkley and the power law models therefore hinges on whether polymer fluids have a yield stress. This is discussed in the following section.

9.4.3 Yield stress

To investigate the possible yield stress of polymer fluids, the Bohlin rheometer (see Figure 9.2) was used to test a high-molecular-weight PHPA mixed at concentrations ranging from 0.2 to 1.0 kg/m^3. Details of this study were reported in Lam and Jefferis (2015a). For the tests, a cone-and-plate measurement geometry (see Figure 9.3) was employed and a series of ramp stress tests were conducted. This test involves applying a gradually increasing shear stress and, at the same time, monitoring the instantaneous viscosity for an inflexion of the instantaneous viscosity–time curve (i.e. the onset of flow). The instantaneous viscosity is the viscosity that is determined from the instantaneous shear stress and shear rate parameters, at which time the fluid may not be in an equilibrium (steady state) flow condition. For each polymer concentration, the shear stress was increased from 0 to 1 Pa over a period of 30, 60 and 120 s. The use of different sweep times effectively altered the shear rate so that the dynamic effect could be taken into consideration. After each test, the yield stress of the fluid was interpreted as the shear stress when the instantaneous viscosity was at its peak. Figure 9.7 shows an example of how the yield stress is determined.

Figure 9.8 shows a summary plot of the yield stress results. It can be seen that the measured yield stresses are very close to zero at the polymer concentration of 0.2 kg/m^3 and then increases approximately linearly with increasing concentration. This can be explained by the fact that a higher concentration means more dissolved polymer molecules in the fluid, and thus greater potential for their interaction. The yield stress values also can be seen to increase with reducing sweep time. This means that dynamic

Figure 9.7 Determination of yield stress from a ramp stress test. (From Lam and Jefferis, 2015a)

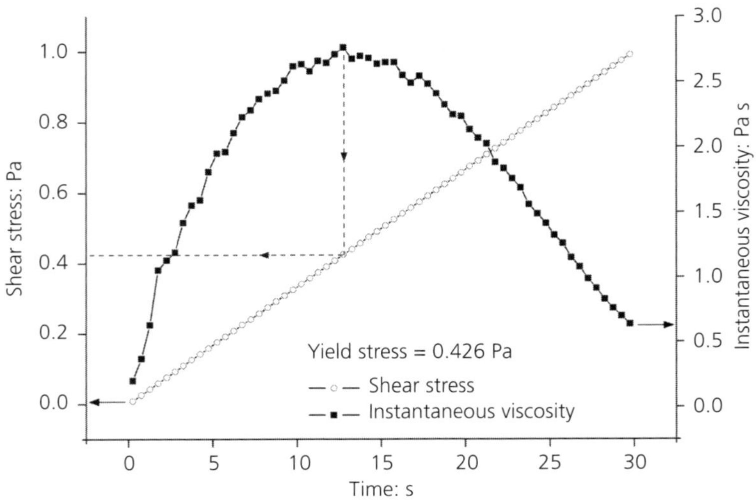

Figure 9.8 Summary of yield stress values for a high-molecular-weight PHPA. (From Lam and Jefferis, 2015a)

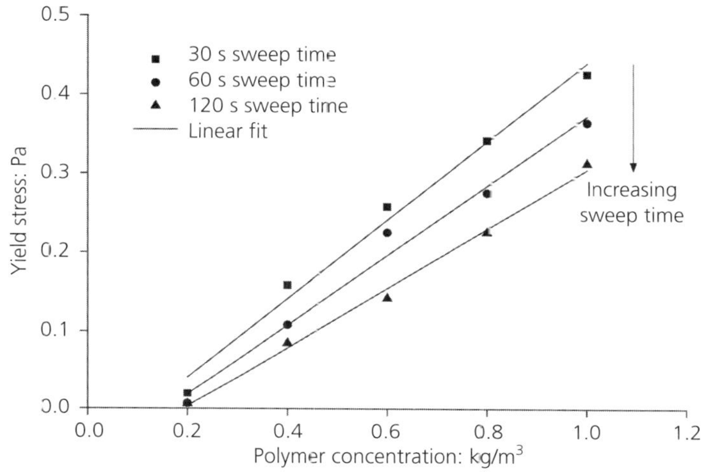

effects play a role in the value of the yield stress measured in this way. If a fluid flows under steady state (equilibrium) conditions, there will be no dynamic effect and thus the yield stress can be expected to be much smaller or even non-existent. Barnes and Walters (1985) present an interesting discussion on whether yield stress actually exists. Nevertheless, since the values of the yield stresses obtained are all very small (<0.5 Pa) even when the dynamic effect is considered, they may be ignored for most practical purposes except perhaps for flow in fine soils or other fine capillary systems.

9.4.4 Effects of shear rate and polymer concentration

To investigate the effects of shear rate and polymer concentration on the viscosity of polymer fluids, further laboratory measurements were carried out, and the results reported by Lam *et al.* (2015). For this investigation, the Bohlin rheometer was used to extend the low shear rate range from the lower limit of the Fann viscometer, from 5.1 s^{-1} down to about 10^{-3} s^{-1}. A Brookfield viscometer with a shear rate range of 0.07–100 s^{-1} was also available, but as its range is well within that of the Bohlin rheometer, the former was only used to confirm the results of the latter. Solutions of a commercial PHPA (Shore Pac) were tested at concentrations ranging from 0.2 to 1.2 kg/m^3. This range was adopted to cover (in fact, slightly exceed) the range recommended by the manufacturer of this polymer. The lower concentrations are recommended for excavating in fine-grained soils, while the higher concentrations are for use with coarse-grained soils. These recommendations are given probably due to the fact that a higher solution viscosity and hence higher polymer concentration is necessary to minimise the bulk fluid loss to coarse/permeable soils.

Figure 9.9 shows the viscosity results plotted against shear rate on a log–log scale, and the Carreau model (Equation 9.6) was used to fit the viscosity data. Since none of the

Figure 9.9 Viscosity curves of a high-molecular-weight PHPA fluid at different concentrations. (From Lam *et al.*, 2015)

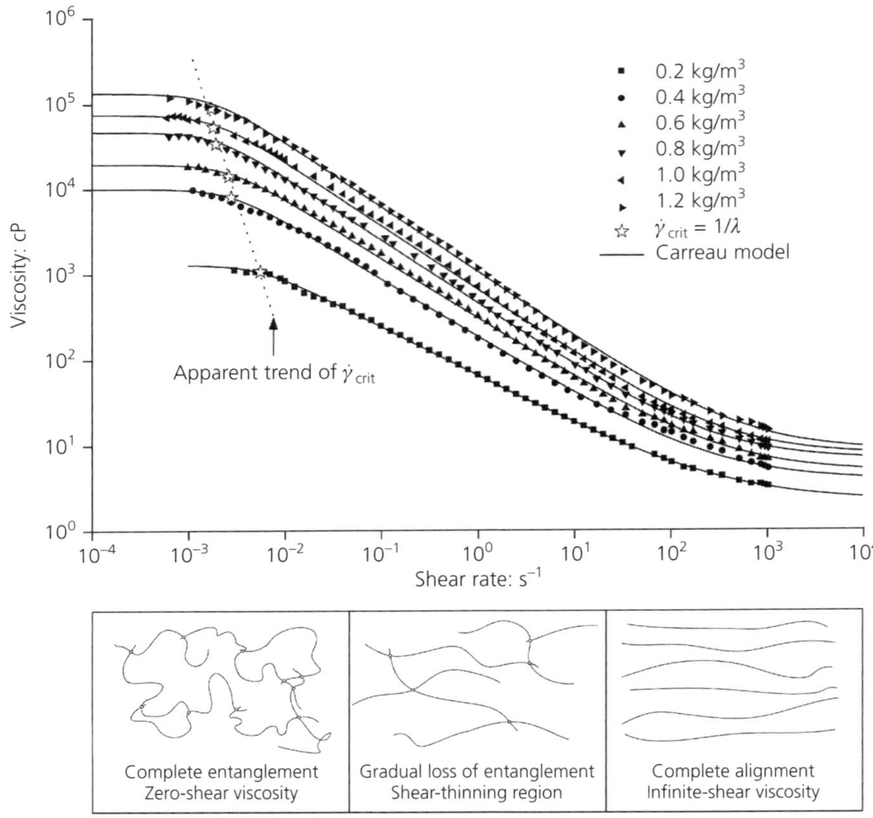

viscosity curves clearly entered the infinite-shear region, the value of the infinite-shear viscosity (η_∞) could not be determined directly from the data, and hence the initial curve-fitting results from a non-linear solver were manually adjusted to give the best overall visual fit. Table 9.1 summarises the Carreau model parameters of the fitted curve for each polymer concentration. The Marsh funnel test results are also given in the table for reference.

These viscosity curves now show a 'bigger picture', and provide considerably more information about the overall flow behaviour of the fluid. Furthermore, it is now obvious that viscosity curves produced by the Fann and Brookfield viscometers (see Figures 6.6 and 6.8) are only parts of the overall flow curves of these fluid solutions. Figure 9.9 shows that the polymer fluid at all concentrations tends towards limiting viscosities at very low (10^{-3} s^{-1}) and at very high (10^3 s^{-1}) shear rates. This behaviour is expected because in the absence of a yield stress the viscosity cannot increase indefinitely at low shear, nor can it go below the viscosity of the solvent (water) at high shear (i.e. 1 cP). It also can

Table 9.1 Summary of Carreau model parameters and Marsh funnel times for PHPA polymer fluids prepared using deionised water

Polymer concentration, c_p: kg/m^3	Infinite-shear viscosity, η_∞: mPa s	Zero-shear viscosity, η_0: mPa s	Flow behaviour index, n	Time constant, λ: s	Critical shear rate, $\dot{\gamma}_{crit}$: s^{-1}	Marsh funnel time: s
0.2	2.2	1 300	0.42	180	0.0056	53
0.4	4	10 100	0.32	360	0.0028	68
0.6	5	19 500	0.3	380	0.0026	80
0.8	7	47 000	0.26	520	0.0019	94
1.0	8	7 000	0.25	550	0.0018	113
1.2	9	135 000	0.24	580	0.0017	125

After Lam *et al.* (2015).

be seen that at 600 rev/min the shear rate of the Fann viscometer (1022 s^{-1}) is high enough that the fluid solutions approach their infinite-shear viscosity.

Before the effect of polymer concentration is discussed, it is useful to explain the processes behind the observed shear-thinning behaviour exhibited by the fluid. At the foot of Figure 9.9 are diagrams showing the schematic arrangement of polymer molecules in three viscosity regions, namely zero-shear viscosity (η_0), shear thinning and infinite-shear viscosity (η_∞). When the polymer fluid is under static or near-static conditions, the long-chain molecules are completely intertwined and entangled as shown in Figure 3.15. Since entanglement resists flow, and viscosity represents the resistance to flow, at this stage the viscosity is at its highest – the Newtonian plateau of η_0. When the fluid is subject to increasing shear, the number of entanglements reduces as the chains orient along the lines of flow. The fluid therefore shows a gradual reduction in viscosity over this middle shear-thinning region. At very high shear rates, all the entanglements disappear, as the flow 'combs' them out – and hence the dependence on molecular weight (which dictates the number of entanglements) also disappears. At this stage, the viscosity is only dependent on the polymer concentration, and any further increase in the shear rate will not have any further effect; the viscosity is therefore at its lowest on the Newtonian plateau of η_∞. It should be noted that the above discussion assumes that the polymer has an intertwined three-dimensional molecular structure under (near) static conditions. This will not be the case if the pH is low and (or) the ionic strength of the solvent is high, because the electrostatic repulsion between the molecules, and hence the network structure, would be weakened or completely lost. For this reason, if the PHPA polymers are dissolved in a solvent with less favourable pH and ionic conditions, the viscosity will be lower than that shown in Figure 9.9.

As regards the effect of polymer concentration, Figure 9.9 shows that when the polymer concentration is increased, the viscosity also increases. This is as expected, as a higher concentration means more polymer molecules in solution, and thus a higher degree of molecular interaction. The critical shear rate ($\dot{\gamma}_{crit}$), which is the shear rate where the viscosity leaves the Newtonian plateau of η_0, tends to decrease as the polymer concentration increases. This means that the fluid becomes more sensitive to shear as the concentration increases, probably due to the increased number of chain entanglements that can be broken by shear. The gradient of the middle regions of the viscosity curves also can be seen to become slightly steeper with increasing polymer concentration. This implies that the fluid becomes more shear thinning again, probably due to the greater potential for entanglements to be 'combed out' by the flow. Numerically, the increase in the shear-thinning behaviour is shown by the decreasing value of the flow behaviour index, n, given in Table 9.1. Similar behaviour for several other polymers was reported by Howsam and Hollamby (1990) using the more basic direct-indicating (Fann) viscometer.

To further examine the effect of polymer concentration on viscosity, Figure 9.10 plots both η_0 and η_∞ against concentration on a log–log scale. It can be seen that both parameters show a power law relationship with concentration, and that they form the lower and upper viscosity boundaries on the plot. It is interesting to see that the slope of the η_0 curve is much steeper than that of the η_∞ curve. This means that a change in polymer concentration has a much greater effect on the low-shear viscosity. This phenomenon can again be explained using the chain entanglement theory, as follows. When the fluid is undisturbed, a higher concentration means that the polymer chains have more opportunities to interact with their neighbours and become entangled. However, at high shear rates these polymer chains are 'combed out' by the flow, so that the effect of extra entanglements becomes negligible. Of course, the simple presence of the extra dissolved material (i.e. the polymer molecules) also has an effect on the viscosity, and this is why there is still a small increase in η_∞ as the concentration increases.

Figure 9.10 The concentration dependence of infinite- and zero-shear viscosities. (From Lam *et al.*, 2015)

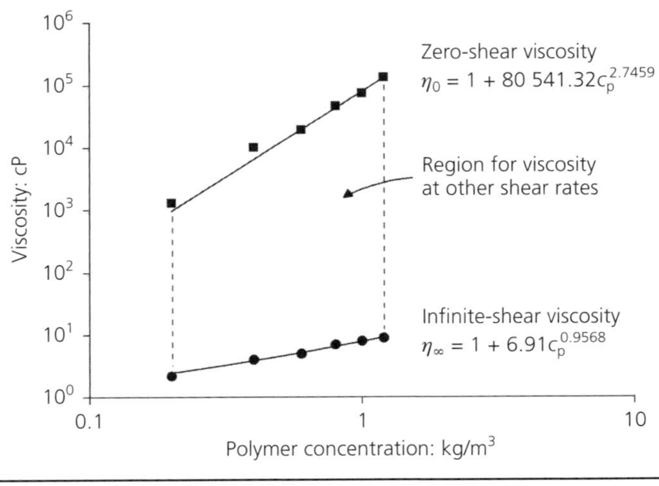

The two curves in Figure 9.10 also show another interesting feature, which is that their slopes remain almost the constant over the entire concentration range from 0.2 to 1.2 kg/m^3. This is worth noting because, in theory, once a critical polymer concentration (c_p^*) has been reached, the slope should change dramatically (e.g. Barnes, 2000). In the molecular network theory, c_p^* is commonly defined as the point at which chain entanglements take place. Williams and Philips (2003) mention that c_p^* is usually the concentration level where the behaviour of polymer fluids changes from Newtonian to non-Newtonian. As the polymer fluid in Figure 9.9 displayed non-Newtonian flow behaviour even at the lowest concentration tested (0.2 kg/m^3), the value of c_p^* for this polymer would appear to be less than 0.2 kg/m^3. The practical implication of this observation is that, over the typical concentration range used for geotechnical construction, fluid viscosity can be predicted using simple power law relationships such as those shown in Figure 9.10 – note these best-fit curves all have their offset fixed at 1 cP, which is the viscosity of water. For example, if a user requires the polymer fluid to have a zero-shear viscosity of 10^5 cP to stabilise an excavation, then from Figure 9.10 it can be seen that the required polymer concentration is about 1 kg/m^3, provided that the same polymer is used and that the mix water used is of similarly good quality. As site water typically contains some dissolved ions, the polymer concentration obtained from this plot should therefore be taken as a minimum required value. If in doubt, it is always good practice to start with a higher polymer concentration, as there are many site factors that could adversely affect the performance of polymer fluids. Factors such as the effects of ionic compounds, ageing, entrained air bubbles and fluid reuse are discussed in the following sections.

9.4.5 Effect of ionic compounds

The viscosity and hence the performance of polymer fluids can be severely degraded by salts in the mix water or the groundwater (e.g. from hard water). This effect is well recognised in the oil industry (Nasr-El-Din and Taylor, 1996). To illustrate the effect of salts in mix water on the performance of commercial polymer products used in geotechnical construction, the viscosity of two selected products was measured over a range of sodium chloride (NaCl) concentrations using an Ubbelohde capillary viscometer (other viscometers could have been used). One of the products was a pure PHPA, while the other was a blended polymer containing PHPA and other chemicals. Deionised water was used to prepare the polymer solutions, and laboratory-grade NaCl was added to achieve a range of salt concentrations. Figure 9.11 plots the viscosity test results as a ratio to the initial viscosities with deionised water. It can be seen that, at above about 100 mg/l of sodium chloride, the PHPA fluid lost about 60% of its viscosity whereas the blended polymer lost only about 40%. However, for both fluids there was little further effect up to 1000 mg/l. This behaviour can be explained by the molecular chain theory together with the schematic diagram given at the foot of Figure 9.11. At zero or low salt concentrations, the negatively charged PHPA molecules strongly repulse one another in the solution. This increases their hydrodynamic volumes, and so results in a large number of chain entanglements. Therefore, the resistance to flow (i.e. the fluid viscosity) is high. As the ionic strength increases, the dissolved cations (Na^+ in this case) shield some of the negative charges of the polymers. This reduces the degree of electrostatic repulsion, thus causing the polymer chains to be able to curl slightly. The viscosity of the fluid therefore decreases. At high salt concentrations (>100 mg/l), further shielding by the dissolved cations has

Figure 9.11 Effect of NaCl on the viscosity of two commercial polymer solutions. (Adapted from Jefferis and Lam, 2013)

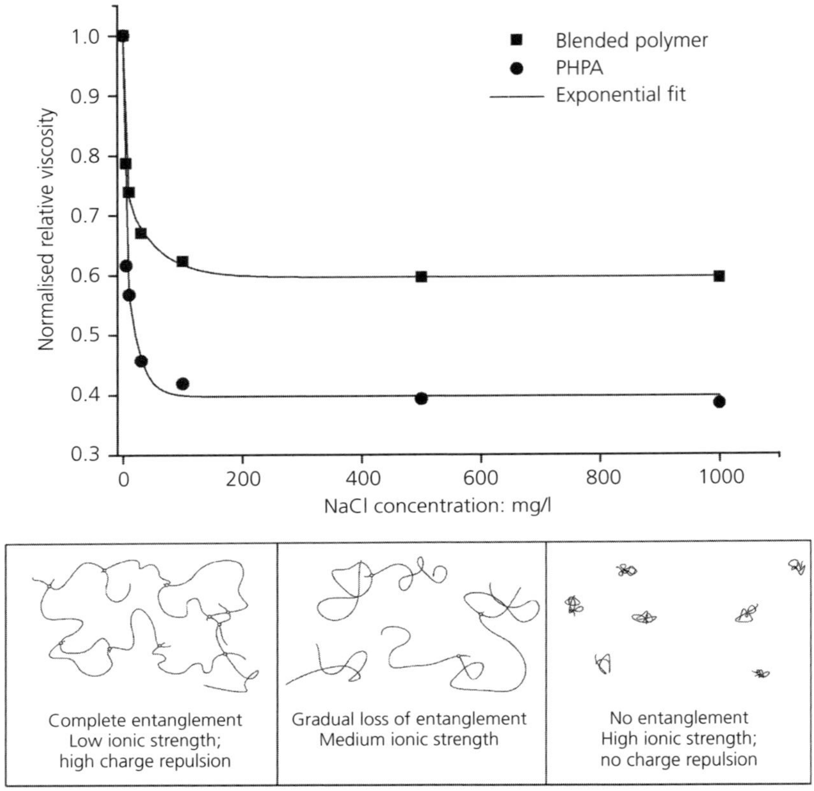

minimal effect, as the polymer chains are curled up. This eliminates the electroviscous effect, so the viscosity is at its lowest. Further addition of salt to the system can have no further effect. Similar test results are given by Nasr-El-Din and Taylor (1996).

Figure 9.11 shows that the blended polymer gives a lower reduction in viscosity. This is probably because some of the polymers in the blend are non-ionic so they are not affected by the presence of added salts. Non-hydrolysed polyacrylamide, hydroxyethyl cellulose and guar gum are all examples of non-ionic water-soluble polymers (see Chapter 3). Blended polymers may therefore be more suitable for use in saline environments. The effects of salts in mix water are recognised by PHPA manufacturers and are compensated by increasing the polymer concentration and raising the solution pH with caustic alkalis, although an increase in pH may give limited benefit. In saline soils there should be regular monitoring of fluid viscosity to check for viscosity loss, as there are case histories of collapses. For example, according to Bustamante *et al.* (1998), during the construction of the Vasco da Gama Bridge, two of the piles had to be redrilled following collapses that were possibly due to fluid contamination (see Section 7.2.2).

Schwarz and Lange (2004) also report a case history of pile bore collapse due to high concentrations of salts at a site in Benin (see Section 7.3.1). Although simple PHPAs can be adversely affected by salts, engineered polymer systems are available that are more tolerant of ionic species in soils, and these should be used in saline grounds (Jefferis and Lam, 2013).

9.4.6 Effect of ageing

In the previous section it has been shown that the viscosity of a polymer fluid can be affected by the ions in the solution. This section discusses viscosity change due to ageing – as distinct from loss of viscosity due to biodegradation. A survey of the relevant literature reveals a wide range of results concerning the viscosity development and ageing characteristics of PHPA fluids. These results may seem contradictory at first sight, but many of them can be explained by the properties of the materials used and/or the mixing methods employed. For instance, both Gardner *et al.* (1978) and Likos *et al.* (2004) state that it took a minimum of 5 h after mixing for the reported polymers to fully develop their viscosity. However, upon examination of their test methods it becomes clear that the results were due to insufficient mixing. For instance, in the case of Likos *et al.* (2004) the polymer fluid was mixed with a large spoon for a mere 5 min, as opposed to the typical over 1 h by aeration on construction sites. Fluids for laboratory work prepared by the authors are typically mixed for 40–50 min (see Section 9.3.2).

In terms of ageing characteristics, Mächtle (1982) saw no ageing effect in their non-ionic polyacrylamide solution, but Henderson and Wheatley (1987) found a gradual reduction in viscosity with time in their 30% hydrolysed PHPA solution. Kulicke and Kniewske (1981) also found that, for the same degree of polymerisation (molecular weight), the higher the degree of hydrolysis, the greater the loss of viscosity with time. The effect of ionic charge on the ageing behaviour of polyacrylamide-based polymers can be explained by the molecular network theory, as follows. For polymers with a high degree of hydrolysis creating charged sites along the polymer chain, the polymer chains are stiffened and entangled with one another (see Figure 3.15), so the fluid has a high viscosity. Given time, some of the entanglement would be lost, as the polymer chains try to return to the more stable flexible state, thus leading to a reduction in viscosity. For polymers that are only slightly hydrolysed or are completely non-ionic, the polymer chains would start as flexible coils with limited interaction with one another. The opportunity for a reduction in chain entanglement is therefore limited, and so is the potential for ageing and viscosity reduction. However, there are also reports describing ageing characteristics that are not readily explicable by material properties or the mixing method. For example, Shyluk and Stow (1969) reported an initial rapid decrease in viscosity followed by a much slower gradual reduction in their polyacrylamide solutions. Bruce and Schwarz (1969) also reported that the viscosity of fresh PHPA solutions decreased by as much as 30% per day. These unusual findings were attributed to chemical reactions (absorption of carbon dioxide) and chain disentanglement.

To obtain further information on the ageing characteristics of PHPA fluids, tests were conducted using the same high-molecular-weight PHPA as was previously tested for yield stress and viscosity (see Sections 9.4.3 and 9.4.4). Since both Henderson and

Wheatley (1987) and Kulicke and Kniewske (1981) found less ageing in polymer fluids prepared with impure water (i.e. not deionised), it was decided to compare the ageing characteristics of fluids mixed with pure deionised water and a hard water. Oxford tap water, which has a moderately high concentration of calcium and magnesium ions equivalent to a total hardness of about 250–280 mg/l (ppm) of calcium carbonate, was chosen as an example of hard water commonly found across south-east England, and indeed many other parts of the world. In accordance with common site practice, soda ash (sodium carbonate) was used to pre-treat the water; a dosage of 0.9 kg/m^3 was found to be sufficient to raise the pH to 10 and to remove about 90% of the water hardness. The remaining 10% was due to magnesium ions that are not precipitated by sodium carbonate. After the water treatment, the granular polymer was added to the mix water at a concentration of 0.8 kg/m^3. Immediately after 40 min of mixing by a laboratory mixer, the solutions were transferred to a 600 ml beaker and then tested using the Brookfield viscometer as described in Section 6.5.4. The Brookfield viscometer is ideally suited to this test since the sample can be sealed by covering the top of the beaker with several layers of plastic film. This allowed continuous measurements of the viscosity over a long period without any significant moisture loss. A moderately high rotational speed of 20 rev/min, which is equivalent to 4.4 s^{-1} with the spindle LV1, was used to keep the measured torque at a reasonable level for both the deionised water and the tap water specimens. Finally, to prevent photo-degradation of the polymer (Caulfield *et al.*, 2002) and the effect of temperature variation, the specimens were wrapped in a layer of aluminium foil and the laboratory room temperature kept at 23°C.

Figure 9.12 shows the ageing test results over a 30-day period. First, it can be seen that the water quality had a significant effect on the viscosity of the fluid. The starting viscosity of the sample mixed with deionised water (190 cP) was four times as high as that of the sample mixed with Oxford tap water (48 cP). This indicates that, despite the soda ash treatment, Oxford tap water was still not a good solvent for the PHPA polymer

Figure 9.12 Ageing behaviour of a high-molecular-weight PHPA. (Adapted from Lam and Jefferis, 2016)

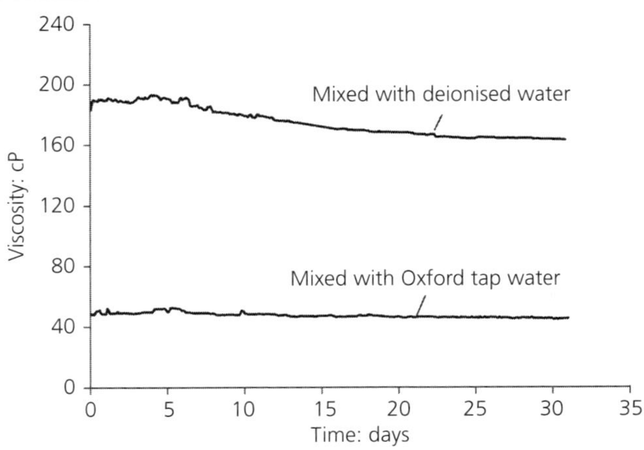

– probably due to the magnesium ions that remained and/or the extra sodium ions that were introduced as a result of the water treatment with soda ash. As discussed above, the cations (positive ions) in solution shield the negative charges of the polymer molecules, thus reducing their electrostatic repulsion and so allowing curling of the polymer chains and hence leading to a decrease in viscosity. In terms of ageing characteristics, it can be seen that none of the samples showed any initial rapid decrease in viscosity as reported by some of the past researchers (Bruce and Schwarz, 1969; Shyluk and Stow, 1969). However, as expected, the specimen mixed with deionised water did show a larger reduction in viscosity than the one mixed with Oxford tap water. This confirmed the findings of Kulicke and Kniewske (1981) that PHPA fluids mixed with pure water age more than fluids mixed with water containing dissolved salts (NaCl). This is because, as explained earlier, polymer chains were fully stretched in the deionised water, and thus there is more potential for them to return to their natural random configuration, whereas the polymer coils in the Oxford tap water were already more compact, and hence fewer conformational changes could take place. Kulicke and Kniewske (1981) put forward the theory that conformational changes of a polymer molecule occur because certain types of intramolecular hydrogen bonds are lost, thus leading to a more stable and compact coil over time. Lastly, it can be seen from Figure 9.12 that none of the samples showed any noticeable ageing during the first 5–6 days. This supports the thesis that the overnight viscosity reduction seen at the Glasgow field trial (see Section 8.2.3) was not caused by ageing but was probably due to the escape of entrained air bubbles. The effect of air bubbles on viscosity is the subject of the following section.

Many natural polymers such as guar and xanthan gums are readily biodegradable, and biocides may be necessary to prevent their rapid degradation in use. Synthetic polymers such as polyacrylamides are more resistant to biodegradation. Further details of the ageing tests discussed in this section are given by Lam and Jefferis (2016).

9.4.7 Effect of entrained air bubbles

In the discussion of the Glasgow case history (see Section 8 2.3), it was noted that the viscosity of freshly prepared polymer fluid was found to reduce overnight in the mixing tank without any obvious reasons. However, a simple laboratory experiment conducted after the field work helped to explain this unexpected fluid behaviour. In the experiment, a high-molecular-weight PHPA was mixed at a concentration of 1.0 kg/m^3 in a 20-litre open-top container by aeration. Other types of polymer or concentrations could also have been used. After mixing, the fluid was simply left standing in the container, and the Marsh funnel time monitored over a 24 h period so as to assess the effect of entrained air bubbles on the viscosity of polymer solutions. Due to the relatively large amount of water (20 litres) required for this test, it was not practical to use deionised or distilled water, so tap water from Manchester was used. Tap water from the Greater Manchester area commonly has a total hardness level of around 10 mg/l of calcium carbonate, so it is classified as 'very soft'. Nevertheless, the water was still pre-treated with soda ash to pH 10 before use in the tests. Since the purpose of this test was to study the effect of air bubbles caused by aeration mixing in the field (e.g. see Figures 5.4 and 5.6), instead of using a laboratory mixer the fluid was mixed in a small open-top container using compressed air. The compressed air was delivered to the fluid via several perforated

Figure 9.13 Reduction in the Marsh funnel time over time

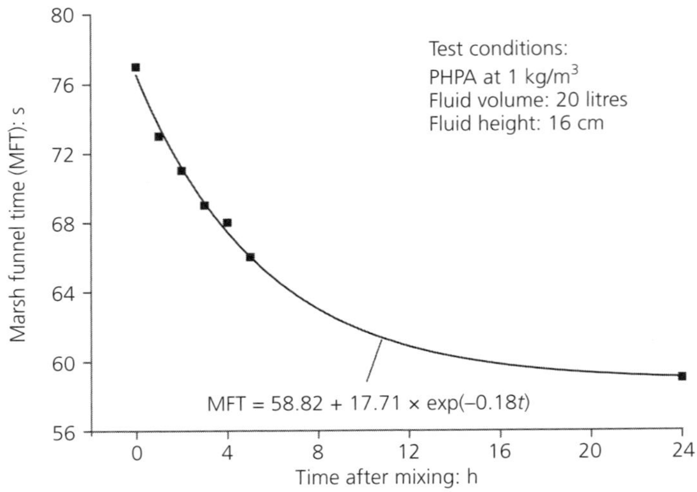

pipes held at the bottom of the container. With the compressed air supply turned on, polymer granules were sprinkled into the water very slowly to achieve a medium dosage of 1 kg/m^3. The addition of polymers should not be hurried as polymer granules have a tendency to clump together immediately after wetting. The aeration mixing process was allowed to run for 45 min before the air supply was turned off.

Figure 9.13 shows the monitoring results. It can be seen that the viscosity dropped from the initial 77 to 59 s in 24 h, which is equivalent to a reduction of 23%. Inspection of the samples showed that the fluid contained numerous air bubbles immediately after mixing, but that the number of bubbles reduced over time to a negligible amount after 24 h. As the effect of ageing only becomes dominant over a much longer period (see Figure 9.12), it can be concluded that the reduction in the Marsh funnel time was caused by the gradual escape of entrained air bubbles in the fluid. This is because the presence of air bubbles can distort the flow path of fluid in the funnel and increase the resistance to flow. This results in an apparent increase in the fluid viscosity. Since the depth of the fluid was only 16 cm in this experiment but is typically of the order of metres in a storage tank on site, the rates of bubble escape and thus viscosity reduction should be lower in the field. Nevertheless, this simple experiment explains why the viscosity of polymer fluid dropped overnight on the Glasgow site under no external influence (see Figure 8.4). The key conclusion from this experiment is that the initial Marsh funnel time of a freshly prepared polymer fluid, if mixed using aeration, can be temporarily increased by the presence of entrained air bubbles in the fluid. Site engineers and operatives should be aware of this possible effect when evaluating the viscosity of fluid in the field.

9.4.8 Effect of repeated reuse

So far, only the properties of clean polymer fluid have been discussed; however, in practice, the fluid in a pile bore or a diaphragm wall trench will contain some suspended soil particles due to the mixing action of the digging tools. Foundation contractors always try to reuse support fluid on site in order to minimise the fluid volume to be handled during construction and their disposal after construction. As a result, the concentration of soil particles in the support fluid tends to increase over time. To gain a better understanding of the rheological properties of polymer support fluids in use, a field study was carried out by the authors during the course of a bored piling project in central London in 2008, and the results reported (Lam and Jefferis, 2015b). This section presents the background of the project and a summary of the monitoring results.

The site was located in the London borough of Southwark where the ground conditions consisted of made ground overlying River Terrace Gravel (sandy gravel), London Clay (stiff to very stiff clay), the Lambeth Group (very stiff clay) and Thanet Sand (very fine dense sand). Thirty-seven bored piles with lengths ranging from 40 to 42 m and diameters ranging from 0.9 to 1.5 m were required for a redevelopment project at the site. The piles were drilled mainly through the London Clay and the Lambeth Group. The total drilled volume was about 2000 m^3. Due to the scale of the project, seven open-top tanks with a combined capacity of 300 m^3 were required to manage the mixing and maintenance of the fluid. The available tank capacity allowed a maximum of two piles to be excavated simultaneously with sufficient reserve polymer for the exchange of any contaminated fluid before concreting, should this be required. Figure 9.14 shows a schematic diagram of the tank layout. As can be seen, the tanks were assigned specific roles: two for mixing and storage, one for primary sedimentation, three for secondary sedimentation and one for fluid reconditioning (polymer replenishment).

The polymer used for this piling project was SlurryPro CDP, manufactured by KB International. The polymer was mixed at a concentration of 0.8 ± 0.1 kg/m^3 in the mixing tank by aeration. Polymer materials were also added to the recycled fluid in the reconditioning fluid tank at a lower concentration of around 0.5 kg/m^3, although the actual dosage was controlled by the condition of the recycled fluid. Throughout the monitoring period (52 days), fluid samples were collected from five different locations in the fluid system, namely the clean fluid tank, the primary sedimentation tank, the secondary sedimentation tank, the reconditioned fluid tank and pile bores. The samples were tested for their density, Marsh funnel time and apparent viscosity over the shear rate range of 5–1022 s^{-1}. A direct-indicating viscometer (as shown in Figure 6.5) was used for the viscosity measurements, and the results interpreted using the two-parameter power-law model.

Figure 9.15 shows the average Marsh funnel time for each of the sampling locations on a bar chart. It can be seen that the average Marsh funnel time was highest when the fluid was clean. The Marsh funnel time then dropped gradually as the fluid went through various stages of the process (indicated by the arrows between the bars) before returning to near the original value after reconditioning by adding further polymer. The reduction in viscosity is believed to have been caused by several factors, including (*a*) for the fresh fluid the gradual escape of entrained air bubbles introduced by air lines used for mixing; (*b*) shear degradation of the polymer molecules by the pumping equipment; (*c*) polymer sorption onto soil particles reducing the effective concentration; (*d*) dilution by soil pore water released by soil dispersion; (*e*) slight dilution by rainwater; and (*f*) sedimentation

Figure 9.14 Schematic diagram of the tank system used for a bored piling project in central London. (From Lam and Jefferis, 2015b)

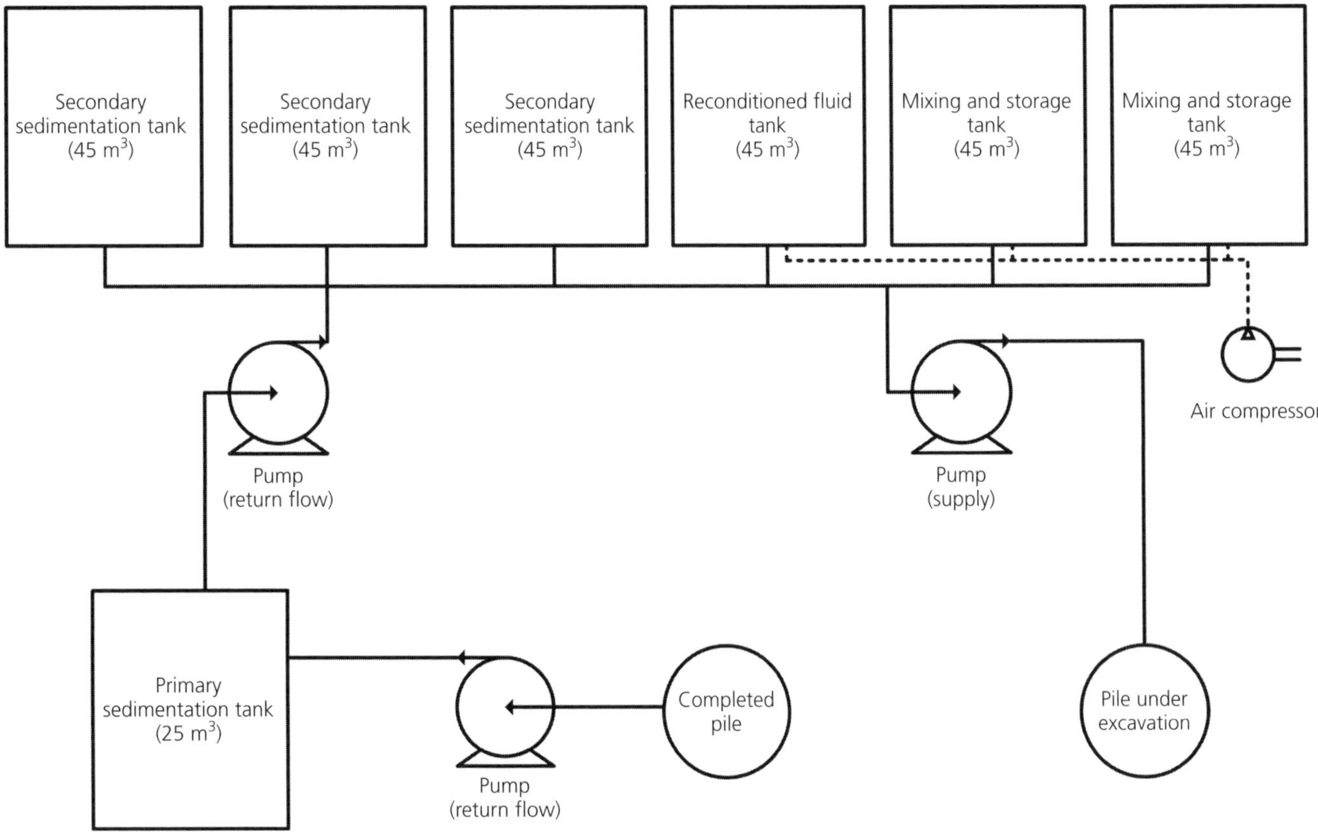

Figure 9.15 The average Marsh funnel time of polymer fluid for each sampling location. (Adapted from Lam and Jefferis, 2015b)

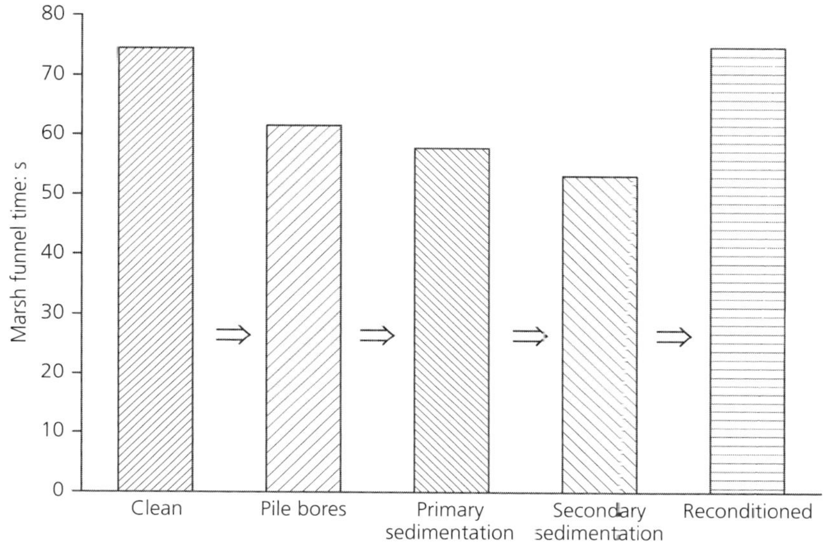

of the coarser soil particles – although these particles do not contribute to the polymer function, they do slightly increase the polymer viscosity. The initial reduction in the Marsh funnel time from the clean fluid tanks to the pile bores was probably due to effects (*a*) to (*d*), whereas the reduction occurring from the pile bores, via the primary sedimentation tank, to the secondary sedimentation tanks was probably due to effects (*b*), (*e*) and (*f*). Note that the tanks were not covered and the work was conducted in the winter, which is a wet season in the UK. An attempt was also made to correlate the Marsh funnel time with fluid density, but no clear trend could be found. This suggests that although the Marsh funnel can be used to monitor the evolution of viscosity during fluid use, digging, sedimentation and polymer replenishment, it is not suitable for the development of a detailed understanding of fluid behaviour.

Further data analyses revealed that the two power law parameters, n and K, show a strong correlation with the density of the used fluid. The correlation results are shown in Figures 9.16 and 9.17. These figures show that the effect of suspended particles (increased density) is to reduce the value of n and increase the value of K. This means that the fluid, although becoming more viscous, was also becoming more shear thinning over the course of the project. To illustrate this concept more clearly, Figure 9.18 shows the viscosity curves of selected fluid samples taken from the pile bores. It can be seen that the overall effect of the suspended particles was to raise the viscosity to up to five times the original value. Since suspended particles had a greater effect on the viscosity at low shear, it made the fluid appear more shear thinning. The reason why the particles had a larger effect at low shear was probably due to them making it more difficult for the polymer chains to disengage from each other, thus creating a synergistic effect on the

Figure 9.16 The flow behaviour index plotted against fluid density. (Adapted from Lam and Jefferis, 2015b)

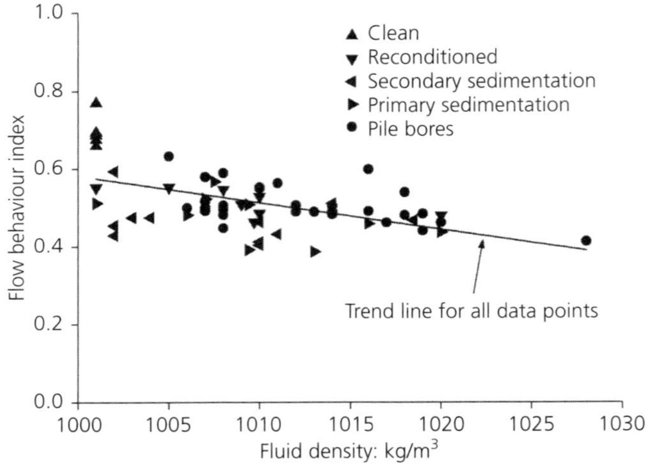

Figure 9.17 The flow consistency index plotted against fluid density. (Adapted from Lam and Jefferis, 2015b)

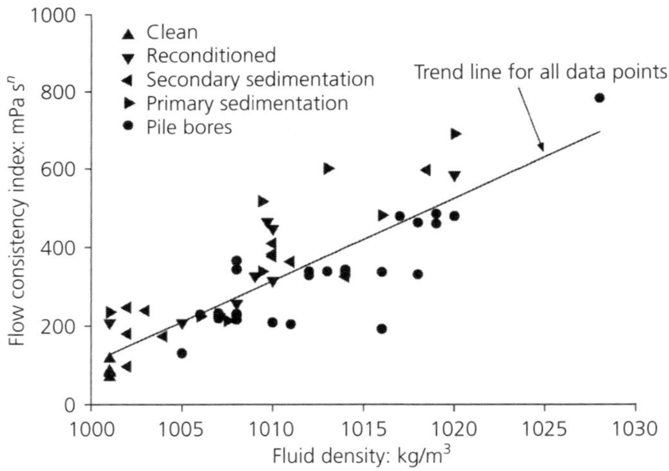

viscosity. At high shear, when the effect of chain entanglement became non-existent, this extra effect therefore also disappeared.

9.5. Transient viscoelastic properties

9.5.1 Background

The shear-dependent fluid behaviour considered thus far is actually only a subset of the types of non-Newtonian behaviour observed in polymer fluids. In the previous

Figure 9.18 Selected viscosity curves of polymer fluid samples taken from pile bores

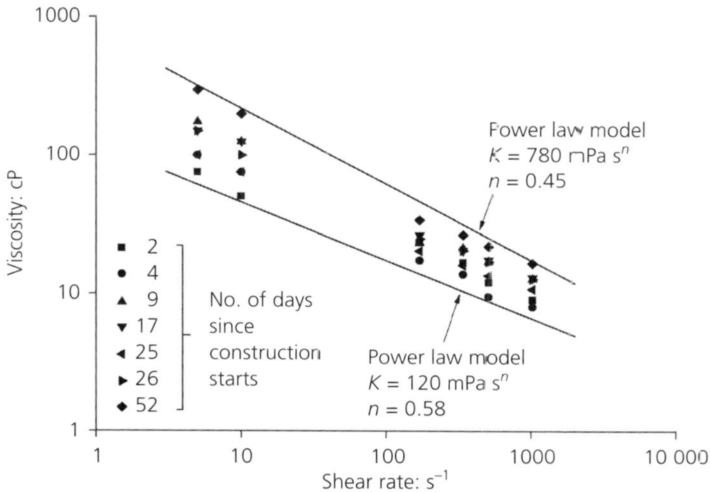

discussions, the fluids were assumed to be inelastic when in fact they exhibit viscoelasticity. Most fluids that we encounter in our daily life actually exhibit demonstrable viscoelasticity. Examples include paint, glue, yogurt, saliva, blood and shampoo, to name just a few. Perfectly elastic solids and perfectly viscous fluids are, in effect, the limiting cases of viscoelasticity, but for most materials it is only these limiting conditions that are observed in everyday life. For example, the elasticity of rock is considerably more obvious than its viscosity, although the latter does exist. The fact that nothing remains still was recognised in biblical times, and this is reflected by the line 'the mountains flowed before the Lord' in the Bible and by the quote 'everything flows' attributed to the ancient Greek philosopher Heraclitus (Reiner, 1964). It is interesting to note that the rheological parameter 'Deborah number' is named after the Prophetess Deborah cited in Judges 5:5 (Reiner, 1964). This parameter is defined as the ratio of the relaxation time (characterising the time it takes for a material to adjust to applied stresses or deformations) to the characteristic timescale of an experiment probing the response of the material.

Previous researchers have shown that fluids such as bentonite slurries, silicone oil and cross-linked polyacrylamide hydrogels can improve the cyclic resistance of soil (Ellis *et al.*, 2000; El Mohtar *et al.*, 2008; Graham and Elton, 2011). It follows that polymer fluids may also have a similar effect, and this may inhibit soil liquefaction at the excavation side walls due to sudden pore pressure surges caused by the excavation tools. The viscoelasticity of a polymer fluid can actually be seen without doing any rheological experiments (Collyer, 1973). For example, anyone who has handled PHPA fluids must have experienced the self-siphoning effect illustrated in Figure 9.19. The fluid in this figure was a high-molecular-weight PHPA fluid prepared at a moderate concentration of 1 kg/m^3. When the fluid was tipped from the beaker, the hanging fluid was able to

Figure 9.19 Self-siphoning effect of a high-molecular-weight PHPA fluid

'pull' more fluid out of the beaker due to its viscoelasticity, resulting in an upward flow over the wall of the beaker. Note that the surface of the liquid was well below the rim of the beaker and yet the stream reached up and flowed over the lip. This effect is a direct indication of the strong but invisible molecular structure of the polymer fluid.

Another manifestation of viscoelasticity can be seen during the mixing of polymer fluids in a laboratory. When water is stirred rapidly in a container, a deep vortex is formed. However, soon after the addition of polymer granules into the water, the vortex will disappear and the fluid will climb up the stirring rod, as shown in Figure 9.20. This phenomenon is known as the Weissenberg effect, after Karl Weissenberg who first reported it. The fluid appears to climb up the rod because when it is sheared along the x axis in the direction of flow, the stresses along the y and z axis are unequal in a viscoelastic fluid. The unbalanced force then pushes the fluid up the rod.

9.5.2 Oscillatory shear: fundamentals

As the viscoelasticity of a material is commonly measured by oscillatory shear experiments, some of the fundamental concepts will be discussed before the test results are presented. In the classical theory of elasticity, Hooke's law states that in a sheared body

Figure 9.20 The Weissenberg effect displayed by a high-molecular-weight PHPA fluid

the shear strain (γ) is directly proportional to the shear stress (τ), hence

$$\tau = G\gamma$$

where G is the shear modulus of the material. If a perfectly elastic (Hookean) solid is subjected to an oscillatory shear stress, the deformation will be instant, and hence the stress and the strain are exactly in phase as shown in Figure 9.21. However, for a perfectly viscous (Newtonian) liquid, the strain will be 90° out of phase, and this is also shown in the figure. For a viscoelastic material, the phase angle (δ) will be in between 0° and 90°. The phase angle is therefore a measure of viscoelasticity. For the interpretation of oscillatory shear test results, phase angle is usually used to separate the measured stiffness (G^*), which is known as the complex modulus, into two components:

$$G' = G^* \cos(\delta) \tag{9.9}$$

$$G' = G^* \sin(\delta) \tag{9.10}$$

where G' is the storage (elastic) modulus and G'' is the loss (viscous) modulus. The assessment of viscoelasticity can therefore be carried out by comparing their relative values. If $G' > G''$, the material is said to be solid-like. Conversely, if $G'' > G'$, the material is liquid-like rheologically.

To assess the viscoelasticity of a material, two types of oscillatory shear test are commonly employed, namely the amplitude sweep test and the frequency sweep test. Figure 9.22 shows the schematic waveforms for these tests. The waveform for a single frequency oscillation is also shown for comparison. It can be seen that in an amplitude

Figure 9.21 Strain response of purely elastic and purely viscous materials when subjected to oscillatory shear

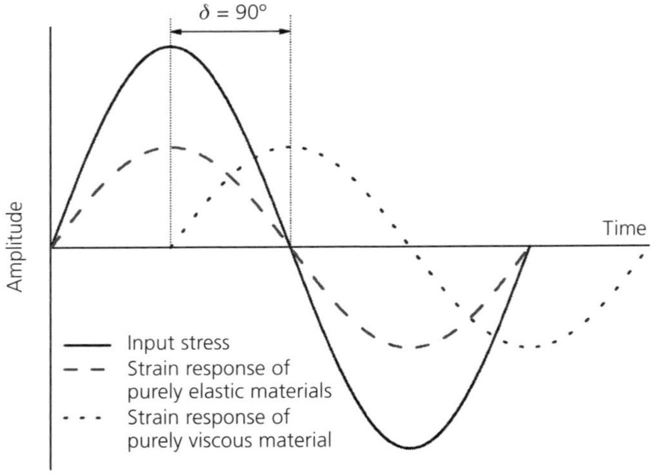

sweep test the amplitude of the wave, and thus the shear strain experienced by the specimen, is gradually increased at a single frequency. On the other hand, in a frequency sweep test the amplitude (shear strain) is fixed over a range of frequencies. These tests allow the user to characterise the viscoelastic properties of the specimen over a wide range of dynamic conditions. The experimental details and the test results for each type of test are given in the following sections.

9.5.3 Oscillatory shear: amplitude sweep tests

As discussed above, a viscoelastic material displays both viscous and elastic properties when undergoing deformation. The amplitude sweep test is a standard rheological test that can be used to determine the strain range over which linear elasticity (Hooke's law) applies. This range is known as the linear viscoelastic region (LVR). Because the stress–strain curve of the material is linear within the LVR, the values of G' and G'' are constant within this region. An understanding of the LVR is a prerequisite for the frequency sweep test since the latter requires a strain amplitude that is within the LVR.

Figure 9.22 Oscillatory test types

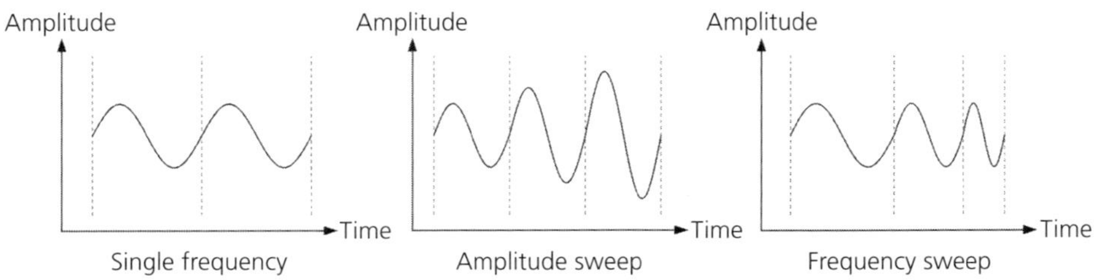

Figure 9.23 Amplitude sweep test results for a high-molecular-weight PHPA fluid. (From Lam *et al.*, 2015)

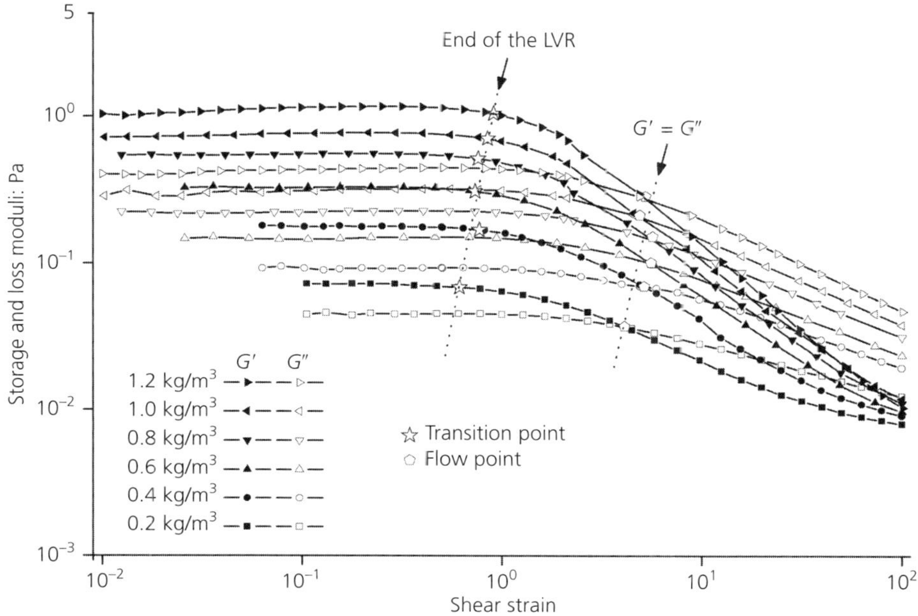

For the series of tests presented here, a Bohlin Gemini rheometer equipped with the parallel-plate geometry was used (see Figure 9.3). The parallel-plate geometry was used as it is the standard measurement geometry for oscillatory tests. The diameter of the top plate was 40 mm and the gap between the plates was 150 μm. A commonly used test frequency of 0.5 Hz was used, but other frequencies could have been used, as in most cases the strain at which non-linearity occurs is independent of frequency. As in other types of oscillatory tests, the results of an amplitude sweep test can be interpreted by assessing the relative values of the two shear moduli or the phase angle.

Figure 9.23 shows the amplitude test results for polymer concentrations ranging from 0.2 to 1.2 kg/m^3, which is the same range used for the viscometry tests discussed in Section 9.4.4. The results show that, at small strains, the elastic (storage) modulus G' is greater than the viscous (loss) modulus G''. Therefore, the polymer fluid is showing predominantly solid-like behaviour. It can be seen that as the strain continues to increase, G' drops faster than G''. This is typical behaviour for viscoelastic materials, as the elastic structure is more prone to damage by excessive deformation. The end of the LVR, denoted as the transition point in the plot, is therefore determined using G', and is defined as when its value has dropped by 5% from the plateau value. It can now be seen that the polymer fluid yields at strain values of just below unity (100%), compared with only 1–2% strain in bentonite fluid as reported by Santagata *et al.* (2008). This means that, when compared with bentonite, polymer fluids can maintain their internal elastic structure over a much larger range of strains. As the strain increases beyond the yield point, G' falls dramatically. The point where G' crosses G'' is denoted the flow point in the plot. After this point, viscous behaviour dominates. It can be seen that the strain values of both the transition and the flow points increase as the polymer concentration increases. This is contrary to what Santagata *et al.* (2008) found for bentonite fluid. Furthermore, the shape of the G'' curves for bentonite fluid is very different from that shown here for polymers, and the value of G' is considerably lower than that observed for bentonite (7.5 Pa for a 4% bentonite fluid (Santagata *et al.*, 2008) compared with the 1.1 Pa maximum shown in Figure 9.23). These differences are perhaps not surprising because polymer fluids are vastly different from their bentonite counterparts both physically (solution versus particulate slurry) and chemically, and also in terms of their typical in-use concentrations, so they can be expected to show different behaviour.

Figure 9.24 plots G' and G'' against polymer concentration (c_p) to show their relationship. It can be seen that the two moduli do not increase linearly with the concentration but instead show an upward quadratic trend. These non-linear relationships may be explained by reference to Figure 9.25, which shows the effect of increasing polymer concentration on the number of chain entanglements. It can be seen that when a new polymer molecule is added to the network it can form more than one additional entanglement, and this ability increases as the number of existing molecules increases. Given that the viscoelastic properties of a polymer fluid are due to the chain entanglements in the solution microstructure, an increase in the polymer concentration therefore has a greater than linear effect on these properties. In terms of the difference between the two curves, it can be seen that G' rises faster than G'' with increasing concentration. This

Figure 9.24 Effect of the polymer concentration on the elastic and viscous moduli. (From Lam *et al.*, 2015)

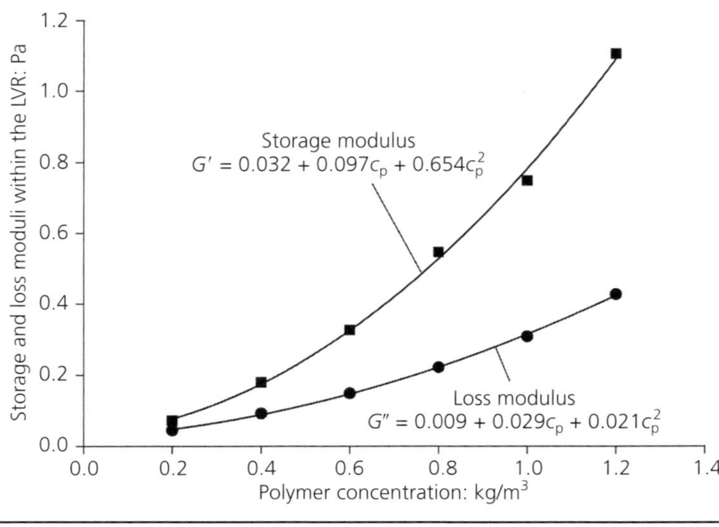

Figure 9.25 Schematic diagram showing the non-linear relationship between the polymer concentration and chain entanglement

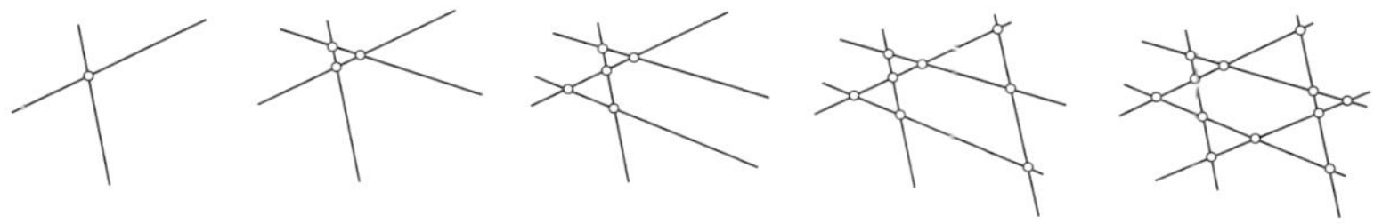

Number of molecules, x	2	3	4	5	6
Number of entanglements, y	1	3	5	8	12

means that the fluid gradually becomes more solid-like as a result of the increasingly stronger molecular structure.

9.5.4 Oscillatory shear: frequency sweep tests

Additional tests were carried out to characterise the response of the polymer fluid at different frequencies (f). As for the amplitude sweep tests, the parallel-plate geometry was used and the gap was set at 150 μm. Based on the results of the amplitude sweep tests, a constant strain value of 0.1 (10%) was chosen as the target strain. At this strain level, the polymer fluid is well within the LVR at all concentrations (see Figure 9.23). To ensure the actual strain value did not deviate from the target value, the 'auto-stress' function of the rheometer was used to allow the control software to automatically adjust the applied stress based on the response of the previous oscillation. Since it was known from preliminary experiments that the stiffness increases with increasing frequency, based on the suggestion of the manufacturer the rheometer was programmed to run from the highest frequency (10 Hz) to the lowest (0.001 Hz), to allow the computer to better control the applied stress.

Since this series of tests involved measurements at very low frequencies, the total testing period was over 4 h. To prevent evaporation of the specimen during the testing period, a vapour hood (also known as a solvent trap or environmental chamber) was used to enclose the specimen. The underside of the vapour hood was also misted with water to saturate the air surrounding the specimen. If this step was not carried out, the periphery of the specimen would have gradually dried out over the test period, leading to erroneous results.

Figure 9.26 shows the frequency sweep test results. First, it can be seen that none of the curves reaches the highest target frequency of 10 Hz, and that the lower the polymer concentration is, the shorter the frequency range becomes. This is due to the effect of inertia. Although the inertia of the rheometer was known and was automatically accounted for by the control software, for low-viscosity samples the inertia of the rheometer could become significantly greater than the true inertia response of the specimen at high frequencies. For this reason, some of the data at high frequencies were unusable and had to be disregarded. The results show that at low frequencies the viscous behaviour is more pronounced ($G'' > G'$), meaning that the fluid is more liquid-like. As the frequency

Figure 9.26 Frequency sweep test result for a high-molecular-weight PHPA fluid. (From Lam *et al.*, 2015)

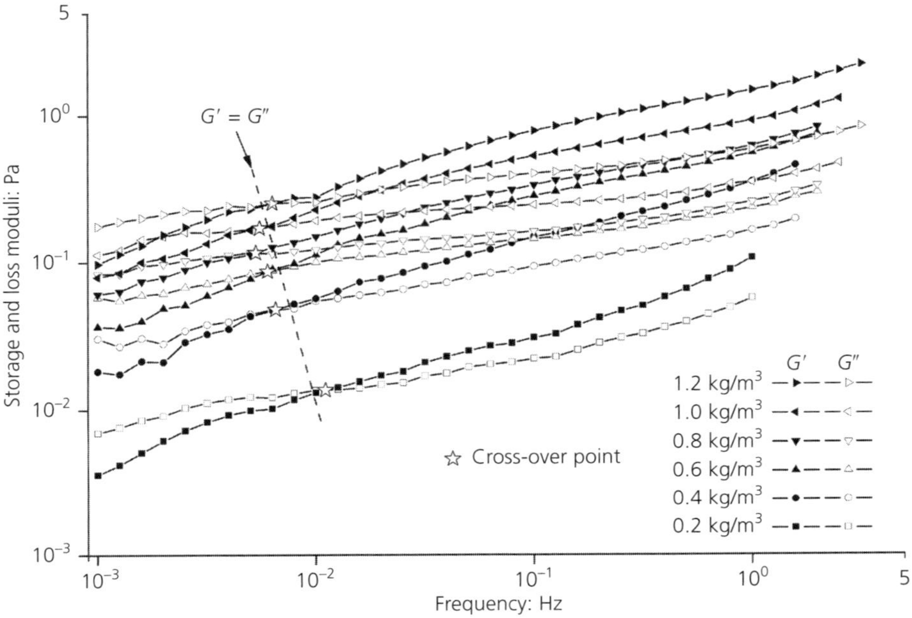

increases, the two curves cross over at about 0.01 Hz. Beyond this point, elastic behaviour clearly dominates, as can be seen by the higher G'. The reason for the observed time-dependent behaviour may be explained by the chain entanglement theory. Since the polymer chains are entangled but constantly sliding over each other under Brownian motion, they form a three-dimensional network structure such as that shown in Figure 3.15. If a stress is suddenly applied to this transient network, initially all the molecules are stretched and the fluid will behave more elastically. However, given time, the molecules can relax and disengage from each other, and the fluid will show more viscous behaviour.

To conclude this chapter, it is worth briefly mentioning the engineering implications of the oscillatory shear test results. As mentioned before, viscoelastic materials such as bentonite slurries and cross-linked polyacrylamide gels have been shown to be able to effectively delay the onset of soil liquefaction (El Mohtar *et al.*, 2008; Graham and Elton, 2011). It follows that polymer fluids may also have this same beneficial effect on soil due to their viscoelastic properties, as shown by the oscillatory shear tests. Such an effect may help prevent local liquefaction of the soil on excavation side walls due to pore pressure surges caused by the digging tools. Further research is required to confirm this, but the oscillatory results presented here should form the basis of such investigation.

REFERENCES

ASTM (American Society for Testing and Materials) (2014) D4016-14. Standard test method for viscosity and gel time of chemical grouts by rotational viscometer (laboratory method). ASTM International, West Conshohocken, PA, USA.

Ata A and O'Neill M (1997) *Characterization of the Effect of POLY-BORE™ Polymer on the Construction and Performance of drilled Shaft Foundations. Phase I: Field Tests.* Department of Civil and Environmental Engineering, University of Houston, Texas, TX, USA.

Ata A and O'Neill M (2000) The physicochemical interaction between PHPA polymer slurry and cement mortar. *Geotechnical Testing Journal* **23(2)**: 225–235.

Barnes HA (2000) *A Handbook of Elementary Rheology. Institute of Non-Newtonian Fluid Mechanics.* University of Wales, Cardiff, UK.

Barnes HA and Walters K (1985) The yield stress myth? *Rheologica Acta* **24(4)**: 323–326.

Beresford JJ, Cashman PM and Hollamby RG (1989) The merits of polymeric fluids as support slurries. *Proceedings of the 3rd International Conference on Deep Foundations, London* (Burland J and Mitchell J (eds)). Balkema, Rotterdam, the Netherlands, vol. 1, pp. 3–10.

Brown DA, Turner JP and Castelli RJ (2010) *Drilled Shafts Construction Procedures and LRFD Design Methods.* Federal Highway Administration, US Department of Transportation, Washington, DC, USA. Report FHWA NHI-10-016.

Bruce C and Schwarz WH (1969) Rheological properties of ionic and nonionic polyacrylamide solutions. *Journal of Polymer Science Part A-2: Polymer Physics* **7(5)**: 909–927.

Bustamante M, Gianeselli L, Boato R and Conedera A (1998) Performance of polymer slurries in large diameter bored pile. *Proceedings of the 3rd International Seminar on Deep Foundations on Bored and Auger Piles (BAP III), Ghent.* Balkema, Rotterdam, the Netherlands, pp. 119–127.

Caenn R, Darley HCH and Gray GR (2011) *Composition and Properties of Drilling and Completion Fluids*, 6th edn. Gulf Professional Publishing, Oxford, UK.

Camp WM, Brown DA and Mayne PW (2002) Construction method effects on axial drilled shaft performance. *Proceedings of International Deep Foundations Congress 2002, Orlando.* American Society of Civil Engineers, Reston, VA, USA, pp. 193–208.

Caulfield MJ, Qiao GG and Solomon DH (2002) Some aspects of the properties and degradation of polyacrylamides. *Chemical Reviews* **102(9)**: 3067–3084.

Collyer AA (1973) Demonstrations with viscoelastic liquids. *Physics Education* **8(2)**: 111–116.

Ellis EA, Soga K, Bransby MF and Sato M (2000) Resonant column testing of sands with different viscosity pore fluids. *Journal of Geotechnical and Geoenvironmental Engineering* **126(1)**: 10–17.

El Mohtar CS, Clarke J, Bobet A, Santagata M, Drnevich V and Johnston C (2008) Cyclic response of a sand with thixotropic pore fluid. *Proceedings of the Geotechnical Earthquake Engineering and Soil Dynamics Congress 2008, Sacramento.* American Society of Civil Engineers, Reston, VA, USA, pp. 1–10.

Ferraris CF and de Larrard F (1998) Modified slump test to measure rheological parameters of fresh concrete. *Cement, Concrete and Aggregates* **20(2)**: 241–247.

FPS (Federation of Piling Specialists) (2006) *Bentonite Support Fluids in Civil Engineering*, 2nd edn. FPS, Beckenham, UK.

Gardner KL, Murphy WR and Geehan TG (1978) Polyacrylamide solution aging. *Journal of Applied Polymer Science* **22(3)**: 881–882.

Graham DS and Elton DJ (2011) A new method of reducing liquefaction susceptibility through pore water modification. *Proceedings of Geo-Frontiers 2011: Advances in Geotechnical Engineering, Dallas*. American Society of Civil Engineers, Reston, VA, USA, pp. 1785–1794.

Henderson JM and Wheatley AD (1987) Factors effecting a loss of flocculation activity of polyacrylamide solutions: shear degradation, cation complexation, and solution aging. *Journal of Applied Polymer Science* **33(2)**: 669–684.

Ho CE and Tan CG (1996) Barrette foundation constructed under polymer slurry support in Old Alluvium. *Proceedings of the 12th Southeast Asian Geotechnical Conference, Kuala Lumpur*. Southeast Asian Geotechnical Society, Pathumthani, Thailand, pp. 379–384.

Holmes MM (1993) Drilled shafts and new slurry technology used instead of driven piles: sets record on the Ohio River. *Foundation Drilling* **32(5)**: 13–15.

Howsam P and Hollamby R (1990) Drilling fluid invasion and permeability impairment in granular formations. *Quarterly Journal of Engineering Geology* **23(2)**: 161–168.

Jefferis SA (2003) Grouts and grouting. In *Advanced Concrete Technology: Processes* (Newman J and Choo BS (eds.)). Butterworth-Heinemann, Oxford, UK, pp. 12/1–12/47.

Jefferis SA and Lam C (2013) Polymer support fluids: use and misuse of innovative fluids in geotechnical works. *Proceedings of the 18th International Conference on Soil Mechanics and Geotechnical Engineering, Paris*. Presses des Ponts, Paris, France, pp. 3219–3222.

Kheng H, Bloomquist D, Townsend FC, Thompson PY and Dobson E (1991) *Slurry Performance in Drilled Shaft Construction. Final Report*. Department of Civil Engineering, University of Florida, Gainesville, FL, USA. (Available from Association of Drilled Shaft Contractors, Irving, TX, USA.)

Kulicke W-M and Kniewske R (1981) Long-term change in conformation of macromolecules in solution, 2. Poly(acrylamide-co-sodium acrylate)s. *Die Makromolekulare Chemie* **182(8)**: 2277–2287.

Lam C and Jefferis SA (2015a) An investigation of possible yield stress and thixotropy in polymer excavation-support fluids. *Proceedings of the XVI European Conference on Soil Mechanics and Geotechnical Engineering, Edinburgh*. Institution of Civil Engineers, London, UK, pp. 3183–3188.

Lam C and Jefferis SA (2015b) Physical properties of polymer support fluids in use and their measurement techniques. *Geotechnical Testing Journal* **38(4)**: 490–500.

Lam C and Jefferis SA (2016) Ageing behaviour of polyacrylamide-based excavation fluids. *Proceedings of the Institution of Civil Engineers – Construction Materials*, **169(6)**: 301–307.

Lam C, Martin PJ and Jefferis SA (2015) Rheological properties of PHPA polymer support fluids. *Journal of Materials in Civil Engineering*, **27(11)**: 04015021.

Likos WJ, Loehr JE and Akunuri K (2004) *Engineering Evaluation of Polymer-based Drilling Fluids for Applications in Missouri Shale*. University of Missouri, Columbia, MO, USA, Report RDT 04-019.

Littlechild B and Plumbridge G (1998) Effects of construction technique on the behaviour of plain bored cast in situ piles constructed under drilling slurry. *Proceedings of the 7th International Conference and Exhibition on Piling and Deep Foundations, Vienna*. Deep Foundations Institute, Hawthorne, NJ, USA, pp. 1.6.1–1.6.8.

Mächtle W (1982) Zur alterung von wäßrigen polyacrylamid-lösungen. *Die Makromolekulare Chemie* **183(10)**: 2515–2525. (In German.)

Majano RE and O'Neill MW (1993) *Effect of Mineral and Polymer Slurries on Perimeter Load Transfer in Drilled Shafts*. Department of Civil and Environmental Engineering, University of Houston, Texas, USA, Report UHCE-93-1.

Majano RE, O'Neill MW and Hassan KM (1994) Perimeter load transfer in model drilled shafts formed under slurry. *Journal of Geotechnical Engineering* **120(12)**: 2136–2154.

Nagashiro W and Tsunoda T (1977) Degradation of polyacrylamide molecules in aqueous solutions by high-speed stirring. *Journal of Applied Polymer Science* **21(4)**: 1149–1153.

Nakano A and Minoura Y (1978) Degradation of aqueous poly(acrylic acid) and its sodium salt solutions by high-speed stirring. *Journal of Applied Polymer Science* **22(8)**: 2207–2215.

Nasr-El-Din HA and Taylor KC (1996) Rheology of water-soluble polymers used for improved oil recovery. In *Advances in Engineering Fluid Mechanics: Multiphase Reactor and Polymerization System Hydrodynamics* (Cheremisinoff NP (ed.)). Gulf Publishing, Houston, TX, USA, pp. 615–668.

O'Neill MW and Reese LC (1999) *Drilled Shafts: Construction Procedures and Design Methods*. Federal Highway Administration, US Department of Transportation, Washington DC, USA, Report FHWA-IF-99-025.

Reiner M (1964) The Deborah number. *Physics Today* **17(1)**: 62.

Santagata M, Dalmazzo D and Santagata E (2008) Deformation behavior of clay–water suspensions from rheological tests. *Proceedings of the 4th International Symposium on the Deformation Characteristics of Geomaterials*, vol. 1. IOS Press, Amsterdam, the Netherlands, pp. 453–459.

Schwarz J and Lange U (2004) Brückengründung mit 70 m tiefen flüssigkeitsgestützten gebohrten Pfählen in Benin/Afrika. *Proceedings of the 19th Christian Veder Kolloquium, Graz*. Graz University of Technology, Graz, Austria, pp. 73–89. (In German.)

Shyluk WP and Stow Jr FS (1969) Aging and loss of flocculation activity of aqueous polyacrylamide solutions. *Journal of Applied Polymer Science* **13(5)**: 1023–1036.

Troughton VM (1992) The design and performance of foundations for the Canary Wharf development in London Docklands. *Géotechnique* **42(3)**: 381–393.

Williams PA and Philips GO (2003) The use of hydrocolloids to improve food texture. In *Texture in Food*, vol. 1. *Semi-solid Foods* (McKenna BM (ed.)). Woodhead Publishing, Abington, Cambridge, UK, pp. 251–274.

Polymer Support Fluids in Civil Engineering
ISBN 978-0-7277-5786-9

http://dx.doi.org/10.1680/psfce.57869.265

Chapter 10
Specifications and associated controls

10.1. Introduction

The key to success for any project involving fluid-supported excavations is for the user to have sufficient working understanding of the fluid properties and control procedures and to apply suitable construction techniques; satisfying a fluid specification alone does not guarantee success. Site practice has been discussed in earlier chapters. This chapter therefore focuses on the parameters that are commonly specified for the control of support fluids, and reviews some existing specifications. This review shows that a one-size-fits-all approach to polymer fluid specification is not appropriate. Rather, specifications should consider the required functions of the foundation elements and any site-specific conditions and constraints. Therefore, when developing a specification it is necessary to consider

- What are the key functions of the foundation element to be constructed: for example, for piles, what are the requirements for end bearing and shaft friction?
- What will be the depth and plan dimensions of each foundation element?
- What are the key fluid properties that must be controlled?
- At what stage of the construction cycle and how often must each of these properties be controlled?
- What are the appropriate test procedures for the control of each of these properties?

As practice varies widely across the world and projects are becoming larger and more complex, specifications should also set out who should do the tests (usually the specialist foundation contractor, i.e. the user of the fluids), on what samples and at what times during each construction cycle, who should witness the tests and who should approve them.

New products will appear in the marketplace driven by innovation, cost, performance and environmental acceptability. The following discussion is therefore intended to be generic and not limited solely to consideration of products that are currently available. It is focused on the rationale for the control tests more than the actual numerical values to be specified for each parameter. As well as new products, new test procedures also can be expected, for example the residual concentration test and developments from it.

10.2. Polymer types

This book is mainly concerned with partially hydrolysed polyacrylamide (PHPA) polymers – but even among PHPAs there can be considerable variation in behaviour and in the optimum operating concentration for any particular combination of ground conditions. Furthermore, individual polymer manufacturers may modify PHPAs and/or add additional materials to their products and/or provide treatment agents for the polymers while in use. Thus, although generic advice can be given, specific products can require particular specifications. This may seem a rather negative position but, on the positive side, a number of generic aspects of specifications can be identified. These include the locations from which test samples should be taken, and the tests to be carried out on these samples – although noting that most of the test procedures are currently common to bentonite testing. Over time we can expect to see the development of polymer-specific tests methods. Discussions of the typical sampling locations and the types of test to be performed are given the following sections.

10.3. Sampling locations

During excavation works under an excavation support fluid, test samples may be taken from one or more of the following locations:

(*a*) At the mixing station used for the preparation of fresh fluid.
(*b*) As supplied to the excavation from storage tanks.
(*c*) In the excavation during digging – depth(s) must be specified.
(*d*) At the end of digging at one or more of the following depths in the excavation: top, middle and near the bottom of the excavation (e.g. at 0.5–1.0 m from the bottom, a sample from the bottom may give a false high reading if it picks up some of the settled soil – this must be managed by separate testing).
(*e*) At one or more of the locations: top, middle and bottom of the excavation prior to concreting to ensure that fluid conditions are appropriate for concreting.
(*f*) During concreting at the treatment/sedimentation tank as the fluid is returned from the excavation.
(*g*) After fluid rejuvenation as the fluid is returned to the storage tanks and is ready for reuse (e.g. after sedimentation of suspended solids and/or addition of fresh polymer and any associated additives).

While there are many test locations that may be considered, especially on research projects, practical considerations will limit the number of locations at which the fluid can be checked on a site during normal operation. Of the above-listed locations, tests that will typically be required, as a minimum, are

(*a*) Viscosity to confirm correct preparation.
(*d*) A full suite of tests on samples from the middle and near the bottom of the excavation, to confirm that the fluid has sufficient active polymer, and/or whether further settlement time is required to ensure that excessive solids will not collect at the base of the excavation, and/or the fluid must be replaced to ensure good base contact. Testing at the top may provide little further information if the fluid from the middle sample is of low solids content.
(*e*) A full suite of tests prior to concreting at the middle and base of the excavation, to confirm that the fluid is satisfactory for concreting. This is especially important when drilling in silty fine sand soils that take a long time to settle and when end

bearing is required. Note that the samples in these tests should be from the same locations as (*d*) but taken after a period of settlement and/or after fluid replacement.

(*f*) Viscosity and pH of the fluid returned from the excavation during concreting, to determine any necessary treatment for polymer rejuvenation.

(*g*) Viscosity and pH to confirm that the treatment identified in (*f*) has been successful and that the fluid is suitable for reuse.

10.4. Test procedures commonly employed

Before considering control values, it is necessary to reprise the tests that are commonly used for the control of excavation support fluids in general and the further tests that may be recommended for polymer fluids. It is also necessary to briefly consider the rationale behind the use of these tests. The commonly used tests for support fluids are

- the Marsh funnel time
- density
- sand content, using the American Petroleum Institute (API) sand content test rig
- filter loss, using the API filter press
- pH

and, rather more rarely,

- rotational viscometry (for the assessment of viscosity and other rheological parameters, and the assessment of active polymer content).

While some of these may not be ideal test procedures for polymer fluids, an advantage of adopting them is that details of the test protocols have been widely published and operatives working with excavation support fluids are likely to have been trained to use them. The basic principles of the tests and the test procedures are set out in Chapter 6. In this section, only matters specific to polymer fluid control will be considered.

10.4.1 The Marsh funnel time

The Marsh funnel is a robust, low-cost instrument originally developed for rough measurements in the field (Marsh, 1931). Although its use for polymer fluids has been opposed (e.g. by Berkovitz and Long, 1995), the present authors accept its use as a field control tool since there is a substantial body of experience with it.

For a test the funnel is filled with 1500 ml of the test fluid. The test result is the time for a specified volume of the test fluid to flow from the funnel, usually 946 ml (=1 US quart) or 1000 ml. This time gives an indication of the fluid viscosity, which can in turn provide a check on the active polymer content of the fluid, provided that the fluid is not too heavily contaminated with soil – a situation to be avoided with polymer fluids.

After use, the Marsh funnel should be rinsed with a polymer breaker (e.g. a dilute oxidising agent) and then washed through with clean water. If not cleaned, remnant polymer fluid can dry in the discharge spigot of the funnel, resulting in partial clogging and false high times. An alternative is to store the funnel in a bucket of dilute polymer breaker and to wash it through with the test fluid prior to use.

For laboratory investigations, a viscometer such as a direct-indicating (Fann-type) viscometer (see Section 6.5.3) or a Brookfield viscometer (see Section 6.5.4) can give more detailed information on polymer fluid flow behaviour than the Marsh funnel.

10.4.2 Density

The density measurement method for polymer fluids should have a resolution of at least 0.005 g/ml. The mud balance (see Section 6.6.2) is a tool developed in the oil industry, and has a very wide measurement range, typically from about 0.8 to 2.8 g/ml, and a repeatability between samples of about 0.01 g/ml – which is also the smallest division on the instrument scale. This measurement range is unnecessarily wide and the resolution too low for civil engineering works (Jefferis, 1977), and especially for polymer fluids. The aerometer (see Section 6.6.3), although a completely different instrument, has a similarly wide measurement range and again a resolution of 0.01 g/ml, and so is no more appropriate than the mud balance. For these reasons, the authors recommend the use of a pycnometer bottle (e.g. a 1-litre spring-top bottle as used by Jefferis and Lam (2017)) and an electronic balance. This combination should give a density resolution of better than 0.005 g/ml. Spring-top bottles have the advantage of a defined volume (defined by the closing cap), low cost, wide availability and that the spring action stops self-siphoning when the bottle is filled to the brim. It may be noted that a density bottle procedure is specified in the *ICE Specification for the Construction of Slurry Trench Cut-off Walls* (ICE, 1999) for the control of cement–bentonite slurries for use in cut-off walls. Details of the density bottle method have been given in Section 6.6.4 and are not repeated here.

The density of fresh polymer fluids will usually be very close to that of water as polymer granules are typically of near water density and concentrations are less than 2 kg/m^3. The polymer concentration in a polymer support fluid therefore cannot be determined from on-site density measurements. Occasional measurements on fresh fluids can serve as a check on the on-site density measurement procedure and to confirm that the baseline density against which other densities should be considered is that of water. For polymer fluids that have been recently prepared by air bubbling, the density can be less than that of water until all the entrained air bubbles have escaped. Entrained air bubbles can also increase the fluid viscosity as measured with the Marsh funnel.

For used polymer fluids, the key parameter assessed by density measurement is the amount of soil suspended in the fluid. Soils, especially fine soils, will sorb polymer and so reduce the active polymer concentration in the fluid. The solids content S (defined here as the dry mass of all soil solids per unit volume of fluid) of a used polymer fluid can be calculated from its density ρ_m (the subscript 'm' denotes mixture) using the formula

$$S = \rho_s\left(\frac{\rho_m - \rho_f}{\rho_s - \rho_f}\right) \tag{10.1}$$

where ρ_f and ρ_s are, respectively, the density of the polymer fluid as mixed and free of entrained air bubbles and the density of the solids in suspension. As an example of the use of Equation 10.1, if $\rho_m = 1.020$ g/ml, $\rho_s = 2.65$ g/ml and $\rho_f = 1.00$ g/ml, then $S = 32$ kg per cubic metre of the fluid. The solids content of a fluid–solid mixture also can be calculated using simple spreadsheet tables (Jefferis, 2014), which are particularly useful for multi-component systems such as blastfurnace slag–cement–bentonite slurries.

10.4.3 Sand content

The sand content test is performed on used support fluids, and determines the fraction of soil coarser than 75 μm – this fraction is taken to be ‘sand’ in the test. The test procedure is essentially the same as for bentonite slurries except that a polymer breaker (typically an oxidising agent such as domestic bleach) must be used prior to the screening step, as intact polymer fluid will not readily flow through the 75 μm screen of the test rig (see Section 6.7).

The result is the bulk volume of sand settled through water in the test thimble as a percentage of the volume of the test sample. A freshly mixed polymer fluid will contain no sand-sized material unless it has been deliberately added as a bulk fluid-loss control additive.

The result of the sand content test cannot be directly related to the solids content assessed by density measurement, as the sand content result is the percentage sand by bulk volume and so depends on the bulk density to which the particular sand settles in the thimble – and this settled density can vary with the sand type and grading. The relationship between the sand content by bulk volume and by dry mass in the fluid is

$$S_m = S_v \rho_d \tag{10.2}$$

where S_m is the sand content as the dry mass of sand per unit volume of fluid, S_v is the volumetric sand content as measured in the test, and ρ_d is the dry bulk density to which the sand settles in the thimble during a test. This dry bulk density tends to be low, perhaps 1200–1600 kg/m^3. Thus, from Equation 10.2, if a used polymer fluid has a volumetric sand content (S_v) of, for example, 1%, this fluid will contain about 12–16 kg of ‘sand’ per cubic metre of soil-laden fluid.

The sand content is a useful parameter, as coarser particles have lower surface-area-to-volume ratios than fine particles, and so their potential for polymer sorption is lower. Consequently, their effect on the active polymer concentration will be less than for finer material. It follows that higher fluid densities may be acceptable if mostly due to a high sand content. Density results therefore should be considered in tandem with sand content results, and higher densities may be permissible when sand contents are higher, provided that the fluid is suitably managed.

A further and important consideration in relation to the sand content is that coarser fractions will settle faster than finer fractions unless the latter have been flocculated. The sand content can thus be an indicator of the potential for sediment build-up at the base of an excavation.

10.4.4 pH

For preference, the pH of polymer fluids should be determined with a calibrated electronic pH meter and the results reported to at least one decimal place. Typically under field conditions, indicator strips are readable only to whole integers, and the results can be operator dependent (see Section 6.3.2). While the authors cannot recommend the use of indicator strips as these do not have sufficient resolution to identify incipient pH problems or resolve actual problems, they recognise that pH strips have been used successfully on many projects. If used, indicator strips must be selected to cover a sufficient pH range, as they will give no warning if the pH to be measured is above or below the range of the test strip. It is suggested that indicator strips should be selected to cover the range from 2 pH units below the specified minimum pH to 1 pH unit above the maximum specified. This may require the use of more than one type of test strip. All indicator strips must be properly stored, in a sealed container away from moisture. The indicator strips can be used only once and cannot be reused once wetted.

10.4.5 Filter loss

For bentonite slurries the filter loss test is carried out in the standard API filter loss cell. The test fluid is pressurised against a filter paper supported on a fine mesh. The standard pressure is 100 psi (689.5 kPa), and the test time is usually 30 min. Fresh, unused PHPA fluids without any additives can show high fluid losses. Typically, no identifiable filter cake will form, and the test cell may empty within a 30 min test time, although the discharge rate from the cell may show some reduction with filtration time. However, used polymer fluids will develop a filter cake, although typically much thinner than that for used bentonite slurries, as polymer fluids must be operated at much lower solids contents than bentonite slurries. The fluid loss of used polymer fluids will depend on the solids suspended in the fluid. Fine soils such as clay will give lower fluid losses than fluids containing only sand-sized material. Currently, no operationally useful information is gained from doing fluid loss tests on fresh fluids in the standard API test cell, but this topic needs to be kept under review, and work is in progress on this.

10.4.6 Rotational viscometry

Direct-indicating viscometers such as the Fann viscometer are sometimes deployed on sites where bentonite slurry is used, especially when it is necessary to monitor the gel strength of the slurry. However, as polymer fluids have effectively no gel strength, there is no need to test for it. The Brookfield viscometer may be used for determinations of the active polymer content of polymer fluids (Lam *et al.*, 2014). However, recent site experience suggests that the Marsh funnel can give a useful indication of the condition of used polymer fluids (see Figure 9.15), provided that there has been and continues to be careful control of the suspended solids content of the polymer fluid.

10.4.7 Monitoring product usage

Although it may seem obvious, if only for cost control, polymer product usage should be monitored during the works by those responsible for the works. Polymers and associated additives must be used at appropriate concentrations. Marked under- or overconsumption should be investigated.

10.5. Excavation practice

While fluid properties tend to be the focus of many specifications, it is important to recognise that site practice is fundamental to achieving a high-quality product. As regards excavation support fluids, this includes the selection of appropriate excavation tools (see Chapter 5), adopting digging procedures that minimise the dispersion of excavated soil into the support fluid (e.g. do not race the digging) and the development of appropriate base-cleaning procedures – soil that has been settled through polymer can be very slippery and difficult to remove. A procedure for base cleaning is to terminate the excavation one or two bites above the founding level, allow the soil to settle through the polymer fluid (for deeper piles this may require an overnight hold) and then remove the settled sludge/slime with the final bites of soil. The removal of settled solids is important not just when end bearing is required. Settled material can be carried up on top of the rising concrete, and if the concrete surface is not level (e.g. it is lower outside a reinforcement cage), then this material may slither down and become entrapped as a defect at the face of the foundation element.

10.6. Review of existing polymer fluid specifications

10.6.1 O'Neill (1991)

O'Neill (1991) presented a general discussion on the use of mineral (bentonite and attapulgite) and polymer fluids for drilled shaft (bored pile) construction. Based on the then current state of knowledge, he proposed an interim specification for polymer fluids, which is shown in Table 10.1. It can be seen that the permitted Marsh funnel times are much lower than those commonly used today for PHPA and blended polymer fluids (see Tables 4.2–4.5). This is probably due to the use of natural rather than synthetic polymers for excavation support at that time. For example, O'Neill noted that from about 1988 the use of natural polymers – developed mainly from enzyme action on natural organic materials such as starches – had become relatively widespread in the USA because of their environmental benefits over bentonite.

Based on this observation, O'Neill also noted that the viscosity of oatmeal-like silt-contaminated fluid was considerably higher. To prevent concrete contamination by such material during casting, O'Neill suggested taking fluid samples from the bottom of the excavation and conducting viscosity tests, in addition to sand content tests. He said that

Table 10.1 Interim specifications recommended by O'Neill (1991) for polymer support fluids

Property	Test method	Recommended range of values	
		When fluid is introduced	Just prior to concreting
Viscosity	Marsh funnel	26–29 s	26–35 s
Sand content	API sand content kit	≤1%	≤1%
pH	pH indicator strip	9–12	9–12

if the measured viscosity exceeds the upper limit of 35 s, the presence of 'oatmeal' is indicated. For today's PHPA polymer fluids that are designed to be operated at near-water densities but higher viscosities, the presence of unacceptable levels of suspended solids can be identified by density measurements. A potential consequence of higher levels of suspended solids is the accumulation of soft sediment at the bottom of an excavation. This can be checked with a weighted tape measure and a base hardness grading scale, as described by Berry (2009) and ICE (2017) or with devices such as the miniature drilled-shaft inspection device mentioned by Camp *et al.* (2002). Since the construction of deep foundations is now commonly undertaken using synthetic fluids that typically have much higher Marsh funnel times, the recommended time range proposed by O'Neill is no longer applicable, but concerns about the accumulation of soft sediment remain – and appropriate site practice should be adopted to manage this.

O'Neill's interim specifications also included an upper limit on the sand content of 1%, when measured for a fluid sample taken from the bottom of the excavation. Field experiments conducted by O'Neill indicated that all the sand had settled out of a 15 m-deep column of fluid within 30 min of retracting the auger on its last pass during drilling of the borehole. With this information, O'Neill proposed that contractors must let the fluid in the borehole stand without agitation for a minimum of 30 min and then clean the base mechanically or hydraulically with a submersible pump or by air lift. The 1% sand content requirement was to ensure that effectively all sand had settled before the contractor was allowed to place the reinforcing steel cage and begin concreting.

This recommendation of O'Neill was put to test by the authors during the two UK trials described in Chapter 8. In both cases, the sand content near the bottom of the pile bores was greater than 1% immediately after excavation. In the Glasgow trial, where the subsoil was silty fine sand, a waiting time of 30 min was clearly insufficient as it took 16 h for the sedimentation process to be largely complete and the sand content to drop to about 0.5% (see Figures 8.5 and 8.6). In the East London trial, where both clay and fine sand were present, a waiting time of 30 min was sufficient for the sand content to drop to between 0.1% and 0.2%. At both of these sites the viscosity of the fluid was much above O'Neill's 35 s maximum (synthetic polymers were used with funnel times over 50 s), and the depths of the pile bores were between 18 and 27 m. In O'Neill's case, a settlement time of 30 min was sufficient because the depth of the pile bore at 15 m was relatively limited and, more importantly, the viscosity of the fluid was low by today's standards (i.e. current practice using synthetic polymers rather than natural polymers), so the suspended soil particles in the fluid will have settled more rapidly. From these investigations, it can be concluded that the time required for settlement will depend on the types of soil to be excavated, the viscosity of the fluid used and the depth of the excavation. Contractors are therefore recommended to undertake trials at the start of a project, determining the sand content at mid-depth and near the base of a trial excavation (and possibly also base hardness tests) as a function of time after the end of excavation so that the required settlement time can be established.

A maximum acceptable sand content also has to be established, and although the two UK trials and O'Neill's field experiments have shown that a 1% sand content limit led to successful construction, less successful cases have been reported. For example, Ho and Tan (1996) reported that a 17 m-long barrette had a poor base response, despite a sand content of only 0.2–1.0% (see Section 7.4.4). Ho and Tan ascribed the poor base response to the accumulation of soil particles at the base of the trench during a 4.5 h waiting period between the completion of desanding and concreting. The Marsh funnel time for the fluid was reported as between 28 and 32 s. This case history shows that satisfying a requirement for 1% sand content at the base of an excavation will not necessarily guarantee a clean base. The total amount of solids in the fluid column and the fraction of these solids that can settle after the end of excavation and before concreting must be considered. If samples are taken from both the middle and near the bottom of an excavation, useful information on the rate of solids settlement can be obtained.

Key points relating to the solids content of a polymer fluid are

- If a fluid contains $x\%$ settleable solids by bulk volume (as measured in the sand content test) and if all these solids settle, sediment to at least $x\%$ of the excavation depth will accumulate at the base ('at least', because sediment settled through the support fluid in an excavation may be looser than in a sand content test where settlement is through water).
- Typically, the +75 μm solids in the polymer fluid (i.e. the 'sand' of the sand content test) are taken as settleable. However, some finer material will settle if times are long or the fines are flocculated (e.g. by cleaning agents added to the fluid). It is therefore important to consider the fluid density (an indicator of total solids content) as well as the sand content.
- For low-viscosity fluids and shallower excavations, effectively all settleable solids may be removed within a modest holding period.
- For higher-viscosity fluids and/or deeper excavations, an overnight holding period may be appropriate to ensure that the rate of settlement of any remaining solids is such that no significant sediment can accumulate after the final base-cleaning operations and before concreting.
- Field trials of settlement behaviour are recommended unless there is pre-existing site-specific experience.

10.6.2 Berkovitz and Long (1995)

Berkovitz and Long (1995), writing from the client's point of view, discussed some of the early issues relating to the use of polymer excavation support fluids. They noted that product-specific manufacturer's specifications had been used on several US state department of transport projects. Although this resulted in a number of what appeared to be successful projects, in others the outcomes were less successful even with the on-site supervision by the polymer manufacturer's representative. Berkovitz and Long counselled that this should serve as a caution to those who believe that an end-product specification, and/or a requirement for on-site supervision by a manufacturer's representative will ensure successful construction. The present authors are not surprised that there were some problem projects at that time, as polymer fluid technology and practice was still in its infancy in the early to mid-1990s. At that time, rather little was known about the behaviour of polymer fluids and the necessary management techniques. The present authors also agree with Berkovitz and Long that simply meeting a manufacturer's product specification cannot guarantee successful construction. Specifications must respect the site situation, and vice versa.

As regards the required fluid properties, Berkovitz and Long noted that some polymer manufacturers recognised that fluids with higher viscosity were essential for excavation support in soils of coarser particle size, although not all manufacturers adopted this practice. As can be seen in Tables 4.2–4.5, polymer manufacturers commonly recommend higher viscosities when soil particle (pore) sizes increase, so the inconsistency noted by Berkovitz and Long is no longer in issue. They also commented on the recommendation by polymer manufacturers that the Marsh funnel should be used for viscosity control; they opposed its use on the basis that 'the polymer chains prohibit valid flow through the funnel', and proposed that other appropriate methods should be used. The present authors are of the opinion that the Marsh funnel is useful for the on-site assessment of the condition of properly managed polymer fluids. However, as noted in Chapter 9, the Marsh funnel test is not suitable for the characterisation of rheological behaviour, as it does not return a viscosity value let alone the shear-rate dependent viscosity data that are required for detailed analyses of fluid flow behaviour. In fact, it is the collective action of the polymer chains that gives the fluid a higher viscosity and hence a higher funnel time than that for water, so the issue raised by Berkovitz and Long is actually a required feature of polymer fluids rather than a measurement problem. In concluding their discussion, Berkovitz and Long made several suggestions, one of which was, if possible, the development of a generic specification for all polymer fluids in all types of soils and conditions. This chapter provides some generic advice, but the present authors are of the opinion that a one-size-fits-all approach to polymer fluid specification is not appropriate, especially if such a specification is for use across many countries, as site practice can be very different across the world. Specification limits must respect site practice and, if drafted to encompass all possible practices, may become overly permissive and a source of problems on sites where a narrower specification of properties is necessary.

10.6.3 Tuen Mun Station, Hong Kong

An early bored piling project in Hong Kong using synthetic polymer fluids was described in Section 7.4.2. The project-specific fluid specification is shown in Table 10.2. As can be seen, the control parameters included the conventional parameters: the Marsh funnel

Table 10.2 Specifications for a bored piling project in Hong Kong using synthetic polymer fluids

Description	Before excavation		Before concrete placement	
	pH	Viscosity	Density	Sand content
Instrument	Indicator strip	Marsh funnel	Mud balance	API sand content kit
Value range	9.5–11.5	>50 s	≤1.1 g/cm^3	≤3.0%

Lam and Jefferis (2014).

time, density, sand content and pH. It is interesting that only the viscosity and pH were controlled before excavation, and that only the density and sand content were checked prior to concreting. As the density and sand content of polymer fluids are usually very low before excavation starts (i.e. water density and near-zero sand content even for reused polymer fluids), the omission of these tests from the before-excavation specification is of no great significance if the fresh fluid has been appropriately prepared and/or the used fluid properly rejuvenated. However, in the authors' opinion, the viscosity and pH of the fluid should also have been tested at the end of the excavation of each pile to assess any loss of viscosity due to fluid contamination by the subsoil or the groundwater and the reduction in pH that can occur during digging. These data are essential for proper rejuvenation of used fluids. Lastly, it is interesting to note that the used fluids were specified to have a density of no more than 1.1 g/cm^3 and a sand content of no greater than 3.0%. These values are very high by current standards, especially the density, which at 1.1 g/cm^3 would permit a total solids content of about 160 kg/m^3. A few case histories of unsuccessful projects due to polymer contamination by soil/groundwater were discussed in Chapter 7.

10.6.4 Polymer manufacturers

A survey of the literature has shown that although all polymer manufacturers provide guidance on material usage (see Tables 4.1–4.4), only a few propose fluid specifications for quality control purposes. Tables 10.3 and 10.4 list the specifications of polymer fluids from two manufacturers, namely KB International (SlurryPro CDP) and CETCO (Shore Pac). Both specifications require tests for the Marsh funnel time, density, sand content and pH. The specified limiting values are quite similar for the two manufacturers except for the Marsh funnel time, where there is a marked difference: KB International suggests a minimum value of 50 s with no upper limit, while CETCO recommends a time of between 35 and 135 s.

Table 10.3 Specifications for SlurryPro CDP polymer fluid

Property	Test method	Requirement
Density	Mud balance	$\leq$1.035 g/cm^3 (see note)
Viscosity	Marsh funnel	>50 s
pH	pH indicator strip with a minimum of four colour bands	11–12
Sand content (from half a foot above the base of excavation)	API sand content kit	$\leq$1.0%

KB International (2007).
Under normal operating conditions, the manufacturer also suggests a density range of between 1.01 and 1.04 kg/m^3, unless the fluid-weighting agent WeightIt and the fluid-loss-reduction additive SandSeal has been used to densify the fluid.

Table 10.4 Specifications for Shore Pac polymer fluid

Property	Test method	Requirement
Density (just prior to placing concrete)	Mud balance	$\leq$64 lb/ft^3 ($\leq$1.025 g/cm^3)
Viscosity	Marsh funnel	35–135 s
pH	pH meter or indicator strip	8–10
Sand content (just prior to placing concrete)	API sand content kit and bleach	$\leq$1.0%

CETCO (2012).

As regards the limiting values for other parameters, it can be seen from Tables 10.3 and 10.4 that both manufacturers specify a sand content of less than 1%, which is unchanged from the recommendation of O'Neill (1991). For the fluid density, CETCO and KB International suggest upper limits of 1.025 and 1.035 g/ml, respectively. These density requirements are markedly more stringent than the <1.1 g/ml used for the Tuen Mun project in Hong Kong (see Table 10.2), although more relaxed than the <1.02 g/ml typical figure given in the third edition of the *ICE Specification for Piles and Embedded Retaining Walls* (ICE, 2017) – hereafter referred to as SPERW 2017. Neither of the manufacturers presents a scientific basis for these density limits, although it is likely that they were selected to limit the effect of polymer sorption, as is the figure of <1.02 g/ml in SPERW 2017. As discussed in Section 6.8, a polymer fluid is an active chemical system, so it is always advisable to keep the density of the fluid (i.e. the amount of suspended solids) as low as possible in order to limit polymer loss by sorption onto the suspended particles. However, the authors have shown that the depletion of active polymer may still occur at a low soil concentration, so it is preferable to directly measure the amount of active polymer in the fluid (Lam *et al.*, 2014); but, to date, the authors are not aware of the adoption of this procedure on a site, perhaps because of the cost of the necessary equipment and the time required for each test. Currently the Marsh funnel time is the only parameter used to assess, albeit indirectly, polymer concentration in the field.

10.6.5 Public authorities

In the USA, at the time of writing this book, only the American Association of State Highway and Transportation Officials has published a specification for polymer fluids (AASHTO, 2010). The AASHTO specification is also referenced in the latest version of the US Federal Highway Association drilled shaft manual (Brown *et al.*, 2010). Table 10.5 presents the limiting values from the AASHTO specification. It can be seen that the requirements are very similar to those suggested by CETCO (see Table 10.4), with only small differences in the limiting pH and viscosity values. Again, no arguments are presented to justify the limiting values. It seems that, not unreasonably, the AASHTO specification was heavily influenced by specifications from polymer manufacturers.

In continental Europe, there appears to be a complete lack of industry guidance and specification. The European standards on the construction of bored piles and diaphragm

Table 10.5 AASHTO specifications for polymer fluids for use in drilled shaft construction

Property	Test method	Requirement
Density (immediately prior to concrete placement)	Mud balance	≤64 lb/ft^3 (≤1.025 g/cm^3)
Viscosity	Marsh funnel and cup	32–135 s
pH	Glass electrode pH meter or indicator strip	8–11.5
Sand content (immediately prior to concrete placement)	API sand content kit	≤1.0%

AASHTO (2010).

walls EN 1536 and EN 1538 (BSI, 2010a, 2010b) do not provide any guidance on the use of polymer fluids, although they are mentioned as possible excavation support fluids.

In the UK, the second edition of the *ICE Specification for Piling and Embedded Walls* (ICE, 2007) was silent on the required properties of polymer fluids, although bentonite slurries were covered in detail. However, the third edition, SPERW 2017, does include some typical compliance values for polymer fluids and specifications for the locations from which samples should be taken for testing. The requirements of SPERW 2017 and its recommended control approach are considered in the following section. SPERW 2017, when taken together with the commentary provided in this chapter, represent the authors' current position with regard to polymer fluid specification.

10.6.6 *ICE Specification for Piling and Embedded Retaining Walls* (ICE, 2017)

Tables 10.6 and 10.7 show the recommended test locations and typical polymer fluid control values developed from Tables B20.1 and C20.2 of SPERW 2017. As shown in Table 10.6, the control parameters are the Marsh funnel time, density, sand content and pH. There is also provision for filter loss tests to determine the filter cake thickness, but to date the authors are not aware of any published field data on filter cake thickness.

Before considering the individual test parameters, it is important to note that the 'during digging' test location is new to SPERW 2017. It is not a test location that appears in other specifications. The authors are not aware of any corpus of 'during digging' test results for either polymer fluids or bentonite slurries. The figures included in Table 10.7 are best estimates of compliance values – values that can be expected to evolve as data become available. Tests during digging are not advised as mandatory, and should be undertaken only if problems occur or are expected. For the other test locations, each of the proposed limiting values is discussed below.

10.6.7 Rationale for the compliance values

10.6.7.1 The Marsh funnel time

The Marsh funnel flow time gives an indication of the fluid viscosity, which itself is a function of the active polymer concentration and the suspended solids content of the

Table 10.6 Polymer fluid test schedule adapted from SPERW 2017

Property to be measured	Test method and apparatus	Freshly mixed	Ready for reuse	Sample from excavation during digging	Sample from excavation prior to concreting
Density: g/ml	Pycnometer bottle and electronic balance	M	M	PS	M
Fluid loss at 30 min: ml	Low-temperature fluid loss test equipment	n/a	PS*	PS*	PS*
Filter cake thickness at 30 min: mm	Low-temperature fluid loss test equipment	n/a	PS*	PS*	PS*
Marsh funnel time: s	Marsh funnel	M	M	PS	M
Sand content: % (by bulk volume)	API sand content kit	n/a	M	PS	M
pH	Electrical pH meter to BS 3445; range pH 7–14	M	M	PS	M

ICE (2017).
M – mandatory test location.
PS – only if required in the project specification.
PS* – polymer fluids containing some excavated soil will develop a filter cake, typically thin because of the low specified maximum solids content. Currently, these tests are perhaps more useful for research than field application.
n/a – not applicable. For the freshly mixed fluid the fluid loss test is not applicable, as pure polymer fluids show very high fluid losses (typically emptying the cell within the 30 min test time) and they build no filter cake, and so tests provide no useful information. Modified polymer systems may be developed that show filter loss control and/or cake formation. Tests on such systems may provide useful control information, and the manufacturer's advice should be sought.

fluid. The purpose of a Marsh funnel test is therefore

- to ensure that the polymer fluid has sufficient viscosity to minimise bulk fluid loss to the ground and especially to coarse soils so that the stabilising pressure can be maintained near the face of the excavation
- to ensure that there is sufficient active polymer to inhibit soil dispersion and prevent the polymer fluid becoming a soil–water slurry.

In Table 10.7, a Marsh funnel time of more than 90 s is proposed for polymers when freshly mixed and prior to use. This value, and indeed all the Marsh funnel times given in Table 10.7, will depend on the particular polymer used and may be adjusted with input from the manufacturer. However, it is important to recognise that the polymer concentration will drop during use, and that the initial and working concentrations must both be sufficient to offer a reasonable margin of safety against the fluid flipping from a polymer fluid to a soil–water slurry. It is very difficult to recover a flipped soil–water slurry as dispersion of soil greatly increases the soil surface area exposed to polymer and thus the amount of polymer that must be added to satisfy the greatly increased number of sites for polymer sorption.

With regard to the limiting viscosity values, ideally these should be derived from rigorous research considering soil properties, support fluid loss to the ground and excavation stability, but at the time of writing such research has yet to be undertaken. In the absence

Table 10.7 Typical compliance values for support fluids prepared from polyacrylamide-based polymers for used in bored piles as specified in SPERW 2017

Property to be measured	Compliance values			
	Freshly mixed	Ready for reuse	During digging	Prior to concreting
Density: g/ml	<1.01	<1.02	<1.05	<1.02
Fluid loss at 30 min: ml	n/a	n/a	n/a	n/a
Filter cake thickness at 30 min: mm	*	*	*	*
Marsh funnel flow time: s	>90	60–120	60–120	90–120
Sand content: % (by bulk volume)	n/a	<2	<5	<1
pH	9–11	9–11.5	9–11.5	9–11.5

ICE (2017).
* If the filter cake thickness is specified, it should be for a fixed volume of fluid (400 ml), as all fluid may drain from the cell during the test.
n/a – not applicable.
For polymer fluids with low solids contents, the Marsh funnel time is an indicator of the active polymer concentration in the fluid (polymer sorbed onto soil in suspension will not contribute to solution viscosity). The maximum Marsh funnel time of up to 120 s for 'ready for reuse fluid' is a typical value. It may be revised if appropriate experience exists or if the polymer manufacturer recommends other values.
The minimum Marsh funnel time of 60 s for digging fluid is again a typical value. A higher value may be adopted if recommended.

of fuller information, the authors are in agreement with the Marsh funnel time range of between 60 and 120 s as specified in SPERW 2017. According to the results presented in Chapter 9, this time range roughly corresponds to a viscosity range of between 40 and 300 cP (mPa s) if measured at a shear rate of 10 s^{-1}, which can be achieved by operating a Fann-type direct-indicating viscometer at 6 rev/min.

Although some of the case histories discussed in Chapter 7 have shown that fluids with viscosity values lying outside the SPERW 2017 specification limits have successfully stabilised pile bores, other cases have shown that collapses and poor-performance issues are a risk. For example, on the Vasco da Gama Bridge project (see Section 7.2.2) most of the piles were successfully completed using a polymer fluid with a Marsh funnel time of only 40 s, but one pile bore did collapse at a location near salt flats due to contamination of the fluid by the saline soil/groundwater. A minimum of 60 s therefore seems to be an appropriate lower bound even when drilling in clay soils, as local failures can still occur if seams of coarse soils are present, as shown in the Guinea-Bissau case history (see Section 7.3.2). At the other extreme, the Munich case history (see Section 7.2.4) has shown that fluids with too high a Marsh funnel time (>200 s) can cause problems during concrete placement, so it seems prudent to select an intermediate figure. An upper limit at 120 s has been found effective – a viscosity that should be sufficient for most situations, including highly permeable sand-gravel mixtures (as on the Munich site). If fluid loss is a problem, polymer-compatible clogging agents can be employed.

10.6.7.2 Density

First of all, it should be recognised that although a higher fluid density can provide the benefit of slightly increased excavation stability due to the increased hydrostatic pressure, a low, near-water, density is always preferred for polymer fluids. This is because fluids that contain a significant amount of solids are much more likely to cause problems due to the depletion of active polymer in the system and the potential for the accumulation of sediment at the base of the excavation. In SPERW 2017 and in the authors' approach to specification, limits are set on density to control these harmful effects. Put simply, the reason for density control is to prevent the polymer fluid turning into a soil–water slurry. Ideally, the density test should be subsidiary to the residual concentration test, as density indicates the amount of absorbent (soil) while the residual polymer concentration test measures the amount of absorbate (active polymer) left in the system. However, the authors recognise that the current residual polymer test has not been widely adopted on sites because of the cost of the equipment (centrifuge and viscometer) and the time taken to do a test. However, it remains a powerful technique for laboratory investigations of the behaviour of polymer fluids when loaded with soil.

SPERW 2017 requires the polymer fluid density to be measured using a pycnometer (density bottle) and an electronic balance (see Section 6.6.4). This procedure should offer a resolution of 0.005 g/ml or better if carried out carefully (Jefferis and Lam, 2017; Lam 2016). As previously discussed, the density of a fresh, unused polymer fluid will be effectively that of water (i.e. 1.00 g/ml at 4°C). In SPERW 2017, the upper limit of 1.01 g/ml has been set, recognising that, even with a pycnometer and an electronic balance, under site conditions it is difficult to measure density to a precision significantly better than

0.005 g/ml. Measurement of the density of fresh fluid sets a baseline for the fluid density and provides a check on the accuracy of the density measurement procedure. Measurements of the fresh fluid density can be limited to weekly or longer intervals, and certainly are not required on every batch of fluid prepared.

SPERW 2017 suggests a maximum density of 1.02 g/ml for fluid for reuse and prior to concreting. A polymer fluid with a density higher than this is at risk of flipping or having already 'flipped' to a soil–water slurry. Recovery of such a slurry will be expensive in polymer consumption. The chosen value of 1.02 g/ml follows from the results of Jefferis and Lam (2013) and Lam *et al.* (2014). These papers report results that show that the amount of active polymer can be significantly reduced if the clay (bentonite and London Clay) concentration reaches 30 kg/m^3. At this soil concentration, the density of the fluid will be about 1.019 g/ml (calculated from Equation 10.1 using a typical soil particle density of 2.65 g/ml). The authors' site experience has shown that it is perfectly possible to control the fluid density to below 1.020 g/ml in a wide range of ground conditions as long as the used fluid is properly cleaned by sedimentation with the help of the manufacturer's treatment products if appropriate and rejuvenated by fresh polymer addition. Details of the relevant site procedures can be found in Section 5.8.

The suggested maximum density value of 1.05 g/ml during excavation is supported by field data reported by Lam and Jefferis (2015). Densities of up to 1030 kg/m^3 (1.03 g/ml) were recorded for polymer fluid samples taken from the bottom of pile bores at a site in central London, where the soil conditions were stiff clay over fine sand (see Section 9.4.8). However, as described in the Glasgow case history (see Section 8.2 and Figure 8.5), a density of 1.05 g/ml was exceeded in those silty soils. Monitoring of the fluid density during excavation is therefore recommended for excavations in silty grounds to confirm that the as-mixed polymer fluid concentration is adequate to limit soil dispersion into the polymer fluid. This monitoring will be necessary only until a regular working pattern has been established.

10.6.7.3 Sand content

Fundamentally, the purpose of the sand content test is to limit the amount of suspended coarse soil particles in the fluid so as to reduce the likelihood of sediment accumulating at the base of the excavation and, in combination with density measurements, to assess the potential for polymer loss by sorption onto soil. A maximum content of 5% is suggested for polymer fluids during digging so as to limit the risk of 'soft toes'. This value is supported by field data collected during the Glasgow project (see Section 8.2 and Figure 8.5), where a maximum sand content of 5% was recorded for fluid samples taken from the bottom of pile bores. The upper limit of 1% prior to concreting is the same as that proposed by O'Neill (1991) and some polymer manufacturers (see Tables 10.1, 10.3 and 10.4). However, it must be emphasised that satisfying this requirement alone cannot guarantee the absence of soft sediment, as the build-up of sediment depends on many factors other than the sand content, including the total solids content, the fluid viscosity, the depth of the excavation (the height of the fluid column through which solids must fall and thus the time over which significant amounts of solids may 'rain' down on the excavation base), any use of treatment additives, and the duration of the sedimentation period after final base cleaning. For example, Ho and Tan (1996) reported (see Section 7.4.4) that soft toes still formed at the base of a test barrette when the sand content was less than 1%. In this case, the fluid viscosity was very low (a Marsh funnel time of 28–32 s) and the waiting time was 4.5 h. As previously suggested, to ensure a clean base, the contractor can monitor the reduction in the sand content over time in a trial or working pile bore to obtain the actual settling rate data (see the Glasgow case history in Section 8.2).

10.6.7.4 pH

For PHPA fluids, a moderately high pH is essential to ensure dissociation of the polymer molecules in solution (see Section 3.5), but too high a pH (>11.5) can promote hydrolysis and degradation of the polymer. The pH is specified to be between 9 and 11 when polymer fluid is first prepared, a range similar to that suggested by polymer manufacturers, although manufacturers may provide other values for their products. A pH of up to 11.5 is permitted in use and for reuse, as cement contamination can lead to an increase, although such contamination should be avoided as the calcium ions from the cement can damage PHPAs.

10.6.7.5 Residual polymer concentration

The residual polymer concentration is not included in SPERW 2017, but is discussed here for completeness. It is a new control parameter – introduced to prevent the depletion of active polymers due to sorption while in use. To prevent the used fluid from becoming a soil–water slurry, contractors should regularly add new polymer material to recycled fluids and remove suspended solids by sedimentation. These procedures have been discussed in Section 5.8. The residual concentration test allows the user to measure directly the amount of active polymer left in the system. The present test method requires the use of a Brookfield viscometer equipped with a low-shear adaptor (see Section 6.8), although other procedures may be developed. The present method only works with pure, unblended polymer fluids; that is, fluids with only a single active polymer.

The current state of knowledge does not allow a minimum residual polymer concentration to be determined with confidence. In the oil industry, Kadaster *et al.* (1992) proposed an active PHPA concentration of at least 2.9 kg/m^3 (1.0 lb per barrel) in a bentonite-based fluid system to prevent hole enlargement while drilling in water-sensitive shale. In a pure PHPA fluid system, as is common in civil engineering, lower concentrations are used (often in the range 1–2 kg/m^3, although this will depend on the polymer) and achieve excellent results. Likos *et al.* (2005) conducted a series of slake durability tests using Super Mud Dry manufactured by PDS. From the results, they concluded that the slake durability index (Id) increased (improved) slightly with increasing polymer concentration until about 0.5 g/l, which is the concentration recommended by the manufacturer when drilling in clay and shale. Polymers are believed to have similar beneficial effects in reducing the swelling of clays, but further research is required to establish the interplay between soil mineralogy, polymer chemistry and concentration. In the absence of further information, the authors advise a minimum active polymer concentration of 0.3 kg/m^3 (0.03% by mass) for used PHPA fluids. Satisfying this residual concentration requirement should help users to maintain the chemical reactivity of the fluid and so prevent 'flipping' of the fluid to a soil–water slurry. However, it may not be sufficient to ensure hole stability and/or limit bulk fluid loss to open grounds – matters that must be considered separately.

10.7. Concluding remarks

SPERW 2017 provides a set of typical compliance values for high-molecular-weight (>10 million g/mol) PHPAs. It must be stressed that these values are for PHPA solutions and are not necessarily applicable to blended polymer systems, as the chemistry and behaviour of such systems can be different. Furthermore, the SPERW 2017 values were drafted for bored piles (drilled shafts), although they may also provide a baseline for diaphragm wall excavation by grab. They are not for diaphragm wall or barrette excavation by hydromill, as blended lower-molecular-weight polymers are likely to be used in these applications.

The authors recognise that, as experience with polymer fluids develops, the numerical values of the control parameters may be refined and become more project-specific. However, they believe that the underlying rationale for the selection of the control parameters will remain and provide a framework for future specification development. As more experience is gained with the use of polymer fluids, major developments are to be expected. There are opportunities to be seized by all the parties involved.

REFERENCES

AASHTO (American Association of State Highway and Transportation Officials) (2010) *LRFD Bridge Construction Specifications*, 3rd edn. AASHTO Washington, DC, USA.

Berkovitz BC and Long CS (1995) Use of polymer slurries for drilled shaft construction. *Proceedings of the 31st Symposium on Engineering Geology and Geotechnical Engineering, Logan, Utah*. Idaho State University, Pocatello, ID, USA, pp. 17–23.

Berry AG (2009) Method for cleaning and checking the base of diaphragm wall panels. *Ground Engineering* **42(11)**: 29–31.

Brown DA, Turner JP and Castelli RJ (2010) *Drilled Shafts Construction Procedures and LRFD Design Methods*. Federal Highway Administration, US Department of Transportation, Washington, DC, USA, Report FHWA NHI-10-016.

BSI (British Standards Institution) (2010a) BS EN 1536. Execution of special geotechnical work. Bored piles. BSI, London, UK.

BSI (2010b) BS EN 1538. Execution of special geotechnical work. Diaphragm walls. BSI, London, UK.

Camp WM, Brown DA and Mayne PW (2002) Construction method effects on axial drilled shaft performance. *Proceedings of International Deep Foundations Congress 2002, Orlando*. American Society of Civil Engineers, Reston, VA, USA, pp. 193–208.

CETCO (2012) *Drilling and Mixing Guide – Use of Shore Pac Polymer Slurry in Slurry-Displaced Foundation Construction*. CETC, Birkenhead, UK.

Ho CE and Tan CG (1996) Barrette foundation constructed under polymer slurry support in Old Alluvium. *Proceedings of the 12th Southeast Asian Geotechnical Conference, Kuala Lumpur*. Southeast Asian Geotechnical Society, Pathumthani, Thailand, pp. 379–384.

ICE (Institution of Civil Engineers) (1999) *Specification for the Construction of Slurry Trench Cut-off Walls*. ICE, London, UK.

ICE (2007) *ICE Specification for Piling and Embedded Retaining Walls*, 2nd edn. Thomas Telford, London.

ICE (2017) *ICE Specification for Piling and Embedded Retaining Walls*, 3rd edn. ICE, London, UK.

Jefferis SA (1977) Contribution to the discussion of 'Design and control of bentonite/clay suspensions and concrete in diaphragm wall construction'. *A Review of Diaphragm Walls, London, 1976*. Institution of Civil Engineers, London, UK, pp. 88–89.

Jefferis SA (2014) Grouts and Slurries. In *Construction Materials Reference Book* (Doran DK (ed.)). Thomas Telford, London, UK, pp. 165–194.

Jefferis SA and Lam C (2013) Polymer support fluids: use and misuse of innovative fluids in geotechnical works. *Proceedings of the 18th International Conference on Soil Mechanics and Geotechnical Engineering, Paris*. Presses des Ponts, Paris, France, pp. 3219–3222.

Jefferis SA and Lam C (2017) Using density to determine the solids content of construction slurries. *Proceedings of the 19th International Conference on Soil Mechanics and Geotechnical Engineering, Seoul*, South Korea, pp. 1695–1698.

Kadaster AG, Guild GJ, Hanni GL and Schmidt DD (1992) Field applications of PHPA muds. *SPE Drilling Engineering* **7(3)**: 191–199.

KB International (2007) *SlurryPro: Surface Plant, Excavation Tools & Technique, and Product Applications – Recommendations for Bored Piles*. KB International, Chattanooga, TN, USA.

Lam C (2016) Evaluation of density-measurement methods for construction slurries. *Geotechnical Testing Journal*, **39(3)**: 507–514

Lam C and Jefferis SA (2014) The use of polymer solutions for deep excavations: lessons from Far Eastern experience. *HKIE Transactions* **21(4)**: 262–271.

Lam C and Jefferis SA (2015) Physical properties of polymer support fluids in use and their measurement techniques. *Geotechnical Testing Journal* **38(4)**: 490–500.

Lam C, Martin, PJ, Jefferis SA and Goodhue Jr KG (2014) Determination of residual concentration of active polymer in a polymeric support fluid. *Geotechnical Testing Journal* **37(1)**: 46–59.

Likos WJ, Akunuri K and Loehr JE (2005) Performance of PHPA polymer slurries for applications in Missouri shale. *Proceedings of Sessions of Geo-Frontiers 2005 Congress, Austin, Texas*. American Society of Civil Engineers, Reston, VA, USA, pp. 665–676.

Marsh HN (1931) Properties and treatment of rotary mud. *Transactions of the AIME* **92(1)**: 234–251.

O'Neill MW (1991) Effects of mineral and polymer slurries on drilled shafts. *Proceedings of the 16th DFI Annual Members' Conference, Chicago*. Deep Foundations Institute, Hawthorne, NJ, USA, pp. 141–154.

Polymer Support Fluids in Civil Engineering
ISBN 978-0-7277-5785-9

http://dx.doi.org/10.1680/psfce.57869.285

Index

AASHTO (American Association of State Highway and Transportation Officials), 1–2, 3, 276, 277
ACCU-VIS, 80
acid dissociation constant (*K*a), 42
additives
  to AquaVIS, 63
  in case studies
    Glasgow, 189, 190, 193, 195–196, 201
    Stratford, 207–208, 211
  for cleaning, 105–106, 195–196
  to CoreSMART, 86
  to Enhanced SlurryPro CDP, 78–79
  to G3, 82
  Highway 17 Replacement Bridge (South Carolina), 175, 176
  to HydroCUT, 85
  to POLY-BORE, 70–71
  in recycling and rejuvenation for reuse, 107, 108
  to Shore Pac, 79–81
  to SlurryPro CDP, 71–76, 207–208, 211
  to SlurryShield and SlurrySMART, 83
  to Super Mud, 68–70
aeration, 91
  case studies
    Glasgow, 190, 191, 199
    Stratford, 207, 212
  rheology and, 234, 245, 247, 248, 249
  see also air bubbles
aerometer, 134–135, 268
Africa, 157–159
ageing and viscosity, 245–247
agriculture, bentonite slurry, 23
air bubbles, entrained, 199, 234, 243, 247, 248, 249, 259, 268
air lifting, 105, 107
  case studies, 163, 211
air-operated double-diaphragm pumps, 96, 97
Alabama, Auburn University National Geotechnical Experimentation Site at Spring Villa, 173–174
Alberta, oil sands, 181
AlfaBond, 32
alluvial sand (Glasgow), particle size, 189, 193
American Association of State Highway and Transportation Officials (AASHTO), 1–2, 3, 276, 277
American Petroleum Institute (API) filter loss test, 139, 267, 270
Americas, 170–180
amplitude sweep test, 255–256, 256, 259
anhydroglucose units, 35
anionic (negative) charge, 32
API (American Petroleum Institute) filter loss test, 139, 267, 270
AquaSed, 98
Aquasorb, 68–69
Aquavis, 63
aquifer, 204
  artesian, 100–102
artesian aquifer, 100–102
Asia, 159–170
Auburn University National Geotechnical Experimentation Site at Spring Villa (Alabama), 173–174
auger
  continuous-flight, 148, 153, 173, 189, 207
  rotary, 159
  twin-flight, 103, 193, 209

back-analysis in load testing, Stratford, 221–226, 226
backhoe, 98
Bangkok Elevated Road and Train System (BERTS), 163–167
barrette, 167–168, 273, 282

base (of excavation)
cleaning/cleanliness, 12–13, 105–107, 271
case studies, 159–160, 192–198, 209–211
minimising soil softening at, 7–10
soft, 10
Benin, Lake Nokoué, 157, 158
bentonite, 17–28
calcium *see* calcium bentonite
case studies, 146–181
constraints on use of, 17–28
granular, 69–70, 81
history, 3–4
polymers mixed with, 164, 165
slurry, 10, 18, 19, 22, 24–25
sodium, 17, 23
Stratford (London) project, 207, 208–209, 211, 212, 213, 214, 215, 221, 225, 226
viscosity and effects of, compared with Telmarch, 60
Berkbent 163, 207, 208, 213
Berkovitz and Long specifications, 273–274
bicarbonate ions (HCO_3^{2-}), 115, 118
Bingham plastic model, 127, 130–131, 231, 236, 237
Binh Loi Bridge (Vietnam), 168–169
biodegradable natural polymers, 31, 59–64, 171, 247
biological oxygen demand (BOD), 63, 109
bleach (sodium hypochlorite), 107, 109
sand content estimation, 137, 138, 269
Slurry Buster, 81
blended polymer, 83–86
rheological properties, 243, 244
bonding to reinforcement, 12
bored pile *see* pile
Brazil, 181
British Columbia, Golden Ears Bridge over Fraser River, 180–181
British standards, polyacrylamide, 40
Brookfield viscometer, 121, 128–131, 140, 141, 232, 233, 239, 240, 246, 268, 270, 282

calcium bentonite
reserves, 17, 18
sodium-activated, 18, 208
calcium ions (Ca^{2+}) and water hardness, 118–120
Canada, 180–181
Canary Wharf, 147–148, 229
Cape Fear River bridge (North Carolina), 172–173
CAPWAP (Case Pile Wave Analysis Program), 199
carbohydrates, 32–33
carbonate ions (CO_3^{2-}), 115, 116, 118, 119
carboxymethyl cellulose (CMC), 29, 31, 36–37, 38, 45
commercial systems, 61, 63, 64, 83
carboxymethyl hydroxyethyl cellulose (CMHEC), sodium, 38
Carreau model, 230–231, 240, 241
Case Pile Wave Analysis Program (CAPWAP), 199
case studies (around the world), 145–185
detailed analysis, 187–228
cations (positively charged ions), 32
viscosity and, 243–244
CDP *see* SlurryPro CDP
cellulose (and derivatives), 34, 35–38
cement contamination *see* concrete
CEMSET in back-analysis in load testing, 223–226
centrifugal pumps, 70, 92, 95, 96, 110
case studies, 190, 191, 199, 209, 212
CETCO
Shore Pac *see* Shore Pac
Slurry Buster and Slurry Buster Dry, 81
chain entanglements, 242, 243, 245, 252, 258, 260
Channel Tunnel Rail Link, 22, 202, 204–205
charge (ionic), 32
positive *see* cations
chemical oxygen demand, 109
chemical reaction with fresh concrete, 11–12
Chin's method of back-analysis in load testing, Stratford, 221–223, 224
chlorine (residual), treatment, 79
Choa Chu Kang (Singapore), 168
Christopher Bond Bridge over Missouri River (Kansas City), 172
CIRIA (Construction Industry Research and Information Association), 3
clay
London Clay, 7, 141, 204, 249, 281
pile performance in, 174–175
smectite, 18
cleanliness (and cleaning), 105–107
additives for cleaning, 105–106, 195–196
case studies, 193, 195–196, 209–211

flat-sided cleaning bucket, 103, 193, 211
pile base *see* base
pile shaft, at concrete–soil interface, 12
self-cleaning, 96–100
in situ cleaning, 10–11, 211
CMC *see* carboxymethyl cellulose
commercial polymer systems, 58–88
compliance values, 277, 279, 283
concentration (of polymer fluids)
residual, 138–142, 280, 282–283
viscosity and effects of, 239–240
concrete (and cement)
bonding to reinforcement, 12
clean soil–concrete interface at pile shaft, 12
contamination, 150, 282
Glasgow case study, 192–193
Telmarch effects on, 60
fresh/wet, 11, 197, 201
intermixing of fluid with *see* intermixing
non-reaction of fluid with, 11
slump test, 229
solid, enabled without inclusions, 11
concreting
case studies
Glasgow, 197–198
Singapore, 168
Stratford, 214–216
Vietnam, 169
cleaning before, 105
reducing amount of suspended particles before, 10–11
specifications and associated controls, 255, 267, 271, 272, 278, 279, 281
tremie pipes and, 10, 11, 95, 107, 169, 201, 215
conductivity, electrical, 113–114
cone-and-plate geometry, 233, 238
cone penetration test (CPT) in case studies
Benin, 157
UK, 145, 205
Vietnam, 169
Construction Industry Research and Information Association (CIRIA), 3
continuous-flight auger, 148, 153, 173, 189, 207
controls *see* specifications and associated controls
Cooper Marl, 175–179
copolymers, 31
CoreSMART, 86
cross-hole (ultra)sonic logging, 198–199, 217, 218, 219

Deborah number, 253
De-Chlor, 79
deep foundations around the world *see* case studies
Delhi metro stations, 170
density, 131–137, 263–269, 280–281
in case studies
Glasgow, 193–195
Stratford, 212, 213
repeated reuse effects, 251
specifications, 269, 275, 276, 278, 279, 280–281
density bottle, 135–137
departments of transportation (DOTs – USA), 170, 172
desanding, 99
dewatering, 204–205
bentonite sludge, 23
diaphragm pumps, 95, 96, 199
diaphragm walls, 3, 23, 26, 61, 83, 196, 248–249
Brazil, 181
India, trench collapse, 169–170
Taiwan (Taipei City), 162
USA, 171
digging bucket, vented, 103, 209
Dingpu Station (Taiwan), 162
direct-indicating viscometer, 123–128
disaccharides, 33
dispersion of soil/excavated materials, 26
bentonite fluids and, 26
preventing, 10
disposal *see* used (waste) fluids
dissociation of polymers in solution, 41–45, 115, 116
dissolved solid *see* solution
double-diaphragm pumps, air-operated, 96, 97
drilled shaft – *see* bored pile
drilling
history, 2–4
horizontal directional, 63, 166
oil or gas *see* oil well (and gas) drilling
reverse-circulation, 159
rotary coring, 65–66
drilling fluid vs support fluid, 2
dry, keeping excavated materials dry by preventing dispersion, 10

dry powder form, 33–34, 41
Dubai, 100, 101
 Metro, 64
dynamic loading testing, Glasgow, 150, 199, 200, 201

East Kent-Ashford to Cheriton section of the Channel Tunnel Rail Link, 22
Edmonton (Alberta), oil sands, 181
elasticity, 253, 254, 255
 see also viscoelasticity
electrical conductivity, 113–114
electron microscope, scanning *see* scanning electron microscope
embedded retaining walls, ICE specification (SPERW 2017), 220, 277, 279, 280, 281, 282, 283
emulsions, liquid, 33–34
Enhanced SlurryPro CDP, 78–79
EnhancIt 100 and EnhancIt 200, 78–79
environmental impact, bentonite, 25
environmental regulations, 107
 bentonite, 22–25
environmental scanning electron microscope (ESEM), 51–52
equipment and instrumentation (site), 89–111
 rheology, 232–233
 Stratford (London), 206–207
ethylene, 32
Europe, 145–157
 public authorities, 276–277
excavation, 1–16
 base *see* base
 in case studies, procedures/techniques
  Glasgow, 192–198
  Stratford, 209–211
 cleaning *see* cleanliness
 critical functions of support fluids during, 7–10
 practice, 271
 site *see* site
 in soils with high groundwater pressures, 100–102
 tools, 102–105
EZ-MUD/EZ MUD GOLD/EZ-MUD PLUS/EZ-MUD DP, 71

Fann viscometer, 121, 123, 124, 214, 232, 236, 239, 240, 241, 242, 268, 270, 280
farming, bentonite slurry, 23
field *see* site
filter cake layer, 12, 21, 113, 164, 169, 174, 201, 270
 thickness, 277, 278, 279
filter loss test (API), 139, 267, 270
flat-sided cleaning bucket, 103, 193, 211
Fleming's method of back-analysis, 223–226
flipping of fluid, 139, 279, 281, 283
Florida, 192
 Miami, 178–179
flow (and flow behaviour), 229–232
 Newtonian, 120–121, 230
 non-Newtonian, 121, 126, 229–232
 PHPA, 241
Fort McMurray (Alberta), oil sands, 181
foundations
 deep *see* case studies
 loading, 10, 12–13
Fraser River, Golden Ears Bridge (British Columbia), 180–181
frequency sweep test, 255–256, 256, 259–260

G3, 82–83
galactose, 33
gas *see* methane gas; oil well (and gas) drilling
gelling (and thixotropy), 21, 105, 173
geometry
 cone-and-plate, 233, 238
 loading tests, 223, 225
 oscillatory tests, 257
 parallel-plate, 233, 257, 269
 viscosity measurement, 230, 231
Geomud-15, 66–67
 case studies, 151, 152, 157, 158, 159
Germany, Munich, 155–157, 280
Glasgow, 21, 150, 187–201, 209, 247, 248, 272, 281
glucose, 33
 dehydrated (anhydroglucose units), 35
Golden Ears Bridge over Fraser River (British Columbia), 180–181
Granular Seal, 69–70, 81
ground conditions in case studies
 Glasgow, 187–189, 199–201
 Stratford, 202–206
 Texas A&M University National Geotechnical Experimentation Site, 180
groundwater
 in case studies, 180
  Glasgow, 188–189

Stratford, 204–205
high pressures, excavations in soils with, 100–102
mechanism of stabilisation with, 5
see also aquifer
GS 550, 64–66
GS-TP (PREMIUM), 62–63, 171
guar gum, 34, 38, 64, 267
Guinea-Bissau, 159

handling, fluid, 9, 11–12
hardness of water, 118–120
Henderson–Hasselbalch equation, 42, 43, 45
Herschel–Bulkley model, 214, 231–232, 236, 237–238
high speed rail
Naples–Rome section of TAV project, 152–155
Taiwan, 159–172
Highway 17 Replacement Bridge (South Carolina), 175–178
historical perspectives, 2–4, 57–59
Ho Chi Min City, 168–169
homopolymers, 31
Hong Kong, 95, 98–99
bentonite disposal, 24
Tuen Mun Station, 162–163, 274–275
Hooke's law, 254–255, 256
horizontal directional drilling, 63, 166
Houston (University of), National Geotechnical Experimentation Site, 179–180
HPAM *see* partially hydrolysed polyacrylamide
hydration requirements of bentonite fluids, 18–21
HydroCUT, 83–86
hydrogen bonds, 41, 247
cellulose, 36
partially hydrolysed polyacrylamide, 44
hydrogen ions (H^+), 114, 115, 116
hydrolysis, degree of, 49–50
hydrometer, 134
hydroxyethyl cellulose (HEC), 37–38
hydroxyl ions (OH^-), 115, 116, 119
hypochlorite *see* bleach

ICE Specification for Piling and Embedded Retaining Walls (SPERW 2017), 220, 277, 279, 280, 281, 282, 283
Impac, 72, 75–76
impeller, 234–235
in situ *see* site
inclusions, enabling solid concrete without, 11
India, 169–170
indicator, pH, 50
strips, 116, 117, 118, 270
infinite shear viscosity, 240, 241, 242
Insta-Clear Dry, 81
InstaFreeze, 76, 78, 83, 175, 176
instrumentation *see* equipment and instrumentation
integrity *see* structural integrity
interim specifications (O'Neill), 271–272
intermixing of wet concrete and support fluid (and its avoidance), 11, 12, 13
case studies, 148, 197, 215, 216
inverse (invert) emulsions, 34
ion–dipole attractions, partially hydrolysed polyacrylamide, 44
ionic charge, 32
ionic compounds, 243–245
ionisation (dissociation) of polymers in solution, 41–45, 115, 116
Italy, 152–155

Janpath Road metro station (Delhi), 170
João Landim Bridge, 159

Kansas City, Christopher Bond Bridge, 172
KB International, 71, 275
CoreSMART, 86
HydroCUT, 83–86
KobbleBlok, 72, 76, 80, 175
SandSeal, 75–76, 80, 83, 175
SlurryPro CDF *see* SlurryPro CDP
SlurryShield, 83, 150
Kentucky, Ohio River bridge, 25, 171–172
KobbleBlok, 72, 76, 80, 175

LA-1-D, 72, 78
Stratford (London), 207–208, 208
laboratory tests/measurements
for fluid disposal, 109
rheological properties, 232–235
Lake Nokoé (Benin), 157, 158
Lambeth Group, 26, 147, 202, 203, 204, 205, 209, 249
landfill, bentonite (UK), 23
lignosulfonates, 31

liquid emulsions, 33–34
Lisbon, Vasco da Gama Bridge, 150–152, 280
load–movement/load–settlement behaviour (case studies)
  Germany, 155, 156
  Italy, 152, 153
  Portugal, 151, 152
  UK
    Glasgow, 187
    Norfolk, 145, 146, 147, 151
    Stratford, 221, 223, 226
  USA, 171, 173, 174
loading
  foundation, 10, 12–13
  testing (in case studies)
    Canada, 180
    Germany, Munich, 155–156
    Italian High Speed Train (TAV) project, 152
    Portugal, Vasco da Gama Bridge over Tagus River (Portugal), 151
    Singapore, Choa Chu Kang, 168
    Singapore, Tampines, 167
    Taiwan High Speed Rail project, 159, 160, 162
    Thailand, Bangkok Elevated Road and Train System (BERTS), 164, 165
    UK, Glasgow, 150, 187, 198–199, 201
    UK, Stratford (London), 220–226
    USA, 170, 171, 172, 180
    Vietnam, Binh Loi Bridge, 168–169
    see also load–movement/load–settlement behaviour
London
  Southwark, 248–252, 281
  Canary Wharf, 147–150, 229
  Stratford, 201–226, 220, 232, 272
London Clay, 7, 141, 204, 249, 281
Low-shear viscosity, 242

Macro-Fill, 80
magnesium ions (Mg^{2+}) and water hardness, 118–120
Mandi House metro stations(Delhi), 170
mannose, 33
manufacturers of polymers, 275–276
marine piles, 150
Marsh funnel (and Marsh funnel time/test), 121–128, 236, 241, 248, 249, 251, 267–268
  in case studies worldwide, 151, 152, 153, 154, 156, 157, 158, 159, 165, 167, 170, 171, 172, 175, 176
    in detailed analyses, 190, 191, 193–195, 199, 212–214
  compliance values and, 277–280
  polyacrylamide-based systems
    Geomud-15, 66–67
    GS 550, 64, 65
    Neat Vis NV-220, 82
    POLY-BORE, 71
    SC MUD, 67–68
    Shore Pac, 79, 80
    SlurryPro CDP, 72
    Super Mud/Super Mud Dry, 69
  SlurrySMART, 85
  in specifications, 267–268, 274, 275, 276, 277, 278
  Telmarch, 59–60
methane gas, 181
metro *see* railways
Miami (Florida), 178–179
MicroBond, 82
microstructure in solution, 51–53
miniature drilled-shaft inspection device (mini-SID), 175, 178
Missouri River, Christopher Bond Bridge (Kansas City), 172
mix water *see* water
mixed polymer–bentonite fluid, 164, 165
mixing, 89–92
  by aeration *see* aeration
  bentonite fluids, 18–21
  in case studies
    Glasgow, 190, 199
    Stratford, 208, 209, 212
  polymer fluids, 89–92
  see also intermixing
molecular weight
  cellulose and derivatives, 37
  polyacrylamides, 45–49, 64
    partially-hydrolysed, 71, 82
  polyethylene, 32
monitoring product usage, 270–271
monosaccharides, 33
MPA *see* SlurryPro MPA
mud balance, 132–133
Multi Seal, 80
Munich, 155, 280

Naples–Rome section of High Speed Train (TAV) project, 152–155
National Geotechnical Experimentation Site (NGES)
  Auburn University (Spring Villa, Alabama), 173–174
  Texas A&M University, 180
  University of Houston, 179–180
natural polymers, 4, 29, 31, 32–33, 45, 247
  biodegradable, 31, 59–64, 171, 247
Neat Vis NV-220, 81–82
negative (anionic) charge, 32
Newtonian fluids, 120–121, 230
nomenclature and terminology *see* terminology and nomenclature
non-Newtonian fluids, 121, 126, 229–232
Norfolk (UK), 145–146
North Carolina, Cape Fear River bridge, 172–173
Novagel polymers, 157, 158

Ohio River bridge, Kentucky, 25, 171–172
oil well (and gas) drilling, 35, 121, 164, 181
  carboxymethyl hydroxyethyl cellulose (CEHEC), 38
  mud balance, 132, 268
  partially hydrolysed polyacrylamides, 41, 282
  solid weighing agents, 76
  viscometers, 123, 125
  xanthan gum, 39–40
oil-in-water emulsions, 34
O'Neill interim specifications, 271–272
OP-4 and OP-7, 61
oscillatory shear, 254–260
oxidising agents, 107, 267, 269
oxygen demand
  biological (BOD), 63, 109
  chemical (COD), 109

PAM *see* partially hydrolysed polyacrylamide
parallel-plate geometry, 233, 257, 269
partially hydrolysed polyacrylamide (PAM/PHPA/HPAM), 2, 40–41, 43–44, 45–53
  case studies worldwide, 45–53, 150, 151, 153, 154, 155, 157, 162, 172, 173, 178, 180
    detailed analyses, 192, 199, 207, 208
  commercial systems, 64, 66, 67, 68, 70, 71, 80, 82
  field testing, 125, 126, 131, 139, 141, 142
  properties, 45–53, 125, 126, 131, 139, 141, 142
  rheological properties, 231, 235, 238, 239, 240, 241, 243, 244, 245–247, 253, 254, 255, 257, 260
  specifications and associated controls, 265–266, 270, 271, 272, 282, 283
  viscosity measurements, 125
particle
  size, alluvial sand in Glasgow, 189, 193
  suspended *see* suspended soil particles
penetration tests, *see also* cone penetration test; standard penetration test
performance (fluid)
  bentonites, perceived reduction, 18
  maintained during course of project, 11
pH, 41, 42–45, 50, 114–118, 270
  case studies, 207, 212
  measurement, 113–118, 270
    combined with electrical conductivity measurement, 114, 117
    techniques, 116–118
  in specifications and associated controls, 267, 270, 275, 276, 277, 278, 282
  viscosity and, 241
PHPA *see* partially hydrolysed polyacrylamide
physicochemical properties of fluids, 113–144
  in case studies
    Glasgow, 189–192
    Stratford, 212–214, 226
  flipping, 139, 279, 281, 283
  rheology, 229–263
  specifications and associated controls, 265–284
pile/piling, 98, 99
  auger *see* auger
  base *see* base
  case studies *see* case studies
  ICE specification (SPERW 2017), 220, 277, 279, 280, 281, 282, 283
  integrity *see* structural integrity
  pumping support fluids into, 6
  shaft *see* shaft
  stiffness *see* stiffness
  test *see* test piles
pK_a, 42
plastic viscosity (PV), 127, 229, 232, 236, 237
polyacrylamides, 34, 40–41, 45–53, 64–83
  commercial systems, 64–83

polyacrylamides (*continued*)
  key properties, 45–53
  partially hydrolysed *see* partially hydrolysed polyacrylamide
POLY-BORE, 70–71
  case studies, 172, 180
polyethylene, 32
polymers and polymer fluids (general aspects), 29–55
  blended *see* blended polymer
  commercial systems, 58–88
  concentration *see* concentration
  definition/concept, 2, 29–30
  Glasgow case study
    cleaning, 195–196
    test procedures and properties after excavation, 193–195
  manufacturers, 275–276
  monitoring product usage, 270–271
  natural *see* natural polymers
  preparation for rheological tests, 234–235
  science, 30–34
  in solution *see* solution
  Stratford (London), effects on concrete quality, 215–216
  synthetic *see* synthetic polymers
  types/classification, 34–41, 265–266
polysaccharides, 33
pore pressure, 205, 253, 260
Portugal, Vasco da Gama Bridge (Lisbon), 150–152, 280
positive charge *see* cations
positive-displacement (PD) pumps, 95–96
powders, 33–34, 41
power law model, 127, 131, 214, 231–232, 236, 237, 238, 242, 243, 249, 251
PREMIUM, 62–63, 171
pressure
  groundwater, excavations in soils with high pressure, 100–102
  pore, 205, 253, 260
producers (manufacturers) of polymers, 275–276
properties *see* physicochemical properties
ProTek 100, 71
  Stratford (London), 207
protons (hydrogen ions; H^+), 114, 115, 116
public authorities, 276–277
pumps/pumping, 91, 92–96
  case studies, 190, 191, 192, 199, 209
  centrifugal *see* centrifugal pumps
  diaphragm pumps, 95, 96, 199

Quick Floc, 70, 81

railways
  Bangkok Elevated Road and Train System (BERTS), 163–167
  Dubai Metro, 64
  Hong Kong Tuen Mun Station, 162–163, 274–275
  India, Delhi metro stations, 170
  Italy, Naples–Rome section of High Speed Train (TAV) project, 152–155
  Taiwan
    Dingpu Station, 162
    high speed rail, 159–162
ramp stress tests, 238
reaction (chemical), fluid with fresh concrete, 11–12
reaction frame (load testing), 220
recycling, 107–109
  self-cleaning of recycled fluid, 97
  see also reuse
reinforcement, bonding to, 12
rejuvenation, 107–109, 266
  see also reuse
reserves of bentonite, 17–18
residual polymer concentration, 138–142, 280, 282–283
resistance in load testing, Stratford, 221–224, 225, 226
resources (reserves) of bentonite, 17–18
retaining walls, embedded, ICE specification (SPERW 2017), 220, 277, 279, 280, 281, 282, 283
reuse, 26, 248–252
  recycling and/or rejuvenation for, 107–109, 266, 275
  repeated, effects, 248–252
reverse-circulation drilling, 159
rheological properties, 229–263
  laboratory techniques, 232–235
  models and their selection, 237–240
rheometer, 232–233
Rome–Naples section of High Speed Train (TAV) project, 152–155
rotary auger, 159
rotary coring, 65–66
rotational viscometry, 267, 270

saccharides, 33
Saigon River, Binh Loi Bridge, 168–169
salts and rheology, 243–245
samples
preparation for rheological tests, 234–235, 249
taking (sampling)
locations, 177, 266–267
sand content and, 137
sand content (and its measurement), 137–138, 269–270
case studies worldwide, 173, 175, 176, 178
in detailed analyses, 189, 193, 193–195, 196, 201, 211, 212, 213
specifications, 271–272, 272–273, 274, 275, 276, 277, 278, 279, 281–282
see also alluvial sand; desanding; Thanet Sand
Sand Sealant, 80–81
SandSeal, 75–76, 80, 83, 175
São Paulo, 181
SC GSTP, 64
SC GUM, 64
SC MUD, 67–68, 155
SC VIS, 64, 155
SC XGUM, 64, 155
scanning electron microscope
of blended polymer systems, 83, 84
environmental (ESEM), 51–52
Scotland, Glasgow, 21, 150, 187–201, 209, 247, 248, 272, 281
screw pump, 95
sediment build-up, 99, 153–155, 170, 270, 281
Glasgow project, 196–197
sedimentation tanks, 97–100
self-cleaning, 96–100
settlement *see* load–movement/load–settlement behaviour
shaft
clean concrete–soil interface at, 12
miniature drilled-shaft inspection device (mini-SID), 175, 178
shear, oscillatory, 254–260
shear degradation, 83, 235, 249
shear rate, 124, 125, 126, 129, 130, 131, 229, 230, 231, 232, 233, 237, 238, 239, 239–243
case studies, 212, 214
constant (SRC), 130
viscosity and effects of, 239–243
see also shear stress–shear rate relationships
shear strength in case studies
UK, Stratford project, 205–206
US
Highway 17 Replacement Bridge, 175
National Geotechnical Experimentation Site located at University of Houston, 179–180
shear stress, 120, 121, 124, 125, 212, 229, 230, 231–232, 233, 238, 255
shear stress–shear rate relationships/plots, 120–121, 231, 232, 236
Stratford (London), 214
shear viscosity, 229 241
infinite, 240, 241, 242
low, 242
zero, 241, 242, 243
Shore Pac, 79–80, 181, 275, 276
Glasgow, 190, 191, 192
rheological properties, 239
Shore Pac XX-Poly, 80
side-wall profiles
India, 158
Taiwan, 159
UK
Norfolk, 147
Stratford (London), 216–217, 220
USA, 171
silos, bentonite, 19, 20
Singapore, 167–168
site (on-site; field; in situ)
cleaning, 10–11 211
conditions in case studies
Glasgow, 187–189, 199–210
Stratford, 202–206
equipment *see* equipment
procedures, 89–111
samples from *see* samples
tests, 113–144, 206
Stratford (London), 201, 209, 210, 212
see also test piles
treatment of fluid on, 22–25
slump test, concrete, 229
slurry *see* bentonite slurry
Slurry Buster and Slurry Buster Dry, 81
SlurryPro CDP, 71–78
additives, 71–76, 207–208, 211
MPA *see* SlurryPro MPA
case studies worldwide, 146, 150, 152, 159, 163, 171, 175

SlurryPro CDP (*continued*)
Glasgow, 189–190, 191, 192, 195
Stratford, 206, 207, 208, 213
Enhanced, 78–79
rheological properties, 237, 249
SlurryShield vs, 83
SlurryPro MPA, 73–74, 75–76, 83, 176
Glasgow, 195
Stratford, 208, 211, 212
SlurryShield, 83, 150
SlurrySMART, 83
smectite clay, 18
soda ash *see* sodium carbonate
sodium-activated calcium bentonite, 18, 208
sodium carbonate (soda ash), 18, 67, 71, 115, 120, 158, 208, 246, 247
dissociation, 43, 115, 116
Enhanced SlurryPro CDP and, 78
Shore Pac and, 79, 190
Super Mud and, 68
sodium carboxymethyl hydroxyethyl cellulose (CMHEC), 38
sodium chloride and rheology, 243–245
sodium form
bentonite, reserves, 17–18
partially hydrolysed polyacrylamide, 43
sodium hypochlorite *see* bleach
soft toe (soft base), 10, 13, 74, 105, 107, 113, 145–147, 150, 167, 168, 192–193
artificial, 145
case studies, 192–193, 196, 201, 212
specifications and associated controls and, 281, 282
soil
case study soil profiles/layers
Glasgow, 188
Stratford, 202, 203, 204
clean concrete–soil interface at pile shaft, 12
dispersion *see* dispersion
with high groundwater pressures, excavation in, 100–102
quality, bentonite (waste) improving, 23–24
softening, minimising, 7
suspended particles of *see* suspended soil particles
solid(s)
content, 23, 273
determination, 50, 269, 273
dissolved *see* solution
solid concrete, enabled without inclusions, 11
solution (dissolved solids), 113–114
dissociation of polymers in, 41–45, 115, 116
microstructure in, 51–53
sonic (ultrasonic) cross-hole logging, 198–199, 217, 218, 219
South Carolina, Highway 176–177
Replacement Bridge, 175–177
Southwark (central London), 248–252, 281
specifications and associated controls, 265–284
review of existing ones, 271–283
SPERW 2017 (*ICE Specification for Piling and Embedded Retaining Walls*), 220, 277, 279, 280, 281, 282, 283
splash plate, 90
Spring Villa (Alabama), Auburn University National Geotechnical Experimentation Site, 173–174
stabilisation, mechanism of, 5
standard penetration test (SPT) in case studies
Benin, 157
Singapore, 167, 168
Taiwan, 159, 160, 161
UK, 205, 245
US, 172, 175, 180
stiffness (pile)
oscillatory shear and, 255, 259
in UK case studies
Glasgow, 215, 216, 225
Norfolk, 147
Stratford, 225
stirrers, 89–91
Stone Stop, 81
storage (and storage tanks), 96–100, 266
viscosity and, 190–191
strain gauges in case studies
Singapore, 167, 168
Thailand, 166
UK, 207
USA, 173
Stratford (East London), 201–226, 229, 232, 272
strength
of concrete, polymer fluid effects on, 215–216
shear *see* shear strength
stress
shear, 120, 121, 124, 125, 212, 229, 230, 231–232, 233, 238, 255
yield, 214, 229, 231–232, 236, 237–238, 238–239, 240, 245

structural integrity, 150
Glasgow project, 198–199
Italian High Speed Train (TAV) project, 152
Stratford project, 216–220
Süd-Chemie products *see* SC
sulfonated lignins, 31
Super Mud, 68–70, 76, 80, 150
Super Mud Dry, 68, 69, 150, 282
support fluids (general aspects)
critical functions, 7–13
at excavation stage, 7–10
fluid handling, 9, 11–12
foundation loading, 10, 12–13
definition/concept of, 1
history, 2–4, 57–59
mode of action, 5–6
terminology, 1–2
surface treatment plant, 21–22
suspended soil particles, 251
measurement, 268–269
reducing amount before concreting, 10–11
sweep test
amplitude, 255–256, 256, 259
frequency, 255–256, 256, 259–260
synthetic polymers, 4, 18, 29
example, 43–44
Glasgow, 187–201
natural and, differences, 31

Tagus River, Vasco da Gama Bridge, 150–152
Taipei City, 160–162
Taiwan, 159–162
Dingpu Station (Taipei Mass Rapid Transit), 162
high speed rail, 159–172
Tampines (Singapore), 167–168
Telmarch, 59–61
terminology and nomenclature
polymers, 30–33
support fluid, 1–2
test(s), 265–284
commonly used, 267–271
field *see* site; test piles
Glasgow case study after excavation, 193–195
laboratory *see* laboratory tests
samples for *see* samples
test piles in case studies
Benin, 157, 158
Germany (Munich), 153, 154
Italian High Speed Train (TAV) project, 152–153
Portugal, Tagus River in Lisbon, 151
Taiwan, 159, 160, 161, 162
UK
Norfolk, 145, 147, 149, 150
Stratford, 202, 204, 206–207, 209, 211, 217, 218, 219, 220, 223, 224, 225
USA, 171, 172
Texas, 179–180
Thailand, 163–167
Thanet Sand, 147, 202, 203, 204, 206, 229, 249
thixotropy (and gelling), 21, 105, 173
time constant, 231, 241
tower crane piles, Glasgow, 190, 191–192
trains *see* railways
trash pumps, 96
treatment plant, surface, 21–22
tremie pipe, 10, 11, 95, 107, 169, 201, 215
Tuen Mun project (Hong Kong), 162–163, 274–275
twin-flight auger, 103, 193, 209

Ubbelohde viscometer, 45–46, 232, 243
UK *see* United Kingdom
ultrasonic (sonic) cross-hole logging, 198–199, 217, 218, 219
United Kingdom (UK), 145–150
bentonite, 3–4
environmental impact, 25
environmental regulations, 23–24
history, 3
public authorities, 277
United States (USA), 170–180
bentonite, 3
environmental impact, 25
environmental regulations, 24–25
history, 3
reserves, 17–18
Ohio River bridge (Kentucky), 25, 171–172
public authorities, 276
University of Houston National Geotechnical Experimentation Site, 179–180
USA *see* United States
usage (of product), monitoring, 270–271
used (waste) fluids, 81, 107, 275
environmental impact, 25
treatment systems/disposal, 81, 99, 109–110
bentonite, 22–25

Vasco da Gama Bridge, 151–152
vented digging bucket, 103, 209
Venturi eductor, 89
Vietnam, 169
vinyl, 31
viscoelasticity, transient, 229, 252–260
viscoplasticity, 231
viscosity (and its measurement – viscometers and viscometry), 45–49, 120–131, 141, 233–260
  ageing effects, 245–247
  concentration effects on, 239–243
  Glasgow project, 190–192, 199
  instruments for measuring
    Brookfield viscometer, 121, 128–131, 140, 141, 232, 233, 239, 240, 246, 268, 270, 282
    direct-indicating viscometer, 123–128
    Fann viscometer, 121, 123, 124, 214, 232, 236, 239, 240, 241, 242, 268, 270, 280
    rotational viscometer, 267, 270
    Ubbelohde viscometer, 45–46, 232, 243
  models, 235–238
  shear *see* shear viscosity
  shear rate effects on, 239–243
  in specifications, 267, 271–272, 274, 275, 277–280
  Stratford project, 214
  Telmarch vs bentonite effects on, 60
  VisKeau effects on, 62
VisKeau, 61–62, 145
vortex mixing, bentonite, 19, 20

walls (of excavation)
  diaphragm *see* diaphragm walls
  embedded retaining, ICE specification (SPERW 2017), 220, 277, 279, 280, 281, 282, 283
  minimising soil softening at, 7–10
  side *see* side-wall profiles
waste and its disposal *see* used fluids
water (mix), 113–120
  disposal and water undertaker, 109
  hardness, 118–120
  historical use, 2–4
  key parameters, 113–120
  pH *see* pH
  Stratford (London), 209–210
  see also aquifer; dewatering; groundwater; oil-in-water emulsions; solution
water-in-oil emulsions, 34
Water Treat, 68
WeightIt, 76–78, 83
Weissenberg effect, 255
world-wide case studies *see* case studies

xanthan gum, 34, 38–40, 267
  case studies, 155, 156, 164
  see also SC XGUM

yield point (YP), 127, 232, 258
yield stress, 214, 229, 231–232, 236, 237–238, 238–239, 240, 245

zero shear viscosity, 241, 242, 243